The British Dam Society

Ensuring reservoir safety into the future

Proceedings of the 15th Conference of the British Dam Society at the University of Warwick from 10–13 September 2008

Edited by Henry Hewlett

thomas telford

Conference organised by the British Dam Society.
www.britishdams.org

Organising committee: Neil Williams (Chairman), Kenny Dempster, Jon Green, Henry Hewlett, Andy Hughes, Mark Morris and Andrew Pepper.

Cover photograph: Carsington Reservoir, courtesy of Severn Trent Water

Published by Thomas Telford Publishing, Thomas Telford Ltd, 1 Heron Quay, London E14 4JD.
www.thomastelford.com

Distributors for Thomas Telford books are
USA: ASCE Press, 1801 Alexander Bell Drive, Reston, VA 20191-4400, USA
Japan: Maruzen Co. Ltd, Book Department, 3–10 Nihonbashi 2-chome, Chuo-ku, Tokyo 103
Australia: DA Books and Journals, 648 Whitehorse Road, Mitcham 3132, Victoria

First Published 2008

Also available from Thomas Telford Publishing
A guide to the Reservoirs Act 1975. DETR and ICE. ISBN 978 07277 2851 7
Reservoir Engineering. Guidelines for practice. E Gosschalk. ISBN 978 07277 3099 2
Risk and uncertainty in dam safety. D N Hartford and G B Baecher. ISBN 978 07277 3270 5
Improvements in reservoir construction, operation and maintenance. British Dam Society. ISBN 978 07277 3470 9

A catalogue record for this book is available from the British Library
ISBN: 978 0 7277 3522 5

Printed and bound in Great Britain by MPG Books, Bodmin, Cornwall

Preface

This book contains the proceedings of the 15th Conference of the British Dam Society, *Ensuring reservoir safety into the future*, held at the University of Warwick in September 2008.

There are 37 papers in the proceedings of this conference covering a wide variety of issues relating to reservoir safety, particularly in the UK. Valuable information on the effect that the significant storm events in the summer of 2007 could have on UK reservoir safety is presented, with particular reference to the incident at Ulley Reservoir where dam failure was narrowly averted. The current status of UK reservoir management and legislation is discussed in several papers. These include the Defra research programme, the safety of small reservoirs not covered by the present legislation, post-incident reporting and proposals for changes to be made to UK reservoir legislation following the Pitt review into the 2007 storm events. Progress on implementation of the requirement in the Water Act 2003 for the preparation of reservoir flood plans is described and there are several papers on dam break analysis and risk assessment.

Internal erosion is seen as an increasing threat to the UK's ageing stock of reservoirs and there are a number of papers on this topic. The repair or replacement of old valves, and the installation of additional drawdown capacity is an area where works are likely to be required in the foreseeable future and various examples of such works are given. Matters associated with the design of spillway gates and tree blockage of spillways are discussed as well as the seismic assessment of intake towers and advances in numerical modelling of embankment dams. Various works to construct new dams and carry out remedial works to refurbish and upgrade existing dams are described and there are two papers discussing the discontinuance of reservoirs.

The conference included the presentation of the biennial Geoffrey Binnie Lecture by Bill Carlyle. The 2008 Lecture, entitled '*The Binnie heritage in dam engineering*' is published in the Society's journal *Dams and Reservoirs*.

Contents

1. Internal erosion

2. Dam break analysis and risk assessment

3. Design, analysis and monitoring

4. 2007 storms and UK reservoir safety issues

5. Construction, remedial works and discontinuance

1. Internal erosion

Locating the leakage route at Torside Reservoir using the Willowstick AquaTrack system.

V.O. KOFOED, Willowstick Technologies, Utah, USA.
K.D. GARDINER, United Utilities, Warrington, UK
A .A. GEORGE, Montgomery Watson Harza, Warrington UK

SYNOPSIS At the 2006 BDS conference in Durham, Val Kofoed and Jerry Montgomery of Willowstick Technologies and Keith Gardiner of United Utilities and Supervising Engineer for Torside Reservoir, introduced the AquaTrack subsurface water-mapping methodology (Kofoed et al 2006) While the technology seemed promising, at that time there was relatively little data to confirm its accuracy. Now there is a wealth of results which provide a clearer picture of this technology's reliability and capacity

This paper will analyse just such results from a seepage-diagnosis project conducted at United Utilities's Torside Dam in Derbyshire. That project investigated the source of water that was leaking into the drawoff tunnel running beneath the dam. As per the AquaTrack procedure, the reservoir was charged with a low-voltage electrical current, which emitted a magnetic signal representing the subsurface water flow. When read and analysed, that signal indicated the path of the water flow into the adit. Following the investigation, UU lowered the reservoir and found a sinkhole where the mapping procedure had predicted.

HISTORY OF PROBLEMS AT TORSIDE RESERVOIR.

Torside is the second in the chain of five reservoirs in the Longdendale valley, constructed between 1847 and 1887, to supply water to the city of Manchester. From its construction there have been problems with the dam. (Bateman, 1884 and Quayle, 2006) On first filling of the reservoir in 1851 the drawoff pipes laid through the dam at foundation level ruptured and in 1854 the Engineer, J F Bateman, reported to Manchester Corporation that the dam had "stretched itself upon its base up and down the valley carrying the pipes with it". After several attempts at repair, Bateman abandoned the drawoffs and drove a tunnel through the rock in the north abutment in which he installed two new drawoff pipes. The original drawoffs were sealed and, fearing that the stretching of the foundations could have compromised the integrity of the central clay core, in 1855 Bateman added a 1.5m to 1.8m thick clay blanket on the upstream face of the dam connected to a new clay filled cut-off trench upstream of the toe.

At the north abutment of the dam, instead of taking the original clay core into the hillside, Bateman excavated what he termed an "arm trench", at right angles to the core, at the foot of the north valley slope, to a depth varying from 9m to 12m backfilled with puddle clay. The fissured rock at the north abutment was covered with a clay blanket, between 0.6 and 1.2m thick, connected to the embankment core to the west, the arm trench to the south and terminating in a puddle-filled wing trench to the east (see Figure 1). The blanket is protected with masonry pitching laid on a thick bed of gravel. The clay blanket, laid later, on the upstream face of the dam was connected to the abutment blanket.

On first filling of the reservoir the abutment blanket was found to be leaking considerable quantities of water into the fissured rock. The first repair was made in 1851 and repeated remedial works, usually associated with the deep puddle filled arm trench, have been needed in this area up until the present day. The lining system was finally completed in 1865, having been delayed by further failures, in 1858 and 1865, that were sealed by placing concrete onto the fissured rock before restoring the clay lining.

Further, similar repairs were made in 1879, 1882 and 1897 and again in 1944, 1946, 1947 and 1949. Between 1949 and 1951 an extensive programme of grouting was carried out using 1:1 sand cement in holes at 1.5m centres up to 38m deep into the rock foundation Grout takes were in the region of 4 tonnes per hole with a maximum of 32 tonnes recorded. There was a further failure of the clay lining in 1963 but since then there had been no recorded leakages in the area until 2004

The abutment blanket and arm trench effectively prevents groundwater from the hillside entering the reservoir so, probably to prevent build up of water pressure causing damage, a deep rubble filled trench was excavated in 1861. An adit or box drain, 550mm by 350mm connects the rubble drain to the draw off tunnel. Water discharges into the tunnel through an opening formed in the masonry lining.

Recent leakage problems.

In 1997 a v-notch weir was installed on the tunnel wall at the discharge point to gauge flows from the rubble drain into the tunnel. A weir plate was also fixed across the width of the downstream tunnel near the portal where the flow could be more easily observed and measured. A rectangular slot

330mm wide and 100mm deep was cut into the plate and, initially, this easily accommodated the leakage flow.

In May 2004 the Supervising Engineer observed that the flow had exceeded the capacity of the slot and was flowing over the full 2m width of the weir plate to a depth of 20mm. This represented about a tenfold increase in the flow rate to around 30l/s. The advice of an AR Panel Engineer was sought and Dr A K Hughes carried out a Statutory Inspection, under the Reservoirs Act 1975, on 17 November 2004. He recommended that "repairs are undertaken to reduce the inflow to the tunnel to at least the levels experienced prior to May 2004" by September 2005.

Photo 1 - Stalactites and stalagmites in the drawoff tunnel.

Prior to 2004 the leakage had been quite steady but a rapid growth of stalactites and stalagmites had been observed in the tunnel. A stalagmite 60mm long had been observed growing on a stainless steel pipe strap that had only been installed 6 months earlier. It was speculated that acid water from the reservoir was dissolving the lime from the cement in the grout injected into the rock in 1950.

It is not known whether a defect in the clay blanket or arm trench was identified when the grouting was being carried out. It is possible that grouting was considered to be a permanent solution, since the rock would now effectively be non-fissured, and that the integrity of the puddle clay ceased to be important.

The reservoir level was lowered to try to identify the level at which the leakage stopped; however this was easier planned than accomplished. The reservoir has a large catchment, only some of which can be diverted, and it is not unusual for the water level to rise by 7 or 8m up to overflow level overnight. Initially, a 3m drawdown would reduce the leakage flow significantly, and this was the case until 2005, however by 2007 a 12m drawdown was required to achieve the same result.

It became evident that the identification and repair of the defect would not be achieved before the due date so Dr Hughes was once more appointed to carry out a Statutory Inspection and made a further recommendation for a repair to be effected by July 2008. It is important to note that at no time was there any evidence of seepage through the dam.

Since lowering of the reservoir to attempt to locate a defect was problematic, Willowstick Inc were commissioned by United Utilities (UU) to carry out a survey using the AquaTrack technique in the winter of 2005. (Kofoed et al 2006) This survey concluded that there were 3 main leakage paths from the reservoir and one from the valley side to the north. However, drawdown to 12m had revealed no visible defects and the reservoir had been full at the time of the survey. Remote detection of leakage at depth though a clay blanket beneath a reservoir had not been attempted before by Willowstick inc, or anyone else, so it was decided that the area should be re-surveyed with the reservoir substantially lowered and using a different antenna configuration. This was done in December 2006 over a 6 day period. However the reservoir rose from 10m below TWL to 3.4m below during the survey and the later readings had once more to be taken by boat.

THE AQUATRACK SURVEY

The AquaTrack technology has been applied twice at the Torside Dam and Reservoir site to map and delineate preferential groundwater flowpaths into the adit and drainage tunnel located beneath the reservoir. The results of both geophysical groundwater investigations showed supportive evidence as to how groundwater was getting into the adit and tunnel. The information presented by the survey results was intended to provide UU with the necessary information to make informed decisions concerning how to accomplish the stated task of minimizing seepage into the adit and tunnel.

There were some differences in magnetic field maps between the Phase I and Phase II surveys, which were performed with different reservoir levels and utilized different antenna / electrode configurations over the same study area. Each antenna / electrode configuration induces current to flow with a particular bias, thus the differences in the relative field strength. It is possible that changes in reservoir level contribute to changes in seepage flow, which would also lead to changes in the magnetic field response from electrical current flowing in the subsurface. Despite the differences however, the technology still revealed the important anomalies in the same locations.

Details of the Phase II survey

The purpose of the Phase II investigation was to re-survey the site to further evaluate subsurface flow and determine the source of water and its path(s) into the rubble drain and the drawoff tunnel. The water level in the reservoir was at a low level exposing a significant amount of north shoreline. The antenna / electrode configuration employed for the Phase II survey allowed energy to flow in either the reservoir (similar to the Phase I investigation) or beneath the reservoir's clay liner, which is unique to the Phase II survey. As with Phase I, there were no leaks observed beneath or through the dam's earthen embankment.

The Phase II work was initiated on 5 December 2006 with the majority of the work being completed over a 6 day period. This project presented a particular challenge in mapping a water leak under a body of water and beneath a clay liner. A significant portion of the field measurements had to be taken along the exposed "wet" clay lined reservoir and/or from a boat. Measurements taken from the boat required the installation of a tie line to prevent the boat from drifting while taking readings and recording the magnetic field data. The work was slow and could only be done when weather conditions (rain and wind) were acceptable.

A horizontal dipole antenna/electrode configuration was employed to energize the leak in the adit for the purpose of conducting the AquaTrack geophysical investigation. The antenna/electrode configuration differed from the Phase I survey in the placement of the return electrode. Rather than putting the return electrode in the reservoir's water body (as was done in the Phase I survey) this investigation placed it in the north hillside in wet soils and grasses at the edge of the shoreline and east of the area of investigation.

The injection electrode was placed downstream of the dam in water discharging from the drawoff tunnel. This antenna / electrode configuration was designed to allow electrical current to concentrate and flow (between the electrodes) in and out of the adit beneath the clay liner.

An AC current, with a specific signature frequency (380 Hertz) was applied to the electrodes. The magnetic field, generated from the large electric circuit consisting of the subsurface study area, antenna wire and strategically placed electrodes, was measured and recorded from the surface of the ground. Measurements were taken along the north and west shorelines and directly over the drainage tunnel, adit, rubble drain trench and puddle arm trench from a boat in a grid pattern. The measurement pattern used in the investigation provided sufficient detail and resolution to adequately evaluate the leak and its flow paths into the adit and drainage tunnel.

The Phase II north shoreline survey consisted of 308 measurement stations. The actual number of measurements taken to fully delineate the anomalies identified required 418 measurement stations. These measurement stations were established on lines spaced roughly 5m apart with measurements taken on each line at roughly 5m intervals, resulting in a 5m by 5m grid pattern covering the entire area of investigation. The readings taken from the boat were taken at 10m intervals and the distance between the lines varied because of the tie-line method that was used to stabilize the boat.

The field data was sent to Willowstick's home office near Salt Lake City, Utah, USA for reduction and interpretation. While this was being performed, field crew members were instructed on a daily basis where to take additional readings in an effort to identify and delineate leakage and flow into the adit.

The results

The results of the Phase II investigation identified and confirmed the same locations of seepage into the adit as did the Phase I work. The Phase II work more specifically identified two main sources of seepage and their flow paths into the adit and the drainage tunnel.

The first source of water, observed to be seeping into the rubble drain, appeared to be flowing from the north hillside (see Figure 1). This water is trapped between the clay blanket liner, clay core of the dam and puddle filled arm trench. This trapped groundwater flows into the rubble drain trench and is conveyed westward via the adit into the drawoff tunnel.

The second source of water identified in the survey data appeared to be flowing from the reservoir through a breach in either the puddle filled arm trench and/or the clay blanket liner.

Point A, in Figure 1, represents the location where water is seeping into the adit. This location is confirmed in both the Phase I and Phase II investigations. Point B represents an area where groundwater from the north hillside is highly concentrated and channelised. This location is above high water level where groundwater may be intercepted from the hillside in an effort to reduce groundwater flow into and beneath the reservoir's clay bottom liner. Point C represents an area where water from the reservoir appears to have broken through the clay blanket liner and/or puddle arm trench.

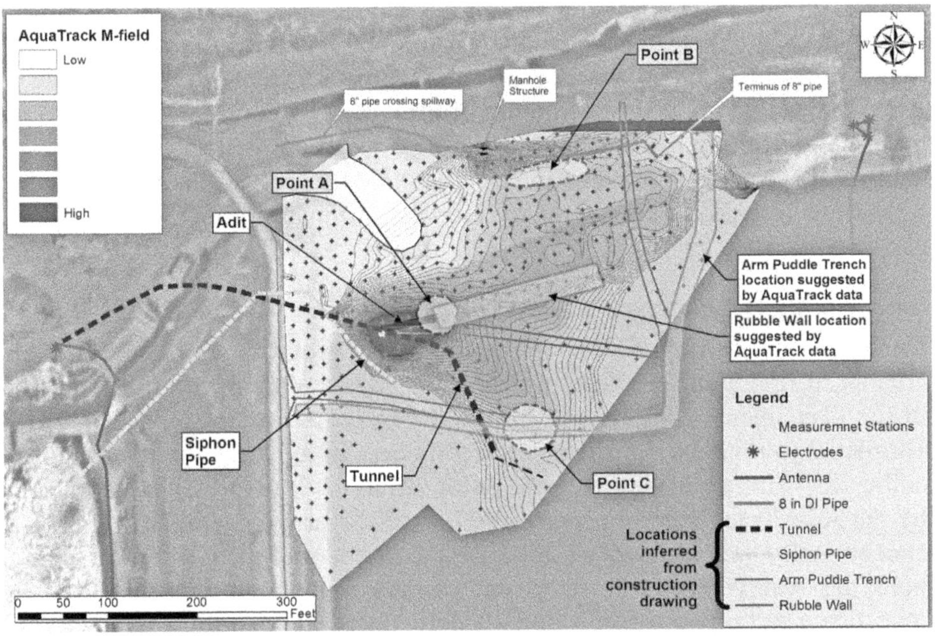

Figure 1 - Plan of the survey area with Phase II Interpretive Results

Survey conclusions

The Phase II AquaTrack survey results provided additional information regarding water movement beneath the reservoir's clay liner and puddle filled arm trench into the adit and a source of leakage into the drainage tunnel. The data obtained from the survey could be used to formulate the next phase of work to identify a remediation strategy to minimize the leakage flow.

The information gathered through the AquaTrack investigation suggested that the survey layout, including antenna/electrode configurations and data grid spacing, were appropriately designed and that the findings of the studies are reliable and accurate in the sense that they seemed to corroborate known information at the site. In general, the AquaTrack survey results were consistent and supported the understanding of existing features in and around the site as noted and shown in the original drawings of the dam. It is believed that the original drawings and sketches (150 years old), when georeferenced to the GPS data provide estimated,but less than entirely accurate, locations of the adit, rubble drain trench and puddle filled arm trench.

The results and final interpretation of the Phase II AquaTrack investigation represents a generalised picture of the subsurface flow. Actual subsurface flow may be more complex than shown because the magnetic field emanating from the cultural features is likely shadowing or over-powering minute groundwater flow paths (e.g., smaller, less permeable features) which may be present, especially given the dynamic conditions of the reservoir (high and low water levels, drainage from the north hillside, etc.). Also, the results are a representation of groundwater conditions observed at the time of completing the fieldwork. It should be noted that different, or additional, groundwater paths may be activated under different hydrologic conditions (e.g., increased level in reservoir, drainage runoff from hillside, dry seasons, etc.).

SITE INVESTIGATION

During the dry spring of 2007 the reservoir was progressively lowered in an attempt to find a surface feature that might indicate the position of a defect in the clay liner. On 23 April 2007 the reservoir was lowered below the 18m level to reveal a 4m diameter, 1.5m deep, hole where the pitching had collapsed. This coincided with the point where the line of the leak predicted by the AquaTrack survey crossed the line of the arm trench. However, the exceptionally wet summer of 2007 meant that it was almost impossible to hold the reservoir at this level and it was September before works to investigate and repair the hole and the defect beneath it could commence.

Draining down the reservoir to this low level had serious consequences for the water yield from the Longdendale valley and meant that the Torside Sailing Club had to abandon its 2007 programme and temporarily relocate to the nearby Winscar reservoir in Yorkshire. The AquaTrack survey provided assurance that such a substantial drawdown was necessary to find the defect giving rise to the leakage.

The physical investigation commenced on 28 September 2007 and was completed by 20 November 2007. The initial excavation exposed a sectional profile around the periphery of the hole comprising pitching overlying ballast bedding above the clay lining. The hole was extended by excavator, within a trench support box, over the plan area of the hole, down to a maximum depth of 5.7m below the surrounding pitching. The excavation quickly confirmed that the hole was located directly above the southern puddle arm trench. The construction of the puddle trench was much the same as that shown in the section taken from a record drawing of repairs carried out in 1882 (see Figure 2).

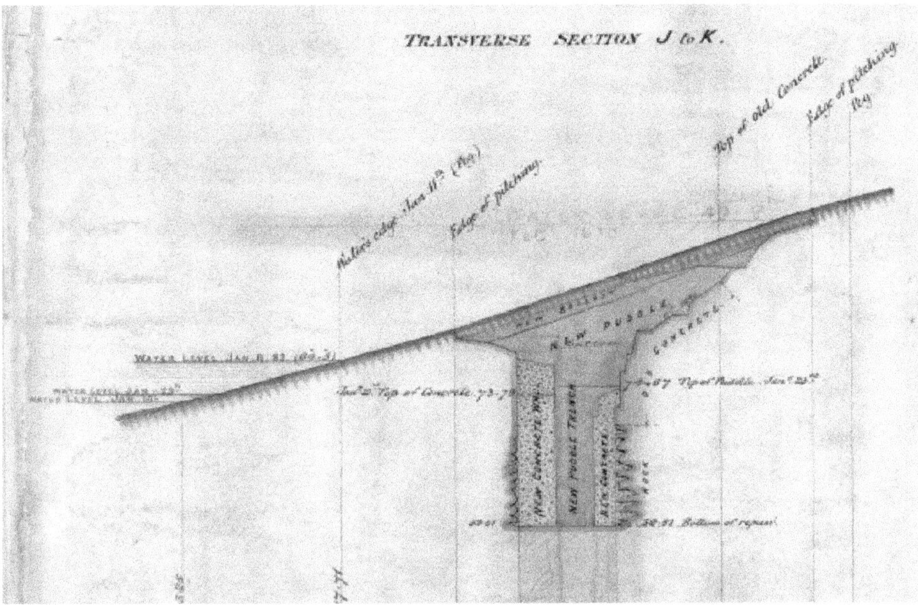

Figure 2 - Section showing the repairs to the arm trench in 1882

Concrete, thought to have been used by Bateman to seal to the rock-head, was uncovered to the northern elevation of the excavation. Exposed at the southern elevation was the top of the arm trench. The material being removed from the hole as the excavation progressed consisted predominantly of the clayey/gravely ballast bedding. This extended down into the arm trench as the construction was exposed. Early evidence indicated that the sides of the trench had been lined with a vertical face of concrete to both the northern and southern elevations.

The concrete lining the south face of the excavation was found to be intact however the north wall contained a large void. As the excavation was deepened, firm intact clay continued to be exposed in the base of the trench. However, the work was continually disrupted owing to inundation from the reservoir. After each inundation event, the contractor removed debris and silty deposits, together with clay that had become soft and slurried, from the base of the trench. On each occasion firm clay was exposed in the base with no evidence to suggest a concrete face to the bedrock on the northern elevation had ever been placed at this particular location. No concrete residue could be seen adhering to the rock nor was any found in the joints. It is probable that this void was once filled with puddle clay laid directly against the rock as part of an even earlier local repair.

The joints in the bed rock were aligned roughly vertical and were generally between 1mm-2mm wide but up to 80mm in places. Very wet, sloppy clay deposits could easily be scraped out from the wide joints by hand. Ground water observed issuing from the fractured rock was initially silty and cloudy but eventually ran clear.

Photo 2 - Excavation at the site of the hole viewed from dam crest. Upstream valve chamber in foreground.

Photo 3 -

View into excavation looking east showing the north face of the arm trench.

The concrete forms a cantilevered overhang below which a void had filled with ballast and clay materials.

Bedrock was encountered behind the overhang

It is postulated that, under pressure from the reservoir, clay was pushed into the fissures in the rock causing settlement and shear cracking in the deep puddle-clay filling to the arm trench. Water seeping through these cracks softened the clay and washed it further into the rock fissures causing further slumping of the clay and the eventual collapse of the masonry pitching. Eventually the flow of the acidic moorland water was sufficient to erode a path through the cement grouted rock until it eventually reached the rubble drain and the draw-off tunnel.

REMEDIATION WORKS

Owing to the continued risk of inundation disrupting the works, a decision was taken to terminate the investigation and implement remedial measures and reinstatement works.

The arm trench, including the void to the rock face behind the concrete facing, was backfilled with in-situ concrete up to the upper elevation of the vertical concrete facing. The existing clay liner, exposed around the periphery of the swallow hole, appeared to be in good condition so a new section of clay was keyed into this making the blanket continuous once more. The repair work was designed by Montgomery Watson Harza (MWH) and supervised by Mr I C Carter, an All Reservoirs Panel Engineer employed by MWH. The Contractor was Eric Wright Construction of Preston, Lancashire.

When the reservoir refilled, which it did in the space of 2 days following very heavy rainfall, the leakage flow was found to have reduced by about 75%. At the time of writing this in the spring of 2008, flow is slightly higher than that measured before May 2004. Plotting of the flow against reservoir head and rainfall may help determine if all of the flow is from the hillside or if there is still some leakage through the clay blanket.

CONCLUSION

The AquaTrack technique is very effective in locating leakage paths below ground and below water. This example of its use showed the technique to be accurate and gave the reservoir owner the assurance that draining the reservoir to the point where there was substantial loss of yield and disruption to the amenity value of the reservoir was justified, while also giving re-assurance to the owner and the Panel Engineers involved that there was no danger to the embankment itself.

The experience at Torside, and at other reservoirs in the Longdendale chain and elsewhere, shows that clay blankets placed directly on fissured or porous foundations are likely to fail at some time. It also indicates that cementitious grouting, although in this case it lasted 50 years, can eventually fail in unfavourable environments.

The authors are grateful to United Utilities for their permission to publish this paper.

REFERENCES

Kofoed, V.O, Montgomery, J and Gardiner, K.D. (2006). *Identifying Leakage Paths in Earthen Embankments*. Proceedings of 14th Conference of the British Dam Society – Thomas Telford Ltd

Quayle, T (2006) *Manchester's Water – The Reservoirs in the Hills.* Tempus.

Quayle, T (editor) *The Bateman Reports 1846 to 1865* – Unpublished.

Bateman, J.F La Trobe (1884) *History and Description of the Manchester Waterworks* - Spon

Assessing the vulnerability of a typical British embankment dam to internal erosion

RODNEY BRIDLE, Dam Safety Ltd, Amersham, UK

SYNOPSIS The vulnerability to internal erosion of a 100-year old typical British embankment dam without filters has been examined using seven methods. Four assessed the potential of the shoulder fill, a low permeability till, to act as a filter and prevent erosion through the core. These were the Sherard grading criteria, the Vaughan permeability criteria, the Delgado permeability criteria, and the Foster & Fell probability-based criterion. Two methods, particularly Fry's, assessed whether sufficient tractive forces would be available to initiate and continue erosion. The seventh was the Interim Guide to Quantitative Risk Assessment for UK Reservoirs, which by comparing the characteristics and performance of the subject dam to that of similar embankment dams makes it possible to assess the probability of failure and whether this is acceptably low. The filter and probability criteria gave only moderately re-assuring results, probably because they are derived from stringent tests to develop conservative design criteria for filters. Methods, such as Fry's, that analyse the mechanics of erosion in the dam seem to best reflect the actual performance of the dam, which has been demonstrated by the QRA to be good when compared to the behavior of many similar dams.

SAFETY INVESTIGATIONS

A 2003 dam safety report recommended safety investigations of the vulnerability of a typical British embankment dam to internal erosion. This was not because there was any obvious cause for concern, but because at the time there was a renewed appreciation of the possibility that internal erosion may pose an unacceptably high threat to the integrity of such dams and their vulnerability to it was not being systematically examined. Such dams do not have filters to protect the core from erosion, and the investigations therefore examined the permeability and grading of the shoulder fill and foundations and the properties of the clay core of the dam with the objective of checking the capacity of the fill to act as a filter and prevent erosion of the core, and the erosive capability of seepage water flowing through fill into and out of cracks in the core, making use of the following:

- Vaughan and Soares (1982) 'perfect' filter relationship between a filter's permeability and its filtering capacity
- Sherard and Dunnigan's (1989) 'critical' filter relationship between a filter's grading and its filtering capacity.
- Delgado et al's (2006) approach linking the percentage of the base soil passing the 0.075 mm sieve to the filter permeability.
- Foster & Fell's (1999, 2001) (approach using logistic regression method in order to replace deterministic no-erosion criteria by a probability-based approach
- Assessments of seepage velocities and forces available to initiate and sustain erosion using Bridle (2007), Fry (2007), Fry & Blais (2006) and Fell & Fry (2007).

This is a work in progress, laboratory investigations have been completed to date, boreholes to check the in-situ properties of the fill have yet to be completed. Many of the findings were reported in the Intermediate Report of the ICOLD European Working Group on internal erosion (Bridle, Delgado and Huber, 2007) and are repeated below.

The Interim Guide to Risk Assessment for UK Reservoirs (Brown & Gosden, 2004), a primary purpose of which is to assess the vulnerability of typical British dams to internal erosion, was published after the 2003 inspection, but an assessment has recently been made and the results are reported below.

THE DAM

The dam was completed in 1904. It is 22.5 m high in earthfill with a puddled clay core 2.7 m wide at the crest with side slopes of 1 to 12 above a clay filled cut-off trench. The upstream slope is 3 to 1 and the downstream slope is 2 to 1. The embankment fill is glacial till.

Photographs (Figures 1, 2) show the puddled clay at the top of the cut-off trench and the site railway that transported the till to the dam embankment, where it was placed on the dam shoulders by side-tipping from the railway wagons. No in-situ tests of the fill's permeability or density have yet been completed, but the photographs indicate that although there is some segregation of the fill, the method of placing may have produced dense fill of varying grading and permeability.

LABORATORY PERMEABILITY TESTING OF GLACIAL TILL SHOULDER FILL

The permeability of the fill in the shoulders of the dam has been established by tests in the large-scale TRL permeameter. This permeameter is 0.3 m square and 1.0 m long. The samples are placed in layers. Water is introduced under pressure at one end and flows along layers of fill, as it would from an opening in a core into the adjoining shoulder fill, and flows out at the other end of the permeameter.

Figure 1: Puddled clay fill placed in cut-off trench up to level of dam foundation

Figure 2: Fill placed on shoulder of the dam by side-tipping from railway wagons

The till is stony with many fines (Figure 3). The samples tested necessarily omitted particles larger than 37.5 mm, about 20% of a typical sample, as Figure 4 shows. The in-situ permeability of the fill has not yet been measured, but the permeability of the truncated fill may be representative of much of the fill in-situ, because the in-situ fill, having been side tipped will be variable, and because there is little difference between the full and truncated grading around 10% passing, which is thought to have most influence on permeability. The permeability varies with density as shown in Figure 5.

Figure 3: Sample of glacial till used for shoulder fill

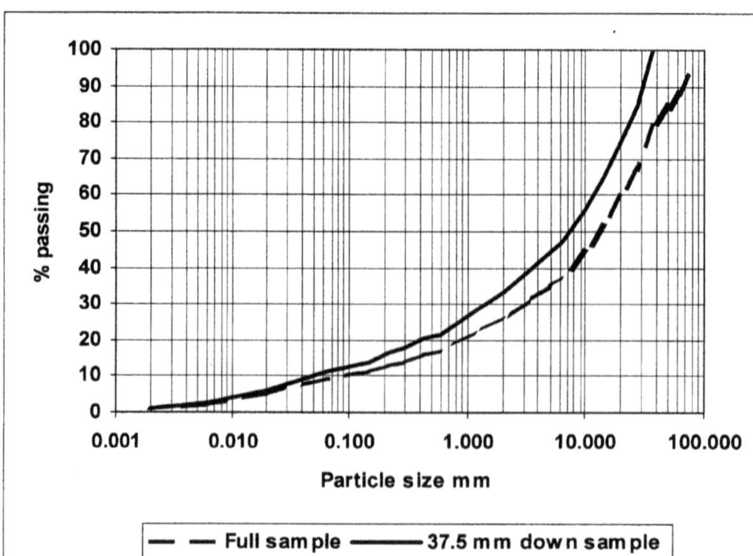

Figure 4: Grading of glacial till used as shoulder fill

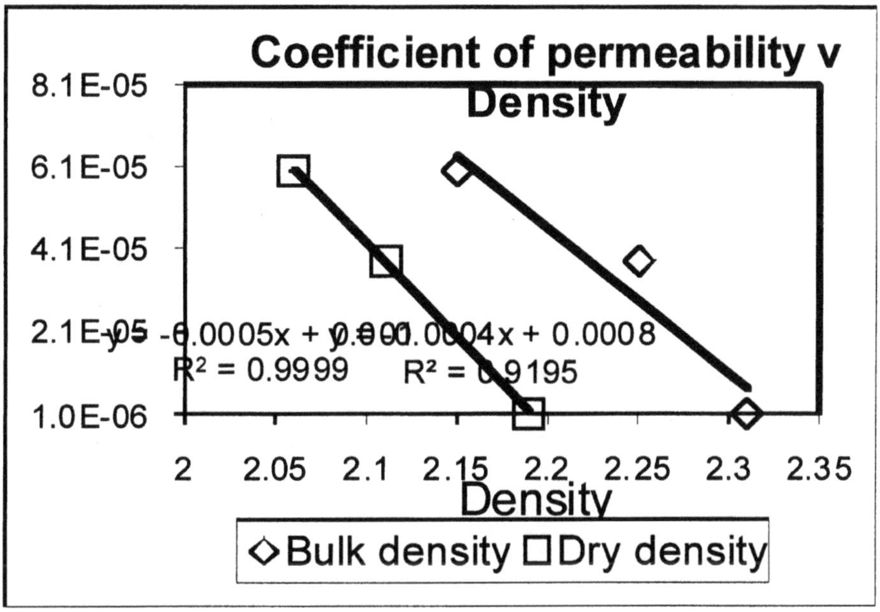

Figure 5: Permeability of glacial till fill as measured in TRL permeameter

Vaughan & Soares (1982) derived the following relationship between permeability and filtering capability:

> $\delta_R = 2.54 * 10^3 (k)^{0.658}$, δ_R in microns, size of floc trapped; k, permeability of filter in m/s

Using the relationship, the fill could trap flocs or particles from 0.3 to 4.2 microns, depending on its density

TESTS ON CORE SAMPLES

The grading of the core was determined from conventional dispersed samples and from undispersed samples. The objective was to determine the smallest floc size present in the core because the Vaughan and Soares 'perfect' filter approach requires that there is no movement of the soil into the filter.

The core samples proved to be very 'murky'(Vaughan & Bridle, 2004), making it impossible to assess the smallest floc size present using Stokes' Law tests because the falling front was not visible through the 'murk' and the velocity at which it was falling could not therefore be measured. The 'murk' was investigated particularly by 'sieving' on very fine cellular micro-sieves and glass fibre filters able to retain particles (or flocs) down to 0.45 microns, smaller than the usual minimum of 2 microns (0.002 mm) measured by pipette in standard soil mechanics laboratory tests. The tests showed that (at this dam, at least) the 'murk' cannot be ignored as colour or other property without physical substance. It was not possible to determine the size of the smallest floc or particle present, but about 7% passed the 0.45 micron 'sieve', as the grading curve on Figure 6 shows.

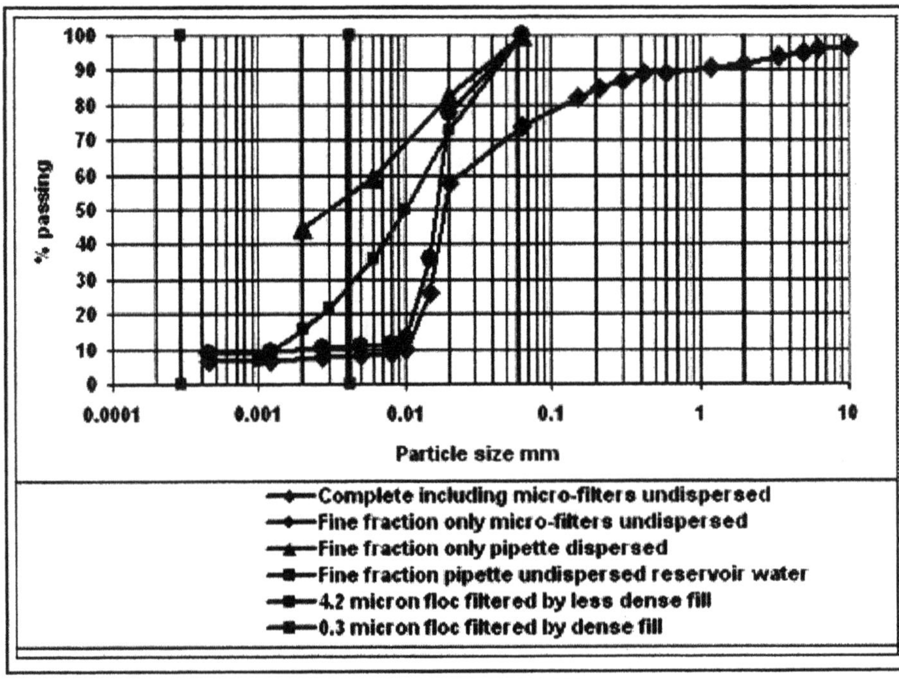

Figure 6: Complete grading of core fill including use of very fine cellular micro sieves and glass fibre filters from 0.063 mm down to 0.00045 mm (0.45microns). Note apparent high clay (0.002 mm) content of dispersed sample (45%), lower clay content (17%), high silt content (83% maximum) of undispersed sample. Liquid limit of core is 35%, plasticity index 19%, on the CL-CI boundary, clay of low to intermediate plasticity.

FILTERING CAPABILITY OF GLACIAL TILL SHOULDER FILL USING VAUGHAN AND SOARES PERMEABILITY CRITERIA

The core may be particularly vulnerable to erosion because, as Figure 6 shows, it may be as much as 83% silt (2 – 60 microns) and therefore erode at a lower seepage velocity than cohesive clay.

The glacial till fill is capable of acting as a filter, trapping, when dense, particles or flocs as small as 0.3 microns. The particle size of the smallest particle present was not determined, but it was proved to be smaller than 0.045 micron and could have been smaller than 0.3 micron, in which case about 7% of the weight of the core could be eroded into the fill, causing some damage. If the fill near the core were less dense, it would trap particles or flocs above 4.2 microns equivalent diameter, but as flocs smaller than these comprise somewhere between 8% and 29% of the weight of the core, the damage could be serious.

FILTERING CAPABILITY OF GLACIAL TILL SHOULDER FILL USING SHERARD GRADING CRITERIA

The results of considering the glacial till fill as a filter to the core following 'critical' filter rules (Sherard & Dunnigan, 1989; USDA, 1986) are shown on Figure 7. The fill would be too coarse in the larger sizes and too fine in the smaller sizes to be the Category 2 filter required. However, the important D_{15} size is 0.4 mm, below the recommended maximum of 0.7 mm and above the recommended minimum of 0.1mm.

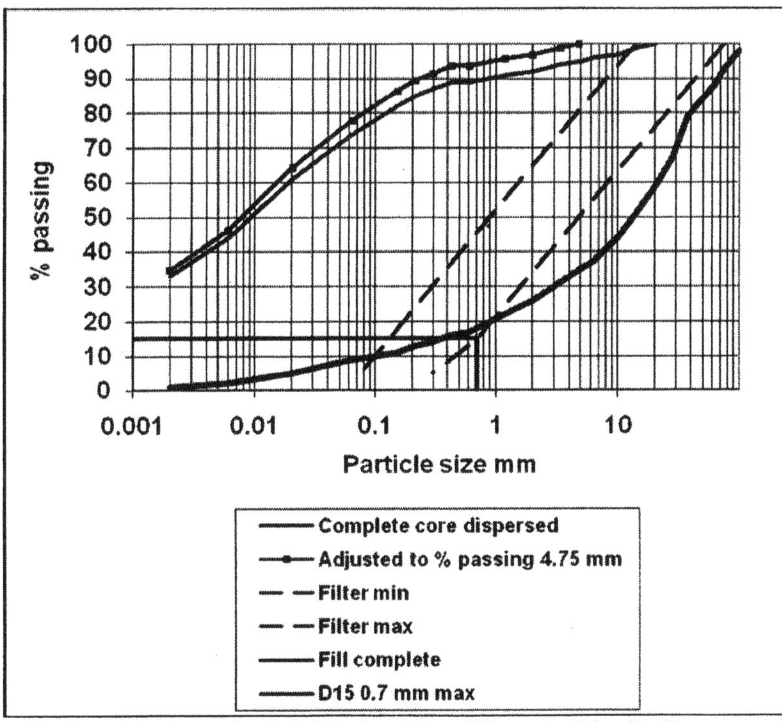

Figure 7: Considering glacial till shoulder fill as 'critical' Category 2 filter to core. Fill too coarse in larger sizes and too fine in smaller sizes, but meets 0.1 mm<D15<0.7 mm criterion

The dispersibility of the core was assessed using the pinhole test. It was a non-dispersive, ND2, soil. The fill's capability as a filter was checked using the no-erosion filter test, which showed that no erosion occurred and that the sides of the 1.0 mm hole through the till had not been eroded. It could be said that this is the conclusive test to demonstrate that the fill provides an adequate filter to the core.

FILTERING CAPABILITY OF GLACIAL TILL SHOULDER FILL USING THE DELGADO PERMEABILITY CRITERIA

Delgado et al (2006) re-examined 340 laboratory tests by others and carried out 348 tests further tests. From the successful tests, those in which the filter trapped the base soil particles and prevented erosion from continuing, he derived the relationship shown on Figure 8, linking the percentage of the base soil passing the 0.075 mm sieve to the filter permeability required to filter it.

It should be noted that the objective was to relate filtering capability to permeability, a hydraulic criterion, as this appears to relate more truly to filter behaviour than geometric grading criteria alone. The 348 Delgado tests show the filtering capacity against a permeability assessed from density through specific regression formulae. Permeabilities were not measured in most of the 340 other tests, and the permeabilities were estimated by applying the Sherard relationship between permeability and D_{15} filter size, as follows:

$$k_{filter} = 0.35*(D_{15})^2\text{, k in cm/s, } D_{15} \text{ in mm}$$

This relationship gives permeability values markedly (nearly three times) higher than the Delgado regression formulae.

Using the Delgado line on Figure 8, the required permeability of the filter to trap the core with 80% (of the below 4.75 mm grading) passing 0.075 mm is 2.0E-02 cm/s (2.0E-04 m/s). The permeability of the fill calculated by the Sherard formula with D_{15} of 0.4 mm (see Figure 7) is 5.6E-02 cm/s, more than twice as high as required and the 'simple' approach suggests that the fill would be too permeable to filter the core.

Using the Delgado regression curve for a compacted filter (Figure 9) shows that the permeability is 2E-02 cm/s (0.02 cm/s), exactly the required value, indicating that the fill would filter the core. The permeability of the fill measured in the permeameter was between 1.2E-04 and 6.0E-03 cm/s, lower than the required value and again indicating that the fill would filter the core.

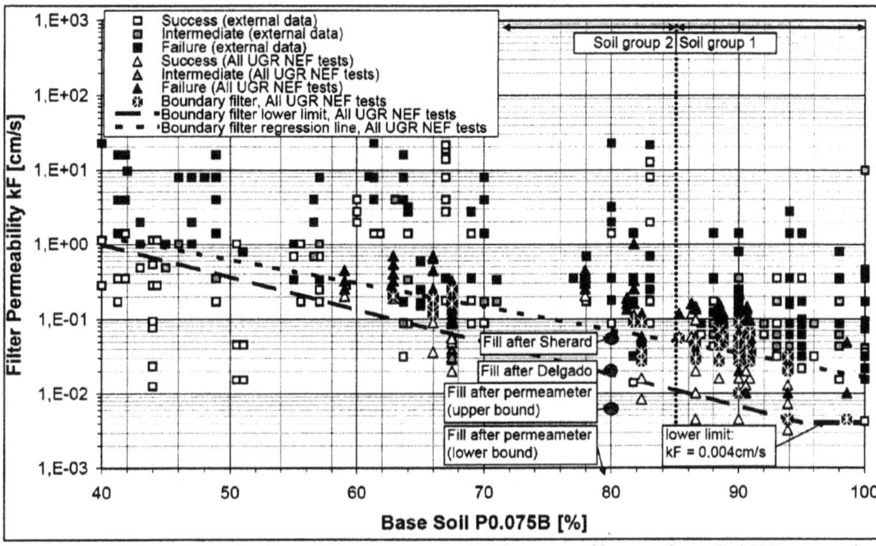

Figure 8: Delgado relationship between % base soil passing 0.075 mm sieve and filter permeability [3]

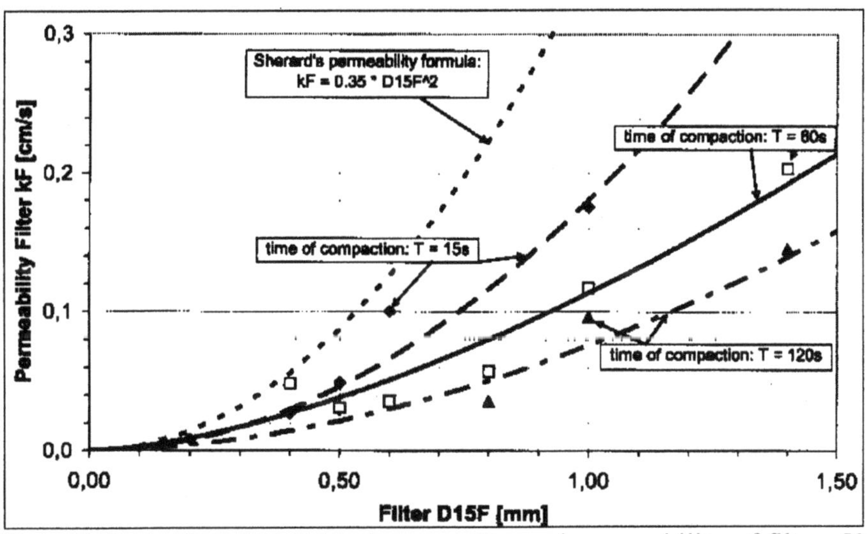

Figure 9: Delgado relationship between D_{15} and permeability of filters [3]

ESTIMATED PROBABILITY OF EROSION USING FOSTER & FELL'S METHOD

Foster & Fell (1999, 2001) use the method of logistic regression, already applied by Honjo & Veneziano (1989), to replace deterministic no-erosion criteria by a probability-based approach. This allows for state-of-erosion criteria being viewed as gradual rather than strict. According to this methodology, the conditional probability of a data point which is characterized by a vector of parameters x_i for showing some defined property C, is defined as follows:

$$P(C \mid x_i) = \frac{e^Z}{1+e^Z},$$

where e is Euler's number and Z is the general function $Z = \omega_0 + \sum_{i=1}^{n} \omega_i x_i$. ω_i are weighting factors which are calibrated using available data. In the present case, C is the "showing of erosion" in filter tests, e.g. the NEF test which is used by Foster & Fell (1999) in order to extend existing data bases.

The core material in the present study shows a $pp_{0.075B}$=80 %, i.e. the percentage of grains of the base soil finer than a particle size of 0.075 mm. According to Foster & Fell (2001), this soil is classified as being in Soil Group 2A, ranging from $pp_{0.075B}$=35 % to $pp_{0.075B}$=85 %, for which no apparent relationship between the base-filter interaction and grain-size characteristics of the base soil can be identified (see also Sherard & Dunnigan, 1989). Therefore, Z in the logistic regression function is only dependent on D_{15F} of the filter. As Foster & Fell (1999) do not give the mathematical definition of Z for soil group 2A, it has to be approximated from the graphical representation in their 1999 paper and the given information is that P(failure|D_{15F}=0.82 mm) = 0.5. This approximation gives the function $Z \approx 3 \cdot (D_{15F} - 0.82)$. Introducing the D_{15F} = 0.2 mm of the dam investigated in this paper, the probability of erosion is calculated as P(erosion|D_{15F}=0.2 mm) = 0.13; and in case of a D_{15F} = 0.4 mm the probability increases to 0.22, in both cases not insignificant but low.

SEEPAGE VELOCITIES TO INITIATE EROSION

The initiator of internal erosion is the energy available from the seeping water to erode soil particles from the walls of seepage channels. Both the upstream and downstream shoulders of the dam are in glacial till fill of low permeability which may limit quantity and velocity of water reaching any flow channels through the core. However, the undispersed 'natural' grading of the core show that it has a high silt content and may therefore behave as non-cohesive silt, very vulnerable to erosion because erosion may be initiated at velocities as low as 0.012 m/s (Bridle, 2007). Very approximate flownet analysis examining flow towards a 10 mm high crack at about one-third height of the core into which water flows from the isotropic upstream shoulder and escapes 'freely' along a shorter flowpath through the isotropic downstream shoulder into a downstream drain showed that velocities could be high enough to initiate erosion in all but the lowest permeability fill (1.2E-06 m/s). However, silt is non-cohesive and could not sustain an open crack if saturated, consequently seepage velocities would be much reduced, probably too low to cause erosion. Clays will resist high velocities, 0.4 to 1.9 m/s, depending on type (Bridle, 2007).

However, calculation of the hydraulic conditions, by Fry (based on Fry & Blais, 2006, and Fry, 2007) leads to the conclusion that 'tolerable' erosion would occur if the permeability of the fill is 6.0E-05 m/s (the highest permeability measured in the laboratory permeameter) or less. The erosion would be tolerable because the hydraulic head in the downstream fill would increase and in consequence the difference of hydraulic head through the core (or the hydraulic gradient) would decrease as the discharge flow increased, in consequence the flow velocity would decrease down to a critical value at which erosion would stop, even though total seepage flows of several litres/second could be occurring. The approach shows that the downstream fill filters the core even if erosion has been initiated.

QUANTITATIVE RISK ASSESSMENT

A Quantitative Risk Assessment (QRA) has been carried out using the Interim Guide to Quantitative Risk Assessment for UK Reservoirs (Brown & Gosden, 2004). Unlike other methods (for example Lacasse, Nadim, Hoeg & Gregersen, 2005) which assign arbitrary probabilities to qualitative descriptions such as 'unlikely', 'likely', etc, to an event train to derive probabilities, the QRA Guide produces approximate, but absolute, assessments of the probability of failure of typical British embankment dams based on a comparison of a subject dam's characteristics and performance to the characteristics and performance of similar dams (Brown & Tedd, 2003). As there are over 1,000 'typical British embankment dams', most more than 100 years old, at least 100,000 dam-years of performance data are available, a few of which were poor performances, recorded as incidents, from which probabilities of failure were projected. The method was developed primarily to examine the vulnerability of embankment dams to internal erosion, as other methods were not available.

The total annual probability of failure (AP) of the subject dam was found to be about 2E-06 (loosely 1 in 500,000-years). It is most vulnerable to internal erosion through the embankment (1.5E-06), condition score 2, resulting from a pessimistic scores of 1 each on settlement (of which there has been none in recent years), and a single episode of rapid refill. Erosion alongside the overflow channel and tunnel is less likely (5E-07) and failure through overtopping is extremely unlikely (1E-07). However, because many lives would be at risk should the dam fail without warning, even this very low overall probability of failure is, to standards recommended by the Health & Safety Executive, only 'tolerable' (in the zone between 'unacceptable' and 'broadly acceptable') as a risk imposed on society. The QRA methodology enables engineers to examine the effects of safety measures to improve safety and whether the costs of doing so are proportionate. In this case small improvements in monitoring and in emptying capacity would likely be sufficient at proportionate cost to further reduce the probability of failure and improve the safety status to be 'broadly acceptable'.

CONCLUSIONS AND RECOMMENDATIONS

Although its performance gives no reason for concern, the objective of the investigations is to assess the dam's ability to resist internal erosion. As it is old, it can be assumed that no large openings exist through which flow could occur at a velocity high enough to initiate erosion and lead to piping and failure. The mechanism envisaged is erosion in a (possibly new) discontinuity through the core, or general erosion if the possibility that the core is silt-sized, as undispersed (natural) samples show, and the question is whether the adjoining glacial till fill would act as a filter, restricting velocity and trapping eroded particles to prevent erosion. The outcomes are summarized in Table 1.

Table 1: Summary of results

Method	Results	How re-assuring?	Remarks
Vaughan & Soares	Filters down to 0.3 - 4.2 microns, but 7-29% dispersed core may be eroded	Not	'Perfect' filter, no movement permitted, thought to be conservative. Natural soils are undispersed, silts more easily eroded than clays
Sherard	D_{15} 0.4 mm, less than critical 0.7 mm, but grading generally too coarse to filter Category 2 core. Fill passed No Erosion Filter test	Moderately	Sherard filters develop by self-filtering, some particles are eroded and build up a filter in-situ. Fill not dispersive.
Delgado	Permeability of fill, assessed from grading and laboratory results, low enough to filter core	Re-assuring	Uses permeability criterion, but less conservative than Vaughan & Soares, perhaps because self-filtering is assumed, as Sherard
Foster & Fell	Probability of occurrence of erosion not insignificant, but low	Moderately	However, scale of 'not insignificant', 'low', not clear

Method	Results	How re-assuring?	Remarks
Bridle	Seepage velocities high enough through 10 mm opening to move silt core in all but lowest permeability fill.	Moderately	Silt core non-cohesive, would not sustain 10 mm opening. Clay would sustain opening, but would not be eroded at velocities generated through it.
Fry	Erosion may initiate, but would be filtered in fill, even with large seepage quantities	Very	Seems to demonstrate what would be expected intuitively in a dam with low permeability, fine fill
QRA	'Tolerable' to 'broadly acceptable' probability of failure	Very	Typical British embankment dams have not shown vulnerability to long term internal erosion, reasons perhaps indicated by results above.

The general conclusion from the results is that methods which analyse the mechanics of erosion in the dam seem to best reflect the actual performance of the dam, which has been demonstrated by the QRA to be likely to be good when compared to the behavior of many similar dams.

The Fry approach and the QRA are very re-assuring about the vulnerability of the dam to internal erosion. The other results are less so but this can be explained by the fact that they are derived from design methods for filters and therefore reflect the very stringent conditions imposed in the experiments used to establish adequately conservative filter design criteria. Filters are usually thin and must operate under high hydraulic gradients and high seepage velocities. Filter design criteria are therefore very conservative. By contrast, because shoulder fill is wide and is commonly, as in this case, similar both upstream and downstream of the core, hydraulic gradients and seepage velocities are reduced thereby reducing the erosive energy of seeping water.

The fill in the subject dam is widely graded and, because it was placed by side tipping, may be segregated with locally coarse and fine variations throughout the shoulders, but because of its geological origins, it is probably unusually uniform when compared to shoulder fill in many typical British dams. Its vulnerability to internal erosion can be examined by simple analyses based on information from laboratory tests on samples collected from trial pits. Borehole investigations are intended to check that the variability of the fill in the dam shoulders is not so great that the findings to date are invalidated.

However, internal erosion can take several forms and be initiated at 'singularities' in fill, of which more are likely to be present in the variable shoulder fill probably present in many 'typical' British dams. Application of the principles set out by Fell & Fry (2007), Fry & Blais (2006) and Fry (2007) are beginning to make it possible to rationally examine and analyse the mechanisms and assess the likelihood of the various types of erosion – backward, crack, suffusion or contact – progressing at vulnerable locations because of the loadings applied through the five stages – initiation, continuation, progression, detection, intervention - that may lead to a breach through internal erosion by enlargement of erosion pipes, by loss of freeboard through crest settlement, by slope instability or by unraveling of the downstream slope. The methodology can be applied to erosion through the body of embankment dams and at interfaces with structures – culverts, spillways- through or on dams.

RECOMMENDATIONS

Internal erosion is thought to be a serious threat to typical British dams, similar to overtopping and instability, for which there is guidance. The QRA Guide is the guidance on internal erosion and should be applied at every embankment dam. If it suggests that the dam is vulnerable to erosion, further investigations, progressively making use of more detailed information collected from existing records, local geology, trial pits, boreholes (and equipment capable of drilling on slopes without scaffolding is now available) to carry out analyses of the vulnerabilities, using methods such as those reported by Fell & Fry (2007), to guide decision-making on whether it is necessary to respond to those vulnerabilities by appropriately targeted monitoring or by safety works.

ACKNOWLEDGMENTS

The author would like to acknowledge, with thanks, the enormous support that the Isle of Man Water Authority has provided for these investigations and for their permission to publish this paper. Thanks, too, to colleagues Richard Young, Peter Horswill, Jean-Jacques Fry, Fernando Delgado, Nils Huber and the Internal Erosion Working Group of the European Club of ICOLD for widening this knowledge of internal erosion, and how to examine and respond to the vulnerability of existing dams to it.

REFERENCES

Bridle R C 2007. Assessing the vulnerability of dams to internal erosion. In *Internal Erosion in Dams and their Foundations, Proceedings of Workshop, Aussois, France April 2005* Fell & Fry (eds), 2007 Taylor & Francis Group, London, ISBN 978-0-415-43724-0 pp 25-33

Bridle R C, Delgado F, Huber N P 2007. Internal erosion: continuation and filtration: current approaches illustrated by a case history. In *Assessment of the Risk of Internal Erosion of Water Retaining Structures: Dams, Dykes and Levees. Intermediate Report of the Working Group of ICOLD* Deutsches Talsperren Komitee, Technical University of Munich, Nr.114 ISBN 1437-3513, ISBN 978-3-9404476-04-3 pp 70-84

Brown A J, Gosden J D 2004. *Interim guide to quantitative risk assessment for UK reservoirs.* Thomas Telford, London. ISBN 0 7277 3267 6.

Brown A J, Tedd P 2003. The annual probability of a dam safety incident at an embankment dam, based on historical data. *International Journal of Hydropower & Dams* 10(2), pp 122-126

Delgado F, Huber N P, Escuder I, De Membrillera M G 2006. Revised criteria for evaluating granular filters in earth and rockfill dams. *Transactions of the 22nd Congress, International Commission on Large Dams, Barcelona, Vol III, Q86 R29, 445-456*

Fell R, Fry J-J 2007. The state of the art of assessing the likelihood of internal erosion of embankment dams, water retaing structures and their foundations. In *Internal Erosion in Dams and their Foundations, Proceedings of Workshop, Aussois, France April 2005* Fell & Fry (eds), 2007 Taylor & Francis Group, London, ISBN 978-0-415-43724-0 pp 1-23

Foster M A; Fell R 1999. Assessing embankment dam filters which do not satisfy design criteria. *Sydney, Australia: The University of New South Wales (Report / UNICIV; R-376). - ISBN 85841-343-4*

Foster M A; Fell R. 2001. Assessing embankment dam filters that do not satisfy design criteria. In: *Journal of Geotechnical and Geo-environmental Engineering, Vol. 127, No. 5, pp. 398-407. - ISSN 1090-0241*

Fry J-J, Blais J-P 2006. Assessment of internal erosion risk of embankment or rockfill dams. *Transactions of the 22nd Congress, International Commission on Large Dams, Barcelona, Vol III, Q86 R69, 1131-1141*

Fry J-J 2007. Current practice and a new view on granular filters. In *Assessment of the Risk of Internal Erosion of Water Retaining Structures: Dams, Dykes and Levees. Intermediate Report of the Working Group of ICOLD* Deutsches Talsperren Komitee, Technical University of Munich, Nr.114 ISBN 1437-3513, ISBN 978-3-9404476-04-3 pp 60-69

Honjo, Y, Veneziano D 1989. Improved Filter Criterion for Cohesionless Soils. In: *Journal of Geotechnical Engineering, Vol. 115, No. 1, pp. 75-94. - ISSN 0733-9410*

Lacasse S, Nadim F, Hoeg K, Gregersen O 2004. Risk assessment in geotechnical engineering: the importance of engineering judgement. In *Advances in geotechnical engineering, the Skempton conference, London, March 2004* Jardine, Potts, Higgins (eds) Thomas Telford, London, ISBN 0 7277 3264 1, Vol 2 pp 856-867

Sherard J L, Dunnigan L P 1989. Critical filters for impervious soils. *ASCE Journal of Geotechnical Engineering Division, January,* Vol 115 (GT7) 927-947

USDA Soil Conservation Service 1986. *Guide for determining the gradation of sand and gravel filters. Soil Mechanics Note No 1 210-VI.* United States Dept of Agriculture, Soil Conservation Service, Engineering Division, Washington

Vaughan P R, Bridle R C 2004. An update on perfect filters. *Long-term benefits and performance of dams. Proceedings 13th conference of the British Dam Society and European Club of ICOLD, University of Canterbury, UK, 22-26 June 2004, Thomas Telford, London, 516-531* (the 'original' paper, corrected in the paper above. Please contact rodney.bridle@damsafety.co.uk for additional copies of the corrected paper if required)

Vaughan P R, Soares H F 1982. Design of filters for clay cores of dams. *ASCE Journal of Geotechnical Engineering Division, January, Vol 108 (GT1) 17-31*

Progress in assessing Internal Erosion

A. J. BROWN, Jacobs, Leatherhead, UK
R. C. BRIDLE, Dam Safety Ltd, Amersham, UK

SYNOPSIS The European Working Group on Internal Erosion presented its Interim Report at the 7th European Conference on dams in Freising in Germany in September 2007, this being the culmination of interactions which had commenced in 2004. This paper summarizes the main technical outcomes from the working group. It then comments on practical implications of progress in understanding internal erosion for dam engineering in the United Kingdom.

INTRODUCTION

The European Working Group on Internal Erosion (EWGIE) was set up in 1993 under the chairmanship of Andrew Charles, with a report of progress to the 2001 European Conference and 2002 BDS Conference (Charles, 2001, 2002). 47 case histories were collected and preliminary conclusions drawn as to the features that increase the vulnerability of dams to internal erosion.

Following a number of serious incidents and failures in Europe due to internal erosion, research funding became available, with chairmanship of the EWGIE passing to Jean Jacques Fry of EDF (Brown and Carter, 2004). An inaugural workshop on definitions and research needs was held at Aussois in April 2005 (Fell and Fry, 2007), with subsequent and ongoing annual workshops. The Interim Report of the European Working group of ICOLD was published at the European Conference on dams in Freising in 2007. It is planned to publish the final report in 2010, at the next European conference.

The focus of the EWGIE is production of technical analytical methods which can be used by engineers to both assess the risk of internal erosion in existing earth structures, and for design of new structures. The presentations and attendees at the workshops are a mix of PhD output and work by practising engineers, most being the engineers for owners of major portfolios of dams.

It is suggested that the level of detail which is appropriate to apply to the assessment of vulnerability to internal erosion should be on a risk based approach. This considered whether the Undertaker has reduced risk as low as reasonably practicable (ALARP), namely whether the measures taken have been proportionate to the reduction in risk that could be achieved. The ALARP decision making process is defined in paragraph 124 and Appendix 3 of "Reducing risks, protecting people" (HSE, 2001) with further guidance given at www.hse.gov.uk/risk/theory/alarp.htm. The Interim Guide to Quantitative Risk Assessment (Brown & Gosden, 2004) provides a screening level of risk assessment, which can be a useful first step.

FRAMEWORK FOR CONSIDERING INTERNAL EROSION

The process of internal erosion is best considered as a series of steps, which can lead from initiation through to progression, as summarized in Table 1, each being applied to the four types of internal erosion shown in Table 2. Further detail on definitions is given in the overview paper in Fell and Fry (2007), being based on the Fell (New South Wales Univ) model of the stages of internal erosion, which itself was based on US Bureau of Reclamation work.

Table 1: Stages of internal erosion

	Phase	Key issues
1	Load conditions	Reservoir level changes. Seismic, Seasonal effects such as sun shrinkage, frost and thawing; External loads such as animal holes, tree roots and human effects such as terrorism or accidents
2	Location	Embankment, appurtenant works, foundation
3	Initiation	Will particles start to move? Four types of internal erosion are differentiated, as shown in Table 3
4	Continuation (Filtration)	Will filters / shoulders stop erosion?
5	Progression	Will the pipe stay open, or collapse? What are the hydraulic conditions, in terms of critical shear stress and rate of erosion? Are there upstream or downstream conditions that would limit progression?
6	Detection	How can the internal erosion be detected, and how far will it have progressed by then?
7	Intervention	What options are there for ways to stop internal erosion, and how effective will they be? E.g. Diaphragm wall? Toe filter?
8	Breach	What is the mechanism(s) which leads to a breach? Four modes are identified a) gross enlargement of pipe b) loss of freeboard (crest settlement) c) slope instability d) unraveling

Table 2: Types of internal erosion

Type	Definition
Concentrated erosion	In soils which are capable of sustaining an open crack, or in the interconnecting voids in a continuous permeable zone. Erosion occurs along the sides of the crack (or voids) where the shear stress (velocity) exceeds the critical value. NB at low flows there may be leakage with no erosion.
Suffosion	Mass erosion in soils which are internally unstable. Fines transported by seepage flow between the larger soil particles.
Backward Erosion (Piping)	Erosion starts at the exit point; a continuous passage is developed by backward erosion when the seepage gradient exceeds the "flotation gradient" of the soil.
Contact Erosion	Erosion at the horizontal boundary of a fine soil overlying a coarse soil, where the fine soil is washed into the coarse soil due to horizontal flow

The Interim Report highlights that for a safety evaluation of internal erosion it is necessary to have data, and a corresponding analytical model, in four the following broad areas:-

a) geometry of the dam,
b) geology of the foundation,
c) geotechnical parameters of the soils comprising the dam and its foundation,
d) hydraulic parameters of permeability and erodibility of soils, and boundary conditions for seepage and flow analysis.

MODES OF FAILURE

In applying the framework for considering internal erosion and thus the vulnerability to internal erosion at a particular dam it has been found helpful to apply the event tree process developed by the US Bureau of Reclamation (Cyganiewicz et al, 2005), as a means of thinking through the processes which determine the vulnerability of a dam to failure by internal erosion. An example of this is given in Table 3, applied in a workshop process of six dam engineers to a cast iron pipe with downstream control in an old dam. This is similar to the event train process in the Interim Guide to QRA (2004). The probability of an answer "Yes" to each step (i.e. the process moving towards failure) can be estimated in a workshop process, and when multiplied together provide an estimate of the overall probability of failure.

The process can also be used to explore potential structural works to reduce the probability of failure, and this is also included in Table 3. This shows that the addition of a upstream valve would have no impact on the annual probability of an incident occurring, but would reduce the probability of the incident progressing to failure (estimated in the workshop as by a factor of 100). Conversely a structural relining of the pipe would reduce both the probability of internal erosion commencing, and the probability of failure (estimated in the workshop as by factors of 40 and 100 respectively).

Table 3: Example of Event tree analysis applied to a bottom outlet pipe laid within the embankment fill

	Stage	Features affecting probability of internal erosion progressing at the subject dam	Potential risk reduction measures
1	Outlet pipe cracks		
2	Initiation	Internal erosion initiates along the outside of the pipe	Reline pipe
3	Continuation – is there an unfiltered exit?	Various exists, some unfiltered	
4	Progression - does a roof form	Assume void can develop along the top, sides or bottom of the pipe	
5	Progression - is the flow unlimited	Flow is limited by the diameter of the bottom outlet pipe i.e. maximum flow through the 280mm pipe is 1000l/s.	Add upstream valve
6	Is early intervention unsuccessful	High level and imported pumps would be used	
7	Will the dam breach process initiate?	The likelihood of breach is reduced by the presence of a rockfill toe	
8	Will heroic intervention be unsuccessful	The rockfill toe should slow the rate of breach development i.e. maximum flow is say 200l/s/m if spread out over 5m width	

TOOLS TO QUANTIFY VULNERABILITY TO INTERNAL EROSION

General

Tools to quantify the vulnerability to internal erosion are still being refined, but basic tools are now available, and are described in papers in the Interim Report of the EWG with highlights given below.

Concentrated erosion

Tools to predict the vulnerability of the watertight element of a dam to the various causes of cracking (hydraulic fracture, differential settlement etc) are reasonably well understood, with Sherard (1986) noting that cracking of the core probably occurs in most embankment dams. In cohesive soils the onset of internal erosion occurs when the shear stress imposed by water flow exceeds the critical shear stress of the soil, determined from the Hole Erosion test or similar. It appears to vary by several orders of magnitude depending on the properties of the core material, from near zero to around

150Pa (Wan and Fell, 2002, 2004). The quoted rate of erosion once this threshold is exceeded similarly varies by several orders of magnitude.

The imposed shear stress τ and flow velocity in the crack can be calculated from standard hydraulic methods and will vary depending on whether flow is laminar or turbulent. Thus the drag force τ on the sides of the crack (or voids) is :

$$\tau = \gamma_w . R_H . i$$

where :

γ_w	:	is the water specific weight (9.8 kN/m^3)
R_H	:	hydraulic radius
i	:	hydraulic gradient

In interconnecting voids R_H is equal to the pore diameter Dp/4, where Dp(m) is issued from the maximal permeability with k (m/s): $Dp = \sqrt{(1E^{-5} * K)}$; whilst in a extended crack of height much greater than width R_H may be taken as half crack width.

Suffosion

Internal erosion of fines within the matrix will occur when the flow velocity in the pores is sufficient to move particles within the pores of the soil.

The mean pore velocity V_{pav} due to seepage flow can be estimated from the Darcy velocity estimated from seepage analysis as follows:

$$V_{pav} = \frac{V}{n\,T}$$

where n is the porosity and T is the Tortuosity (taken as $2/\pi = 0.64$). The relationship between particle size D (mm) and velocity to cause movement may be taken from Equation 20 or Figure 4 of Perzimaier et al (2007), but for pore velocity less than 10^{-2}m/s (particle < 100μm) is approximately $D = 11.5\, V_{pav}^{1.285}$.

The size of particles within the soil which are vulnerable to movement, due to the soil being internally unstable, may be estimated using Kenney and Lau (1985, 1986).

Backward Erosion (Piping)

For detailed analysis the safety factor against piping would be calculated from pore pressures at the toe of the dam taken either from piezometers, or detailed seepage analysis. However, for preliminary analysis average gradients along the whole length of the potential failure path may be used. Critical average gradients for piping were given by Bligh (1910), Lane (1935) and recently by Weijmers and Sellmejier (1993). The latter quote

0.34, 0.28, 0.24 and 0.14 for gravel, coarse, medium and fine sand respectively, all for a uniformity coefficient of 3.

The most likely location for piping is in the foundation, where overlying cohesive soils mean a roof will remain stable. Piping in the embankment is likely to initiate as backsapping and local slope instability at the toe of the dam because any embryonic pipe within the embankment would tend to collapse. A possible exception is in the vicinity of the phreatic surface, where a roof may be sustained over a pipe due to suction forces in the overlying partially saturated soil

Contact erosion

The methodology for estimating the onset of contact erosion is similar to that for concentrated erosion, but with suitable adjustment for pore size and/ or crack width.

SYSTEM RESPONSE

Even where internal erosion initiates it may not continue, and if its does continue it may not lead to failure. For example where the shoulders of a dam are of fine cohesionless materials migration of fines from the shoulders may block any concentrated erosion within the core. Secondly as the flow through a crack in the core increases the head loss in the shoulders will increase, which will reduce the head across the core, potentially to a shear stress (or velocity) below which erosion stops. In assessing internal erosion it is therefore important that all stages of internal erosion are considered, as shown in Table 1, and that the interaction between the various elements of the system are understood.

FILTER DESIGN

It must be appreciated that some migration of particles within the body of a dam under hydraulic loading is likely to occur in all dams, and is part of the bedding in process for new dams. What can be prevented is loss of those particles from the dam, and this is one of the roles of filters; to trap particles on the face subjected to water load. Where the number of particles trapped is large, this leads to clogging on the upstream face. A second role is to provide drainage of water seeping through the dam, such that the phreatic surface is drawn down towards the base of the dam thereby improving the stability of the downstream face and its foundation. Clogging is not an issue for fine filters on the downstream side of the core as it increases impermeability of the water retaining system. However, it could be a problem where it occurred in a filter blanketing the foundation, as it could lead to increased uplift pressures.

The issue of filter design is well covered in the ICOLD Bulletin (1994), with an example of application of this and more recent rules given in Bridle (2008) and not repeated here.

DEBATABLE ISSUES

How variable are UK "puddled clays"?

It is clear that the term puddle relates to the form of construction, and not the properties of the clay. Figure 1 is reproduced from Moffat (2002) and shows clearly the wide variation of plasticity of clays used in UK dams, and that in fact some dams have cores constructed of silt. Non-cohesive silts will be vulnerable to rapid erosion in the event of a concentrated leak.

Are UK soils dispersive?

Recent laboratory testing as part of feasibility studies for dam rehabilitation works, and new flood detention reservoirs is summarized in Table 4 and has showed that some UK clays are dispersive, that is the clay flocs are either naturally dispersed into single clay particles or small clay flocs, or vulnerable to dispersion when the water chemistry changes. This is similar to the findings from Atkinson et al (1990) where tests on four UK puddle clays showed one is dispersive. Dispersive clays are more vulnerable to extremely rapid erosion in the event of concentrated leakage. It is considered that further research is required into this aspect of UK soils.

Figure 1 The range of fine grained soils used in UK "puddle clay core" dams (as Table 3 of Moffat, 2002)

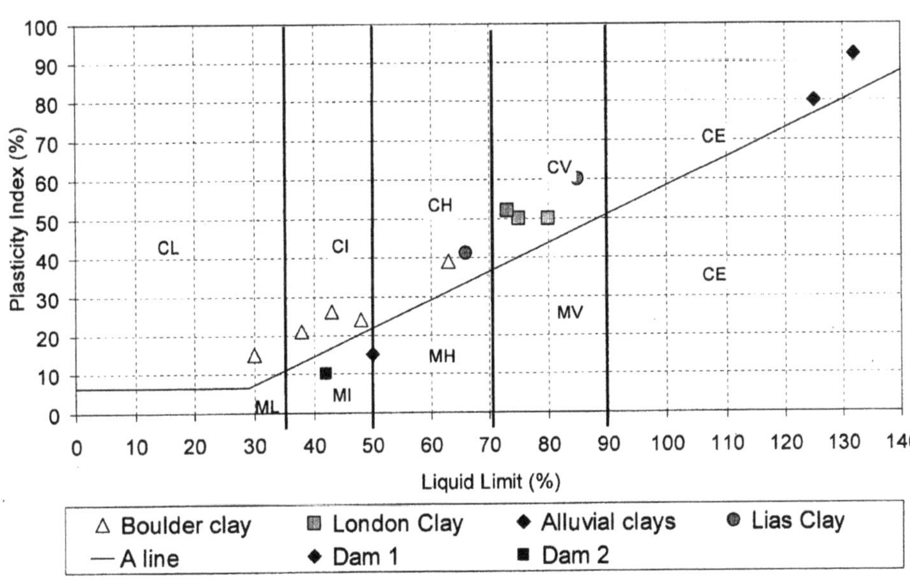

Table 4: Results of dispersion related testing on soils used in UK dams

Dam	LL	PI	%< 75μm	Double Hydrometer	Crumb	Pinhole
A	31	12	62, 69	47%, 42%	Both dispersive (1 x G3, 1 x G4)	Not tested
B	27 to 54	8 to 28	39 to 99	9 tests, 5 >50%, 1x 30-50%	10 tests 1xG4, 2xG2, 7xG1	9 tests, 1xD1, 3xND3 4xND2, 1xND1
C	35	19	76%	29%, 30%	Both Dispersive (G3)	ND2 (non-dispersive)

Notes. Test methods and classification of results described in Head (1982) and BS 1377-5, Section 6. For hydrometer tests 100% is that addition of chemical dispersant does not increase % passing 0.002mm; Sherard et al (1976) suggests that > 50% is classified as dispersive and 30 to 50% as slightly dispersive. The range of potential results for Crumb and Pinhole test are G4 (dispersive) to G1 and D1, D2, ND4 to ND1; ranging from Dispersive o Non-dispersive.

Does the likelihood of internal erosion vary with the time of year?

It sis suggested that the vulnerability of dams to internal erosion will vary seasonally. This is partly because the permeability of a soil varies due to variation of viscosity of water with temperature, water at 20°C having 56% of the viscosity at 0°C. Secondly in summer there is the risk of desiccation of cores which are not protected by a capillary break, which can lead to surface cracks along which concentrated leakage can occur. Figure 7 of Brown (2007) examines the seasonality of a limited number of internal erosion incidents, this data suggesting the frequency of incidents in the summer is up to three times higher than other seasons.

STATUTORY OBLIGATIONS

Currently most Inspection Reports under Section 10 of the Reservoirs Act 1975 do not consider the adequacy of the embankment, in the same way that the adequacy of the spillway to pass the design flood is considered. This is partly because Statutory Instrument 1986 No 468 which sets out the minimum contents of a Section 10 Report does not explicitly require comment on the adequacy of embankment, whereas it does require comment on the following

v – adequacy of the waste weir or overflow
vii – adequacy of the margin between dam level and overflow level
viii – efficiency of the scour pipe or discharge culvert or other means of lowering the ……reservoir

It is of relevant to note that Form E of the 1930 Act required "*...particular reference to (b) leakages, settlement or movement of reservoir banks or walls, movement of surrounding land which might affect stability of* *reservoir"* This is downgraded in the 1975 Act into *"recommendations involving ...the installation of instruments or gauges for measuring leakages, deflections, settlement, uplift, pore pressures or similar matters"*

CONCLUSIONS

The understanding of internal erosion has progressed significantly in recent years following research work in Australia and Europe, with key developments published in Fell and Fry (2007) and the Interim Report of the European Working Group in Freising (2007). This paper has highlighted some of the key issues.

These developments have significant implications for UK dam engineers, in that it is practicable to carry out a basic assessment of vulnerability to, and risk of, internal erosion, It is therefore recommended that the contents of Section 10 reports should include an explicit statement of the adequacy of the dam embankment or other structure retaining the water, similar to that on adequacy of the spillway and outlet. It is recommended that

a) Legislation is changed to add to SI 1986 No 468 Schedule 2 "(ix) a statement as to the adequacy of the dam and its foundation, in terms of ability to retain the reservoir"
b) Guidance on Section 21(5) of the Reservoirs Act should include that the Undertaker should provide "information on the material properties of the dam and its foundation (e.g. geotechnical for earth embankments, concrete and rock for concrete dams), where available"

The level of detail should vary with the consequences of failure, following a risk based approach, and could start with a screening level quantitative assessment of risk. Where more detailed assessment is warranted this would require increased data, but as a minimum owners could carry out a portfolio wide desk study of the information that is available on all of their dams, including construction drawings (although these may be unreliable), site investigation and published geotechnical data on geological deposits in the area which are likely to have been used in dam construction. For some very high consequence dams it may then be appropriate to carry out further site investigation to establish, or confirm, dam construction details, including the filtering ability of the downstream shoulder and erodibility of the core.

REFERENCES

Atkinson JH, Charles JA, Mhach HK (1990) *Examination of erosion resistance of clays in embankment dams.* QJEG 23(2). Pp103-108

Bligh, W.G.(1910) *Practical design of irrigation works,* Second edition, Constable, Edinburgh.

Bridle, R (2008) *Assessing the vulnerability of a typical British embankment dam to internal erosion.* BDS Conference, Warwick.

Brown AJ (2007) *Issues in the management of internal erosion in UK embankment dams.* Paper at Aussois 2005 Published by Taylor & Francis

Brown AJ, Carter I (2004) - European Symposium on Internal Erosion. *Dams and Reservoirs,* Vol. 14 No 2, September

Brown, A.J. and Gosden, J.D., 2004, *Interim Guide to Quantitative Risk Assessment for UK reservoirs.* Thomas Telford.

BSI (1990) BS1377-5 *Methods of test for civil engineering purposes. Part 5* Compressibility, permeability and durability tests.

Charles J A (2002) *Internal erosion in European embankment dams.* BDS Conference, pp378-393

Cyganiewicz JM, Engemoen WO and Redlinger CG (2005), Bureau of Reclamation *Experience with Evaluating Internal Erosion of Embankment Dams*Paper at Aussois 2005 Pub. Taylor & Francis

Fell R& Fry JJ, (2007), *Internal Erosion of Dams and Their Foundations*: Paper at Aussois. Published by Taylor & Francis

Head KH (1982) *Manual of soil laboratory testing.* Volume 2.Pentech Press.

HSE (2001) *Reducing risks, protecting people.* HSE's decision making process. On website www.hse.gov.uk/

ICOLD (1994) *Embankment dams – Granular filters and drains – Review and recommendations.* Bulletin No 95. 256pp

Kenny TC and Lau D (1985) *Internal stability of granular filters.* Can Geotech J 22, 215-225. Discussion. Can Geotech J,1986, 23, 414-423

Lane EW (1935) *Security from under-seepage masonry dams on earth foundations,* Proc. ASCE [republished with discussions in 1935 in Transactions ACSE, Paper no. 1919, Volume 100, pp. 1236-1351].

Moffat AIB (2002) *The Characteristics of UK Puddle Clays - a Review.* British Dam Society Conference

Perzimaier S, Muckenthaler P, Koelewijn AR (2007) *Hydraulic criteria for internal erosion in cohesionless soil.* Paper No 4 in Interim Report of European Working Group on Internal Erosion. Freising, Germany

Sherard (1976) *Identification and nature of dispersive soils.* J ASCE 102(GT4) April

Sherard (1986) *Hydraulic fracturing in embankment dams.* ASCE Jour Geotech Eng. 112(10) Oct pp905-927

Wan CF and Fell R (2002) *Investigation of internal erosion and piping of soils in embankment dams by the slot erosion test and the hole erosion tests.* UNICIV Report R-412. July. ISBN 85841 379 5

Wan CF and Fell R (2004) *Investigation of rate of erosion of soils in embankment dams* ASCE Geotech April 373-380

Weijers JBA and Sellmeijer JB (1993) *A new model to deal with the piping mechanism.* In Brauns, Heibaum and Schuler (Edt) Filters in Geotechnical and Hydraulic Engineering, Rotterdam, Balkema, 345-355

Defra research into Internal Erosion

A. J. BROWN, Jacobs, Leatherhead, UK
J. D. GOSDEN, Jacobs, Croydon, UK

SYNOPSIS. This paper summarises the background to, and the key findings and conclusions from, two Defra research projects on internal erosion. It then discuses the practical implications for dam owners and panel engineers in terms of monitoring and surveillance. The paper concludes with a discussion on current uncertainties on rates of progression of internal erosion, and what future work would be useful to reduce these.

INTRODUCTION

Internal erosion occurs when the soil particles within an embankment dam or its foundation are carried downstream by seepage flow. Available data on reservoir safety incidents in the United Kingdom in the period 1975-2000 are summarised in Table 1. It can be seen that over half of the incidents where emergency drawdown is considered necessary to avert failure have been a result of actual or anticipated internal erosion. Experience internationally is similar, with ICOLD Bulletin No 99 (ICOLD, 1995) noting that 44% of dam failures internationally are due to internal erosion.

Despite the, on average, two emergency drawdown incidents a year due to internal erosion in UK there have been no failures of dams with loss of life due to internal erosion since the Reservoirs (Safety Provisions) Act was implemented in 1930. It is suggested that a significant contributor to this absence of failures includes the effective surveillance and intervention regime, which has provided time to detect developing problems and lower the reservoir before the problem developed to failure. However, the continued pressure for reduction in costs, and thus reducing frequency of surveillance, together with the increasing average age of UK dams, is likely to lead to increasing risk of dam failure from internal erosion.

Following a serious internal erosion incident in January 2002 (Gardiner et al, 2004) a Defra research contract into early detection of internal erosion was awarded to KBR in 2002. The first phase, a feasibility study, was completed in 2004, with the research report available on the Defra website and reported in Brown and Gosden (2004). Funding difficulties meant that the second stage of the contract, measurement of erodibility of UK soils, was cancelled. The contract was therefore completed (after novation to Jacobs) by production of draft guidance on early detection of internal erosion, and a study into the practicability of real time monitoring. This paper summarises the key points from these projects, as an update on the 2004 paper in Dams and Reservoirs (Brown and Gosden, 2004).

Table 1: Data on internal erosion incidents in the United Kingdom in the period 1975-2000

Level of incident	2: Emergency drawdown	3: Precautionary drawdown
Average number of incidents per year (Brown and Gosden, 2004)	3	10
% of UK incidents due to varying causes (Table 6 of Brown and Tedd, 2003)		
• internal erosion	60%	63%
• slope instability	23%	9%
Inferred annual probability of an incident due to internal erosion, for 2500 UK dams	1 in 500 (0.2%)	1 in 160 (0.6%)
Note: The annual probability of failure at less than 1 in 50,000 is much lower because most incidents do not progress to failure		

FRAMEWORK FOR CONSIDERING INTERNAL EROSION

The framework for assessing the likelihood of each of these types of internal erosion is described in a companion paper on the European Working Group on Internal Erosion (Brown and Bridle, 2008) and is not repeated here. This paper focuses instead on the rate of development of internal erosion, and means to detect and manage this.

RATE OF INTERNAL EROSION - EXPERIENCE

Key elements of historic internal erosion incidents which have been recorded in published papers are summarised in Tables 2 and 3. It can be seen that although incidents during the first few years following construction are generally rapid, incidents at dams more than five years old generally develop much more slowly.

A questionnaire to dam owners on the likely time to failure, if no action was taken, was presented as Figure 3 in the 2004 paper and confirmed the wide variation in time to failure, but also that the rate of development in UK dams is generally modest within the body of the dam but much faster when associated with internal erosion along the contact with a structure.

Table 2: Rate of development of internal erosion incidents at embankment dams

Dam, Date of Incident	Available information on rate of development, and type of internal erosion (Reference)
During wear in period	
Warmwithins, UK June 1970	The dam is a 10m high puddle clay core dam built in the 1860s. It failed over a period of 12 hours, along a new 1.5m diameter tunnel lined with concrete segments installed about 1965. (Wickham, 1992)
Balderhead, 1970	On first impounding the first sinkhole was above the filters on the downstream side of the core, followed six weeks later by large holes on the upstream side of the core. The underdrain water had also become turbid. (Kennard, 1972)
Teton, USA June 1976	The dam failed over a period of 2.5 hours. The dam core was built of wind deposited non-plastic to slightly plastic clayey silts, and founded on moderately to intensely jointed volcanic rocks. The time stated is from the first muddy flow being seen to the first whirlpool being seen in the reservoir. Concentrated leak? (Report into failure of Teton dam)
In service	
Upper Rivington Jan 2002	New concentrated leak into culvert. Emergency drawdown such that the reservoir ceased to overflow 48 hours after being detected, and was 7m below top water level six days after ceasing to overflow. (Gardiner et al, 2004)
Scandinavian dams	35 cases of sinkholes and sudden leakage due to suffosion, generally after many years in service. These usually seal themselves after a short time. Settlements, sinkholes and leakage have often occurred at the junction between an embankment dam and a concrete structure. (Nilsson, 2007)

Table 3: Examples of slow ongoing internal erosion in UK dams

Dam	Inferred erosion rate (Reference)
Brent	1.0 litre/ year measured in V notch chamber (Tedd et al, 1998)
Anonymous	1.0 kg/year measured in V notch chamber (private communication from dam owner, 2007)
Lower Slade	3ft diameter swallow hole exposed when upstream pitching lifted. 60 tons of grout injected into open holes into 70 year old dam, where repairs had been carried out 15 years before. Say 1 ton/ year (Kennard, 1972)
Lluest Wen	A horse fell into a hole near the valve tower at Christmas 1969. Emergency declared with evacuation of old and infirm downstream, 50 ton of clay/ cement grout injected into watertight element, which had been previously treated in 1912. Approx 1 ton/year. (Little, 1977)

RATE OF INTERNAL EROSION – PREDICTION AND RISK ASSESSMENT

It is suggested that the level of detail which is appropriate to apply to the assessment of vulnerability to rapid internal erosion should be on a risk based approach. This considers whether the Undertaker has reduced risk to as low as reasonably practicable (ALARP), namely whether the measures taken have been proportionate to the reduction in risk that could be achieved. The ALARP decision making process is defined in paragraph 124 and Appendix 3 of "Reducing risks, protecting people" (HSE, 2001) with further guidance given at www.hse.gov.uk/risk/theory/alarp.htm. The Interim Guide to Quantitative Risk Assessment (Brown & Gosden, 2004) provides a screening level of risk assessment, which can be a useful first step.

The process to assess the vulnerability of a dam to internal erosion is described in Brown & Bridle (2008), and requires data on the geometric, geological, geotechnical and hydraulic properties of a dam. The vulnerability to rapid failure is an extension of this, but also considers the likely rate of failure. This is likely to vary depending on the erodibility of the soil and type of internal erosion, the latter as indicated in Table 4.

Although in theory analytical predictions of the rate of development of internal erosion at concentrated leaks are now possible, currently this analysis is not common in UK. Nevertheless assessments carried out as part of the research contract comprised a sensitivity study of flow and erosion in a 1m high 3m long crack under 10m head which showed that

a) the rate of erosion varied between 0.1 and 1000kg/ day, depending on the erodibility of the soil (Figure 6 of Brown and Gosden (2004); erodibility as defined by Wan and Fell, 2004)
b) Turbidity of less than about 5FTU is not visible to the human eye; so for this combination of crack width and reservoir head, material eroded from a soil with a value of Erosion Rate Index of between 4 and 6 would not be visually detectable

In terms of risk assessment the rate of development can be categorised in terms of the ability to manage the incident and prevent dam failure, and a suggested categorisation is given in Table 5, together with the implications for bottom outlet capacity.

Table 4: Types of internal erosion (after European Working Group)

Type	Definition	Likely time to failure; remarks
Concentrated erosion	In soils which are capable of sustaining an open crack. Erosion occurs along the sides of the crack where the shear stress (velocity) exceeds the critical value. NB at low flows there may be leakage with no erosion.	The rate of erosion is dependant on the erosion resistance of the clay core, and may be limited by the permeability of both shoulders. Where cracks exist in the dam crest (e.g. desiccation, differential settlement) then the critical failure mode may be concentrated erosion during flood conditions
Backward Erosion (Piping)	Erosion starts at the exit point; a continuous passage is developed by backward erosion when the seepage gradient exceeds the "flotation gradient" of the soil.	Can be fast with little warning. Failure is often associated with first filling, or an increase in seepage gradient (for example under flood conditions)
Contact Erosion	Erosion at the horizontal boundary of a fine soil overlying a coarse soil, where the fine soil is washed into the coarse soil due to horizontal flow	Little information. Likely to be similar to concentrated erosion, but accelerated where along the interface between a structure and the embankment
Suffosion	Mass erosion in soils which are internally unstable. Fines transported by seepage flow between the larger sizes of soil	Normally leads to an increasing quantity of seepage as fines erode, but is unlikely to lead to rapid failure

Table 5: Suggested categorisation of speed of failure due to internal erosion

Speed of failure	Time to failure	Relation to Surveillance frequency	Value of bottom outlet for emergency drawdown
Extremely fast	< 1 day	Unlikely to be spotted	Irrelevant
Fast	1 to 7 days	Where there are frequent visits >50% chance of being spotted while there is still time to avert failure	Dependent on the frequency of surveillance visits, and thus the residual time available for intervention
Medium	8 to 90 days	Good chance of being spotted, provided frequent visits are made	Likely to be of most value
Slow	> 90 days	Almost certain to be spotted provided there are routine surveillance visits.	Magnitude of ongoing inflows likely to be the largest influence on the size of the bottom outlet

SURVEILLANCE AND MONITORING

Once an understanding of the risk from internal erosion has been obtained as described above, the type and frequency of the surveillance and monitoring regime can be determined. It would normally include visual inspection as a minimum, and monitoring on higher consequence dams. Current international practice for the frequency of surveillance visits is generally based on the consequence class of the dam, for example on high hazard dams typically being weekly although ANCOLD guidelines (Table 5.2 of 2003) suggest it should be daily. For low consequence dams a typical frequency is monthly.

It is suggested that it would be more rational if the frequency of surveillance were risk based, in that as well as consequence class it also took account of the probability of (vulnerability to) rapid internal erosion. A suggested risk matrix to determine a surveillance and monitoring regime is given in Table 6, whilst a possible starting point for defining the surveillance frequency of dams in service is given in Table 7. The latter is developed from ANCOLD Guidelines on Dam Safety Management (2003). Both are suggested starting points for application and adjustment on a dam specific basis, following the principle that the cost should be proportionate to the benefits obtained.

Table 6: Indicative risk matrix for the surveillance and monitoring regime at a dam

Vulnerability to internal erosion [1], as output from risk analysis	Consequence Class of dam (as defined in Interim Guide to QRA, 2004)				
	A1	A2	B	C	D
High	α	β	β	γ	δ
Medium	β	β	γ	δ	ε
Low	γ	γ	δ	ε	ε

Notes

1. Ranking on the vertical axis could be on the basis of one of the following
   - The tolerability of risk, comprising the Broadly acceptable, ALARP and Unacceptable zones on the FN chart (Sheet 11.3 of Interim Guide to QRA)
   - Likelihood of failure, as a qualitative assessment
   - Speed of failure, as defined in Table 5, with Fast and extremely fast considered together

Table 7: Indicative surveillance frequency for dams in service

Regime (Table 6)	α	β	γ	δ	ε
Routine visual surveillance					
Exterior; including exterior of culverts/ shafts (and interior where no confined space)	Daily to 48 hours	48 hours to twice weekly	Twice weekly to fortnightly	0.5 to 2 months	Two to six months
Interior of culverts/ shafts, where confined space[2]	Weekly to monthly	Monthly to 3-monthly	3-Monthly to annual	Five yearly	Ten yearly
Number of visits a year by Supervising Engineer[3]	4	3	2	1	1

Notes

1. The timing should be calendar time, and thus occur at weekends and bank holidays where these fall in the routine cycle.
2. The timing can be reduced by say up to two classes, where any progressive internal erosion would be visible by inspection at the downstream end
3. Where the dam owner does not have suitable trained and experienced employees then the frequency of Supervising Engineer visits would normally be increased, and they should include an element of training of the dam owner and his staff

REAL TIME MONITORING

Introduction

One of the Defra research contracts was a review of current practice in real time monitoring, both in relation to dams and other industries, leading to production of a Guidance Note. This section summarizes some of the key findings from this review.

When could this be appropriate for UK dams?

Real time monitoring may be appropriate at high consequence dams where it is considered there is a possibility of rapid development of internal erosion, if it initiated. It is emphasised that this should not be a substitute for visual inspection, as visual inspection covers the whole of the dam structure, whereas monitoring only covers the element or part of the structure where the sensors are installed.

Current practice

ICOLD Bulletin 118 (2000) provides case histories and guidance on automated monitoring at dams internationally. In UK real time monitoring of sensors is now standard practice in the water industry as part of the de-manning to reduce operational costs. It is also sometimes necessary to demonstrate compliance with discharge consents or other statutory requirements.

In the UK, British Waterways undertakes real time monitoring of key assets for operational purposes, including some dams, using a system developed in-house. Parameters monitored relate mainly to operational issues of water levels, weir operation, flows at gauging weirs, sluice operations and gate positions. Alarms are incorporated where appropriate, with relevant personnel being notified directly by mobile telephone or pager if a trigger level is reached. A 'reach out' facility, in which remote operation can be undertaken, for example of drawoff sluices, is also available at a limited number of sites. At a few sites the sluice operation is automatic based on reservoir water levels, with various fail-safes included in the system.

The real time monitoring system is internet based and has been in operation since 2004. All software has been developed in-house. In general, remote readings are recorded at 30 second intervals, being averaged to provide a reading over 5-minute intervals. Where mains power is not available, the power source is commonly solar cells and batteries. In order to conserve batteries, stored information is only transmitted to the central server every 24 hours, although the stations can be polled at any time in-between if required. 90% of the information is relayed by mobile telephone technology, with the remaining 10% being transmitted by land line. Data from the last ten years, including information obtained both manually and remotely, can be retrieved upon request. There are no limits to the number of persons who can access the database at any one time. Problems associated with the system include the susceptibility to vandal damage and the rapid rate at which current technology becomes obsolete.

In relation to reservoir safety water levels are recorded automatically at some reservoirs, although water levels noted manually at the remaining reservoirs are also entered into the database. Manually read v-notch, piezometer and flows measured using jugs will soon be entered into the same database. The information can readily be extracted from the database live through the internet, and presented in graphical format which is considered better than Excel format. In summer 2009 SCADA monitored V-notches will be monitored at a remote reservoir with a history of seepage above a small town; the V-notches will have trigger flows set, which will raise an alarm.

A number of UK water companies have each installed real time monitoring of seepage at V notch weirs at one or two dams, generally because of specific issues at those dams. In Canada it is now routine practice by BC Hydro to have real time monitoring of flows and turbidity at the seepage weirs at all of their dams.

Parameters

The parameters which are most useful to provide early warning of internal erosion are seepage flows and turbidity. It is noted that suspended fines in seepage water cannot be seen visually where Formazine Turbidity Unit (FTU) is less than 5, and thus would only be detected by quantitative monitoring. For clay fines an approximate conversion is 2.5mg/l per FTU, such that seepage of 5 litres/ min with a FTU of 5 could correspond to solids of 90g/day

Procurement and maintenance

The practicality, cost and reliability of real time monitoring is significantly dependent on resources available within the Undertaker's organisation to install and maintain any system. Table 8 contrasts current practice at a water treatment works and at British Waterways. As a budget figure the cost of real time monitoring of a simple V notch chamber is likely to be of the order of installation cost of £20k for the civils works and monitoring equipment; an annual maintenance cost of £1k/ year; and £5k every ten years for replacement of sensors and other monitoring equipment. This cost is thus likely to only be proportionate at high consequence dams, where it provides a significant reduction in the probability of failure.

Table 8 Range in variation in strategy for real time equipment

<table>
<tr><th>Issue</th><th>Water Company – Real time monitoring installed at water treatment works</th><th>British Waterways</th></tr>
<tr><td>Equipment installation</td><td>In-house staff</td><td>External contractor</td></tr>
<tr><td>Maintenance cleaning: frequency and by whom</td><td>Twice a week by plant engineers who visit each WTW regularly</td><td rowspan="3">All carried out by external contractor six monthly, and under call-out where problem identified</td></tr>
<tr><td>ICA check i.e. check that it still works correctly</td><td>By instrument technician (the company employs 50 specialist ICA staff)</td></tr>
<tr><td>Recalibration</td><td>Periodically; often in the laboratory at the WTW</td></tr>
<tr><td>Instrument replacement</td><td>On failure, or when technology superseded</td><td>Ten years</td></tr>
<tr><td>Replacement of batteries in power pack</td><td>Not relevant</td><td>Routinely as part of six monthly check</td></tr>
<tr><td>Alarm sent to</td><td rowspan="2">24 hour control centre, manned by in-house staff</td><td>Mobile phone of Duty Engineer</td></tr>
<tr><td>24 hour control centre for receipt of alarms from public</td><td>Contracted out to First Response (West Midlands Ambulance Service)</td></tr>
</table>

Use of data

A key element of assessing the practicality and value of real time monitoring is the use and archiving of the data collected. This will depend on the software used and management of alarms. For water companies and other major owners with control rooms which are manned on a 24/7 basis real time monitoring should be a viable option, although some development work would be required to integrate it with other existing real time systems. For owners who do not have this resource then the system could be developed to provide a SMS text or other form of automatic alarms, but this is likely to be less reliable.

FUTURE RESEARCH

The research projects identified that there is still much work to be done to bring our understanding of internal erosion to the same level as our understanding of other threats to dam safety, such as hydrological assessment. Issues identified for future research are summarised in Table 9.

Table 9: Priorities for future research relating to internal erosion

Description
Laboratory testing of threshold, and rate of internal erosion, of sides of pathway through core, similar to that in Australia. This would be an extension of the pinhole test (BS1377:part 5: 1990 Section 6.2); including changes such that a 6mm hole is used instead of a 1mm hole, and the rate of erosion is measured
Collection and analysis of data on incidents, specifically a) Lessons learnt reports from incidents b) Data collected as part of the database, to allow statistical analysis, on the range of geotechnical properties of UK dams relevant to internal erosion c) Data on the annual probabilities of an incident, and dam failure respectively, both inferred from incidents and estimates in published papers
Materials present in UK dams a) Desk study of existing site investigation reports on SI carried out in the past, and (relevant) published data on geological strata b) Field investigation with associated laboratory testing of core and shoulder to better understand the range of properties for typical UK dams; followed by identification and analysis of the credible failure mechanisms for the system of dam core and supporting shoulders c) Field investigation of construction details of pipes and culverts through dams, where the dams have been discontinued

CONCLUSIONS

The overall objective for the research project was to "provide a cost-effective approach to the early detection of progressive internal erosion in embankment dams". Although there is no simple single tool which can achieve this objective, the research project and other ongoing parallel research in Europe and elsewhere have led to an increased understanding of the vulnerability to, and potential rate of development of, internal erosion. This has highlighted

- The large variability in erodibility of materials used to form the watertight element of dams
- That internal erosion can occur at a slow rate for many years with no apparent surface effect, and with suspended particles in concentrations too low to be detected visually
- The rate of development is likely to vary significantly depending on the type of internal erosion
- The increased likelihood of internal erosion along the contact with structures, relative to internal erosion in the body of the dam

Site specific engineering assessment to inform an assessment of the vulnerability to internal erosion should include

a) The consequences of the failure of the dam, in terms of impact on people and property downstream, and thus what measures would be proportionate to reduce the likelihood of failure to as low as reasonably practicable
b) Determination of the erodibility of the material(s) used to form the watertight barrier
c) An understanding of the system response, as to how the dam shoulders or other elements may affect internal erosion of the watertight element

Suggestions have been made of normal surveillance type and frequency, following a risk based approach. Real time monitoring has been reviewed in relation to current practice, potential costs and issues relating to data and alarm management.

ACKNOWLEDGMENTS

We acknowledge the contribution to this project from British Waterways and a number of water companies who provided valuable assistance in making information and staff available, and also to the European Working Group on Internal Erosion, chaired by Jean Jacques Fry of EDF. We also gratefully acknowledge the time spent by panel engineers and other dam professionals in completing questionnaires at the start of the project.

The project benefited from a Steering Group to advise Defra which comprised at Stage 1, the draft strategy for management of internal erosion, Jim Millmore (chair), David Dutton, Nick Reilly and Andrew Robertshaw.

The work described in this paper was carried out as a research contract for Defra. However, the opinions expressed are solely those of the authors and do not necessarily reflect those of Defra or other who contributed to the project.

REFERENCES

ANCOLD (2003) *ANCOLD Guidelines on Dam Safety Management*

Brown AJ and Gosden JD (2004a) The early Detection of Internal Erosion. *Dams and Reservoirs* 14(1) pp13-20

Brown, A.J. and Gosden, J.D (2004b) *Interim Guide to Quantitative Risk Assessment for UK Reservoirs.* Thomas Telford.

Brown and Bridle (2008) *Progress in assessing Internal Erosion.* BDS Conference

Gardiner K, Hughes, Brown AJ (2004) Lessons from an incident at Upper Rivington reservoir. *Dams and Reservoirs.* September. Pp35-37

HSE (2001) *Reducing risks, protecting people. HSE's decision making process.* On website www.hse.gov.uk/

ICOLD (2000) Bulletin 118 *Automatic dam monitoring systems*

KBR (2004) *Reservoir safety research contract for Defra. Task B : Early detection of internal erosion.* Feasibility Report. On Defra website.

Kennard (1972) Examples of the internal condition of some old earth dams. *Journal of the Inst. of Water Engineers,* v 26, n 3, May, 1972, p 135-154

Little (1977) Investigating old dams. *Quarterly Journal Eng. Geol.* Vol 10. p271-280

Nilsson (2007) *Filters and internal erosion in Swedish dams.* Paper from workshop at Aussois. Published by Taylor & Francis

Tedd P, Dutton DPM, Horton IR (1998) *Investigating internal erosion at Brent Dam.* Proc BDS Conf. pp71-78

Wan CF and Fell R (2004) *Investigation of rate of erosion of soils in embankment dams* ASCE Geotech April pp 373-380

Wickham (1992) Collapse of an earth embankment dam. *Dams and Reservoirs.* October 2(3) pp18-19

Possible indicators of internal erosion prone dams comprising broadly graded materials

H. RÖNNQVIST, Vattenfall Power Consultant AB, Stockholm, Sweden

SYNOPSIS. Embankment dams comprising broadly graded materials of glacial origin in the impervious core (base soil) and protective filter have in the past been identified as being prone to develop sinkholes more frequent than dams composed of materials of other origin. Sinkholes on the crest of a dam are many times an indicator of internal erosion. An internal erosion process can initiate and continue to develop for many reasons, but mainly due to root causes coming from core/filter properties and interaction, possibly affected by the dam design and/or construction related reasons. More than 70 existing embankment dams are reviewed in this paper, predominately Swedish glacial moraine core dams, with objective to see if there are indicators in dams with performance history of internal erosion. The investigation shows that a coarsely graded filter, grading instability of the core and filter, and high susceptibility for filter segregation makes an over-represented combination for dams with performance history of internal erosion. Internal erosion of dams is an important concern from a dam safety and dam engineering perspective.

INTRODUCTION

It was early on recognized that dams constructed from broadly graded filling materials of glacial origin, exhibited signs of internal erosion such as sinkholes to a larger extent than dams comprising other types of materials. Numerous internal erosion incidents in the 1960s and 1970s with piping occurred in dams surprisingly similar in design (Sherard (1979), Vestad (1976), Lafleur *et al.* (1989) and Ripley (1986)); i.e. thin core rockfill dams with core materials of broadly graded materials (often glacial) and widely graded filters. This is confirmed in Foster *et al.* (2000) by statistical analysis that the great majority of cases of piping accidents to central core earth and rockfill dams involves dams with broadly graded core materials of glacial origin with piping into coarse segregated filters.

In a survey carried out on Swedish moraine (till) core dams by Nilsson *et al.* (1999) a total of 84 rock- and earthfill embankment, with objective to investigate the ageing and deterioration processes, showed that a considerable amount of the dams (20 %) reported to have experienced some kind of internal erosion either in the dam's body or its foundation. This is also discussed in Norstedt & Nilsson (1997).

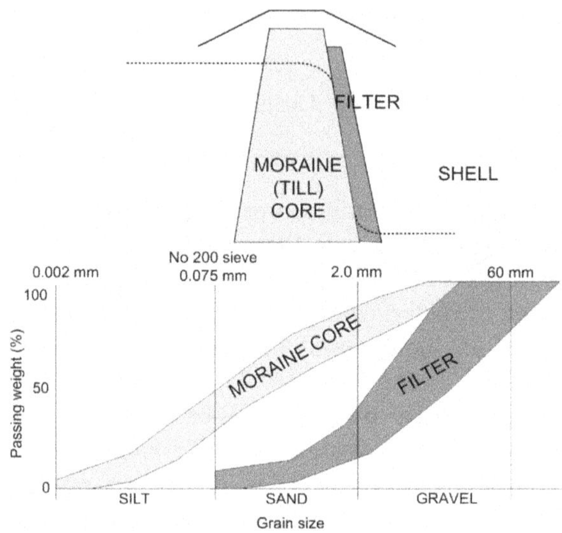

Figure 1. Typical cross-section of moraine core dams and typical sieve curves of widely graded material of glacial origin.

Sinkholes can often be attributed to internal erosion in the dam body or its foundation and can serve as a late indicator of internal erosion if occurring on the dam crest or the upstream face. Other signs may be dirty water (turbidity) and unexpected increases in seepage downstream. The process of internal erosion is usually described by the initiation, continuation and progression phases (Foster & Fell (2001)). Initiated internal erosion, resulting in the loss of impervious core function, can in many cases be traced back to a "root cause" for internal erosion, and the subsequent continuation of the internal erosion is mainly dependent on the protective filter function (Nilsson & Rönnqvist (2004)). Root causes being e.g. internal instability of the core, coarsely graded filters creating an unfiltered interface to the core the filter is intended to protect, or different structural or construction related reasons (e.g. arching effects, low effective stresses or unintentional seepage paths created due to ways of construction or design).

Existing moraine (till) core dams are reviewed in this paper, predominately Swedish moraine core dams, with objective to statistically investigate whether embankment dams that have a performance history of internal erosion have features/properties that are over-represented, properties such as coarsely graded filters for instance. Such common denominators can be used as indicators and provide an idea of which dams that are prone to develop internal erosion. A dam with only one potential indicator may not necessarily affect the potential for internal erosion, but several combined will probably influence a dam's susceptibility for internal erosion and thus give a heads-up during dam safety assessments.

DATA SET OF EXISTING DAMS IN THE STUDY

To date there are 72 glacial moraine core embankment dams comprised in this study (53 % earthfill dams). In the data set of 72 dams there are 13 dams with confirmed internal erosion (whereof rockfill dams are somewhat over-represented with 62 %). The composition and structure of the dams in the study varies slightly, but common features are the core of glacial moraine and a sand-gravel filter of some sort (see examples of gradation curves in Fig. 1). All of the dam's impervious moraine cores are cohesionless with generally low to nil clay content and classifies as Sherard & Dunnigan (1989) soil group 2 base soils (40-85 % finer than No. 200 sieve (0.075 mm) determined from base soils re-graded on the 4.75 mm sieve).

The dam data in this study has been collected by the author, as part of an ongoing review of Swedish moraine core dams, which was started in Rönnqvist (2006) and further developed in Rönnqvist (2007). Data from previous inventories of Swedish dams by Nilsson (1995) and Nilsson *et al.* (1999) has also been used where applicable. The data set is predominately comprised of Swedish dams, except for two dams; a Norwegian glacial moraine core dam with rockfill shoulders that experienced internal erosion (dam data from Vestad (1976)) and a Pennsylvanian dam (US) with materials of glacial origin that suffered extensive sinkhole development at first filling of the reservoir (dam data from Talbot (1991)).

The typical Swedish embankment dam is a central and vertically placed core composed of a broadly graded glacial moraine (till), protected by filters of sand and gravel (in many cases also widely graded) and shell material comprising rockfill or earthfill, see examples of gradation curves in Fig. 1. The glacial moraine (till) in Sweden, and northern Scandinavia, was formed sub-glacially under conditions of hard crystalline bedrock giving it negligible clay content and up to around 40 % fines content (<0.075 mm). Glacial soils depend to an extent of the mother-rock from which it is formed, e.g. British tills with relatively high clay content (up to 25 %) and also Canadian tills in the Western Prairie provinces with high plasticity fines (Milligan (2003), this due to the formation in softer sedimentary bedrock conditions. Tills in eastern Canada are formed in conditions of igneous and metamorphic bedrock giving the till fines with non-plastic characteristics

(Milligan (2003)), which are conditions resembling to that of Scandinavian moraines (tills).

The dams have been categorized in regards of occurrence of internal erosion, in three proposed categories; namely 1-dams, 2-dams and 3-dams, with qualitative definition as described below:

- "1-dams" Embankment dams with *Probable occurrence of internal erosion* – The dam has had incidents with visible sinkholes and settlements on the surface of the dam and leakage with dirty waters (eroded material in suspension) or other observed signs of internal erosion. The presence of internal erosion is documented (e.g. by test pits or drill cores).
- "2-dams" Embankment dams where *Observations may be signs of internal erosion* – Sinkholes and settlements have occurred on the dam, but no leakage with eroded material has been noted.
- "3-dams" Embankment dams with *No observations to indicate internal erosion* – No sinkholes or settlements have occurred on the dam and no leakage with eroded material has been sighted.

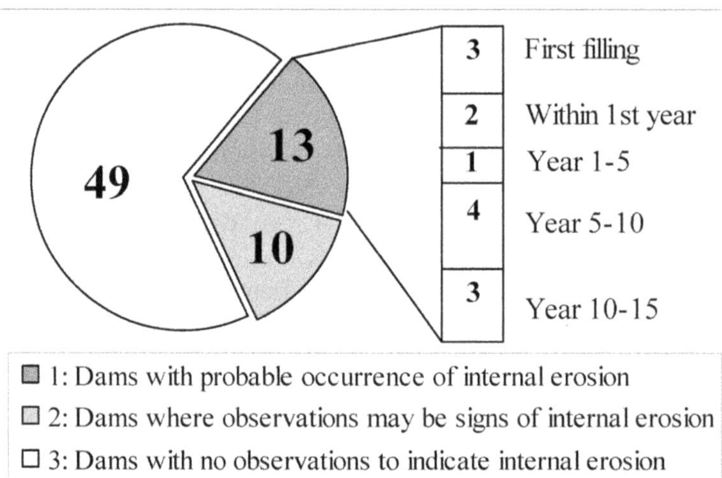

Figure 2. Number of dams and elapsed time until first internal erosion sign.

Category "1-dams" has experienced typical signs of deterioration related to internal erosion and the prerequisite on category "1-dams" is a "confirmed and documented" occurrence of surfaced internal erosion in the dam in question. Category "2-dams" are "borderline cases" that have exhibit signs that can be related to internal erosion but it is not clear-cut, and until further notice the signs of the dam in question is considered not internal erosion related. Category "3-dams" have exhibited no signs of internal erosion, and are regarded as fully functional embankment dams.

There are 13 dams that qualify as "1-dams" (with probable occurrence of internal erosion), see Fig. 2, out of these there are 3 embankment dams where the internal erosion (in the shape of sinkholes and turbidity) surfaced almost immediately after the first raising of the reservoir. But as can be interpreted from Fig. 2 the majority of the "1-dams" experienced signs of internal erosion as long as 5-15 years after the dam's commissioning. Similar findings in Norway where internal erosion have surfaced not only during initial filling of the reservoir, but also first several years after (Höeg (2001)). This is somewhat contrary to the findings of Sherard (1979) that noted that sinkholes in dams with broadly graded materials usually appears in connection to the first filling of the reservoir.

FILTER GRADATION RELATED INDICATORS

The fundamentals of protective filter design is to design the filter's grain size curve so that the filter void openings are small enough to prevent and limit erosion from the base the filter is protecting. The generally accepted filter criteria nowadays in regards to filters for broadly graded base soils are based on the testing in the 1980s by the Soil Conservation Service (SCS) that amounted to the guidelines proposed in Sherard & Dunnigan (1989). A filter designed accordingly is usually expected to be "self-filtering", meaning that some base erosion needs to take place at the filter interface in order for gradually finer particles to be prevented from eroding into the filter. Prior to the SCS investigations the filter requirements were generally less strict and filter design criteria for broadly graded core materials were at the time of the bulk of hydropower dam build-out that occurred in the 1960s and 1970s (at least the Swedish peak construction period) somewhat uncertain. This resulted in many embankment dams constructed from broadly graded materials having filters too coarse to satisfy today's filter requirements.

Filter D15-distribution

The void opening size in filter is dependent on the filter's finer graded particles, and for design purposes the D15 (particle's size at 15 % passing weight) is usually used to describe this void size. Sherard *et al.* (1984a) showed that for granular soils the opening size is given by D15/9, confirmed in Foster (2001) by testing. The filter D15-distribution of in total 72 existing embankment dams is given in fig. 3. Filter D15 is plotted in relation to performance history of internal erosion (categories 1 to 3 where 1-dams being dams with probable occurrence of internal erosion). The "CE-boundary" in fig. 3 is the Foster & Fell (2001) continuing erosion boundary, i.e. DF15>9DB95, that when exceeded a filter is too coarse to allow the eroded core particles to seal the filter.

Dams with confirmed internal erosion (1-dams) generally have coarse filters, since more than 90 % of the 1-dams have filters with D15 coarser or equal to 1.4 mm. The filter D15 coarser or equal to 1.4 mm is one filter characteristic suggested in Rönnqvist (2007) for analysis of potential for internal erosion in moraine core dams. Almost 20 % of the dams apparently without internal erosion (3-dams) also have filter D15 coarser than 1.4 mm. Foster & Fell (2001) found that dams with poor filter performance generally have filters with an average D15 coarser than 1.0 mm as one of general characteristics of dams with poor filter performance. Filter coarseness by itself is not a clear-cut indicator of the potential for internal erosion but provides a pointer. None of the dams with internal erosion (in the data set) satisfy current filter criteria in relation to broadly graded base soils; Sherard & Dunnigan (1989) criterion filter D15 smaller or equal to 0.7 mm).

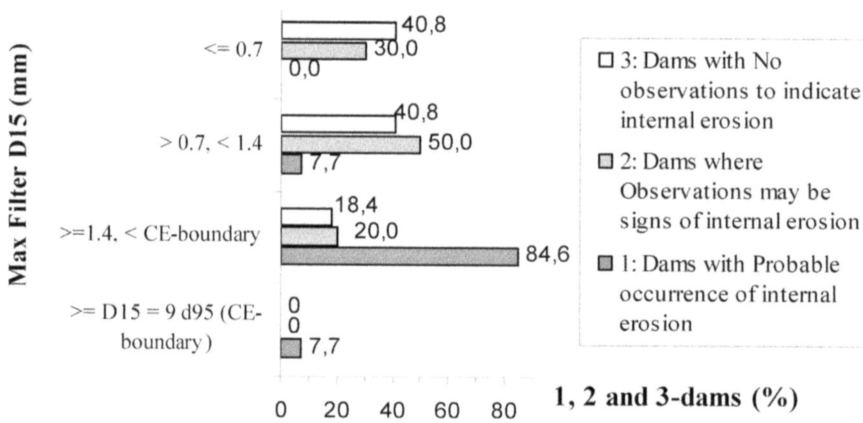

Figure 3. Filter D15-distribution in relation to performance history of internal erosion (dam categories 1-3).

Out of the 72 dams in the data set, 70 are Swedish embankment dams and approximately 68 % of these (regardless history of internal erosion) have filters that do not satisfy current filter criteria. But coarsely graded filters are not unique for Swedish dams, the situation is similar in Norway for instance; Höeg (2001) indicate that out of 122 Norwegian large rockfill dams with broadly graded moraine cores only few, if any, would satisfy today's strict filter criteria.

<u>Filter grading instability</u>

Grading stability can be assessed using the method by Kenney & Lau (1985, 1986). A soil having a smaller volume of fines in relation to volume of voids between coarser load-bearing particles can be suffusive/internally unstable, i.e. finer particles risk passing through the openings between the coarser particles by seepage flow. The boundary between internally stable and unstable material (Kenney & Lau (1985, 1986) shape curve ratio H/F = 1) is determined by the passing weight between particle size D and 4D. The interval of x4 being the size of the predominant constriction in a void network of a filter which is approximately equal to ¼ the size of the small particle making up the filter (Kenney *et al.* (1985)). A material is considered internally unstable if the Kenney & Lau (1985, 1986) H/F ratio is smaller than 1, meaning that the soil assessed as unstable has a grain size curve with inclination flatter than that of the "Fuller curve" (i.e. gradation curve for optimum density).

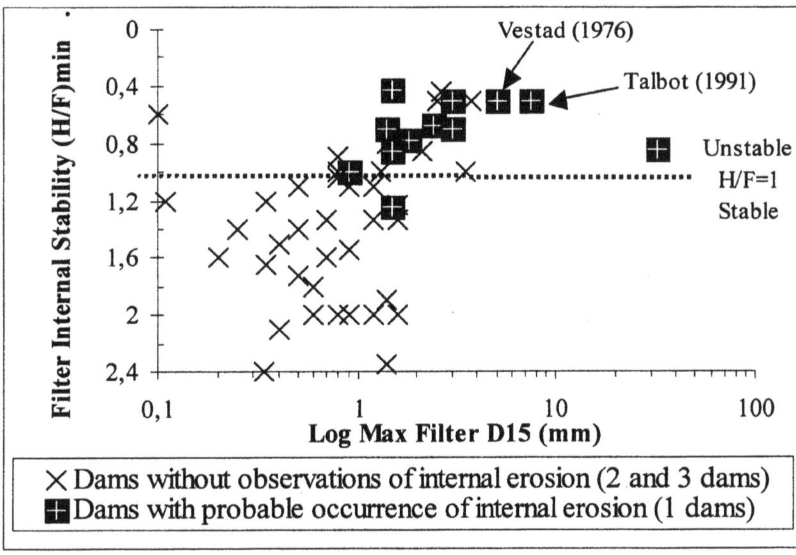

Figure 4. Filter D15 put against filter internal stability in relation to performance history of internal erosion (dam categories 1-3).

By cross-referencing filter instability (assessed according to the method by Kenney & Lau (1985, 1986)) and maximum filter D15 a plot as seen in fig. 4 is achieved. Internal stability against filter coarseness indicates that the coarser D15 of the filter, the more likely the filter is unstable. Furthermore, from Fig. 4, it appears that as the filter is getting coarser and passing the boundary into internal instability, the more likely the dam is a 1-dam (i.e. dam with probable occurrence of internal erosion). The data set in fig. 4 consist of 53 dams (the rest of the dams in fig. 2, 19 dams, have incomplete records). 90 % of dams with internal erosion (1-dams) have an internally unstable filter; 30 % of the dams without internal erosion (2 and 3-dams). The (H/F)min in fig. 4 is the minimum value of the ratio along the sieve curve within the Kenney & Lau (1985) evaluation range of 0-20 % passing weight (applies to widely graded material). In Kenney *et al.* (1985) it is furthermore suggested that for larger values of the coefficient of uniformity (Cu = d60/d10) of the filter the loss of base is increased. It is explained by saying the wider the gradation of the filter, the larger are the pores, and the larger is the amount of base that penetrates the filter before equilibrium.

In fig. 5 the filter Cu of the dams (set of 53 dams, the rest in fig. 2, 19 dams, have incomplete records) is put against the filter's D15 and no clear pattern can be seen regarding dams with internal erosion (1-dams). From fig. 5 it is clear that the majority of the filters are widely graded, with Cu generally exceeding 5-10.

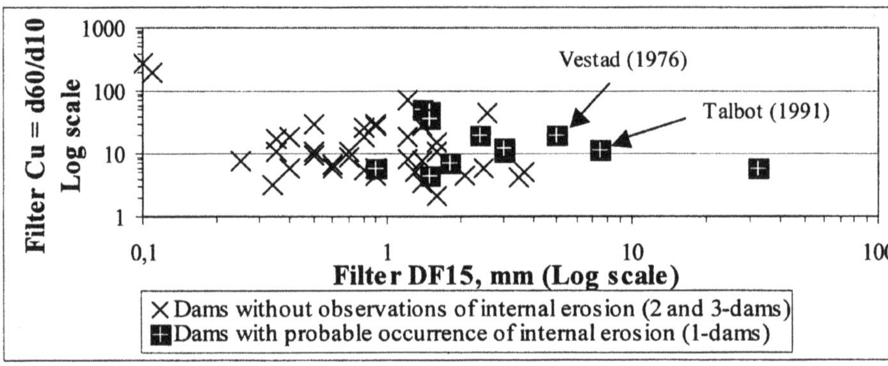

Figure 5. Filter D15 put against coefficient of uniformity (Cu) of the filter in relation to performance history of internal erosion (dam categories 1-3).

FILTER SEGREGATION RELATED INDICATORS

The significance of segregation in filters, especially concerning coarsely and widely graded filters, in relation to internal erosion is strongly emphasised in many references; e.g. Ripley (1986) and Sherard (1979). As pointed out by Milligan (2003) some segregation will probably occur when practicably placing most broadly graded materials, so in the processing, handling, placing, spreading or compaction of widely graded materials there is evidently a risk for segregation, and thus forming streaks or pockets of gravel-sized particles at the core/filter connection. The insufficient finer grained material in the voids can create grounds for an unfiltered interface to the impervious core. Noticeable filter (and core) segregation was observed while reconstructing the Swedish Juktan dam (Kjellberg *et al* (1985)), and also at the filterface of the Swedish Lövön dam (Ericsson&Jender (1998)).

In Ripley (1986), Sherard *et al.* (1984b) and Foster & Fell (2001) there is guidance on how to minimize the susceptibility for segregation in the practical handling of widely graded filter materials. These guidelines are plotted in fig. 6, i.e. Ripley (1986) suggested to limit the maximum particle size in the filter to 20 mm and at least 60 % finer than 4.75 mm in the filter, and Sherard *et al.* (1984b) suggested that the filter should be no coarser than 50 mm and at least 40 % finer than 4.75 mm. Foster & Fell (2001) found based on their research that if maximum particle size in filters are coarser than 75 mm it increases the likelihood for segregation. In fig. 6 the amount coarser than 4.75 mm in the filter is plotted against the filter's maximum particle (max. particle sieved to be correct).

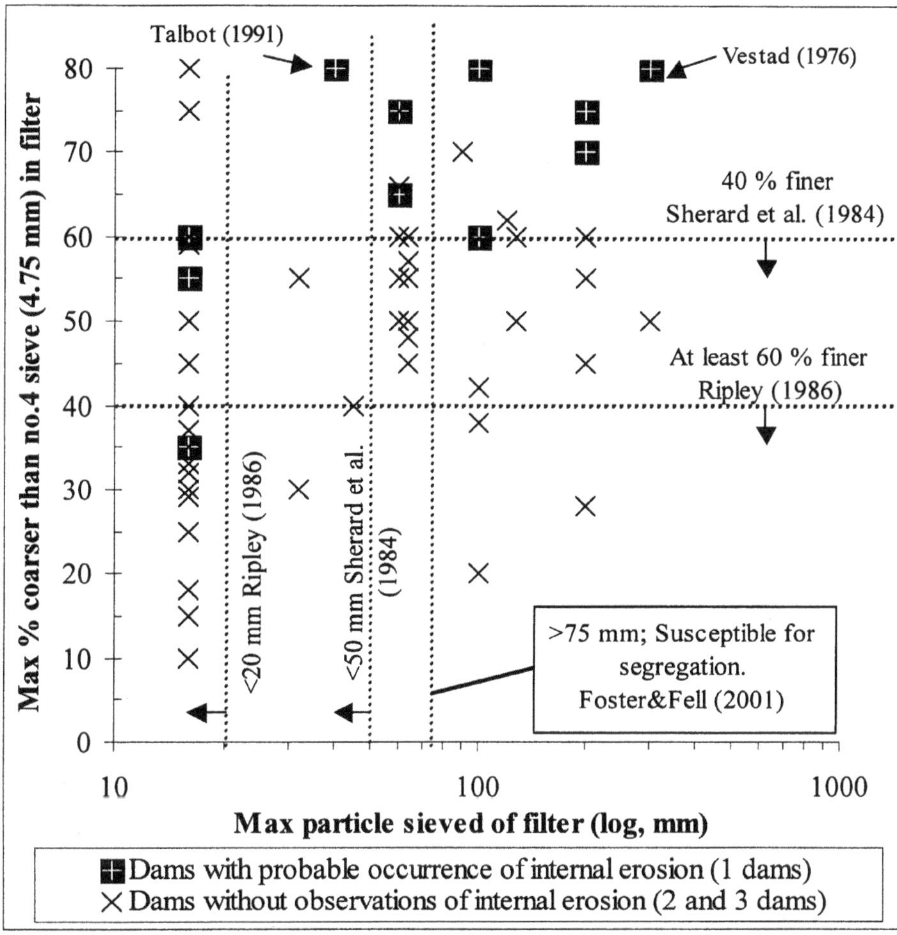

Figure 6. Percentage coarser than no. 4 sieve in filter put against maximum particle sieved in filter in relation to performance history of internal erosion.

There are 72 dams in the set. When using data sets from Nilsson (1995) and Nilsson *et al.* (1999) no complete grain size curves were available, only specific grain sizes (e.g. filter D15, filter D50 and maximum sieved particle). In these cases some degree of judgment were necessary in determining amount coarser than 4.75 mm. It is also important to point out that in the dam safety guidelines at the time of the hydropower build-out in Sweden (peaked in the 1960s and 1970s) no instructions were given regarding maximum particle size in filters nor obligation to document amount of material coarser than maximum sieved. That is why many dams in fig. 6 appears to have maximum particle size of not more than 16-20 mm, which is up to today's standard, but probably the maximum particle size is coarser than that and these values are therefore less reliable.

As can be seen in fig. 6 most filters in the dams are widely graded, having low amount of sand fraction and a large maximum particle size and these properties together increases the likelihood that the filter segregated during placing. Furthermore dams with internal erosion (1-dams) are over-represented in the upper right corner of the plot, where the filter's maximum particle is coarse (exceeding 100 mm) and the amount coarser than 4.75 mm is high (i.e. the sand fraction is lower than 40 % in the filter), which agrees with the research of Foster & Fell (2001) on characteristics of dams with poor filter performance.

CONSTRUCTION AND DESIGN RELATED INDICATORS

Guidelines and trends

Both Ripley (1986) and Foster *et al.* (2005) speaks of the increase in embankment dam undertakings in the 1960s and 1970s, and how this coincided with a period of less stringent construction requirements. During this period it appears that when it comes to dams with cores of broadly graded materials there was an over-reliance in the capacities of the glacial moraine (till) core as being self-filtering in itself, which resulted in the use of wide single filters (of wide gradation) instead of multiple filters and filter/transition components. This agrees with the review made on Swedish dams in Nilsson *et al.* (1999) where it was found that dams constructed around the 1970s were appears to be equipped with coarser graded filters than previously, presumably due to design factors with relaxed filter requirements and poor construction practices at the time.

Table 1. Number of rockfill dams with and without filter-transition component in relation to performance history of internal erosion.

With transition layer(s)	Without transition layer	
17 dams	9 dams	Without internal erosion (2 and 3-dams)
1 dam	7 dams	With internal erosion (1-dams)
18 dams	16 dams	

Filter-transition component

Filter-transition components, between the core's protective filter and the shell material, are generally necessary if filter and drainage functions are to be upheld between a moraine (till) core and the support fill, especially if the shell consists of rockfill. Based on the data set of 72 dams in this study, 34 are rockfill dams. 56 % (18 dams) of these rockfill dams are equipped with one or more transition layer(s) between the filter and the rockfill shell as shown table 1. Only 1 of 8 dams of the rockfill dams with internal erosion (1-dams) is equipped with transition layers, and conversely 44 % (or seven dams of 16) of the dams without transition layers are dams with internal erosion (1-dams), see table 1.

CONCLUSIONS

With objective to investigate possible indicators of internal erosion prone embankment dams of broadly graded soils a review is made in this paper comprising existing embankment dams constructed from broadly graded moraine (till) cores and generally widely graded filters. By back-analyzing performance history of internal erosion and comparing with sets of dams internal erosion free it is possible to suggest some common features of dams that experienced internal erosion (that surfaced in shape of sinkholes or other clear signs). The internal erosion prone dams can in this study statistically be traced back to the use of filters of coarse and unstable gradings with high susceptibility for filter segregation and in the case of rockfill dams there is usually a devoid of filter-transition components. Statistically, based on the data set of dam records of 72 dams in this study, the following core/filter properties have been shown to be over-represented among reviewed dams with a performance history of internal erosion:

- Coarsely graded filter (generally with max. D15 coarser or equal to 1.4 mm).
- Grading instability of the filter.
- Grading instability of the core (base) soil (Rönnqvist (2007)).
- High susceptibility for filter segregation.
- No filter-transition component (applies to rockfill dams).

These properties, when combined, can serve as indicators and give an idea of dams with potential for internal erosion. One factor by itself may not necessarily influence the potential for internal erosion, but an unfavorable combination of the above core/filter properties appears to increase the potential for internal erosion in embankment dams of broadly graded soils. Nevertheless it is important to point out that there can still be dams having the above-mentioned factors that operate without problems connected to internal erosion.

Best practice for protective filters for broadly graded moraine (till) core dams would be to provide the dam with:

- A uniformly graded sand-filter satisfying current filter criteria (e.g. Sherard & Dunnigan (1989) filter D15 finer or equal to 0.7 mm).

- A filter with internally stable grading. (Internal stability assessed by using the Kenney & Lau (1985, 1986) method).
- A filter with low to nil susceptibility to segregate. The guidelines for this are less clear, although Ripley (1986) suggested an upper limit of particle size of 18 mm and high amount of sand fraction (at least 60% finer than 4.75 mm).
- A cohesionless filter unable to sustain a crack. Although Ripley (1986) suggested a clean sand filter with preferably 0 % passing the no. 100 sieve (0.150 mm), but max. 2 % minus the no. 200 sieve (0.075 mm), the Vaughan & Soares (1982) "bucket test" can also be used to detect cohesion in filters.
- Sufficient filter-transition components to the shell material that the current filter criteria require.
- A filter that will not degrade or change in gradation with time (Fell *et al.* (2005)).

ACKNOWLEDGMENTS

The research presented in this paper was carried out as part of the "Swedish Hydropower Centre – SVC". SVC has been established by the Swedish Energy Agency, Elforsk and Svenska Kraftnät together with Luleå University of Technology, the Royal Institute of Technology, Chalmers University of Technology and Uppsala University. Further information can be found on www.SVC.nu. Additional financial support has been received from Vattenfall AB Vattenkraft.

REFERENCES

Ericsson, J., Jender, M. (1998) *Documentation and investigation of damages in Lövön dam*, Master's thesis 1998:335, Luleå Tekniska Universitet, December. (In Swedish). (English summary).

Fell, R., MacGregor, P., Stapledon, D., Bell, G. (2005) *Geotechnical Engineering of Dams*, Taylor and Francis.

Foster, M.A., Fell, R. (2001) *Assessing Embankment Dam Filters that do not satisfy Design Criteria*, J. of Geotech. and Geoenv. Engrg, Vol. 127, no. 4, May, pp. 398-407.

Foster, M., Fell, R., Spannagle, M. (2000) *The Statistics of Embankment dam failures and accidents*, Can. Geotech. J., vol. 37, pp. 1000-1024.

Höeg, K. (2001) *Embankment dam engineering, safety evaluation and upgrading*, proc. of 15th ICSMGE, Istanbul, Turkey, Aug.

Kenney, T.C., Chahal, R., Chiu E., Ofoegbu, G.,I., Omange, G. N., Ume, C. A. (1985) *Controlling Constriction Sizes of Granular Filters*, Can. Geotech. J. vol 22, pp. 32-43.

Kenney, T.C., Lau, D. (1985) *Internal stability of granular filters*, Canadian Geotechnical Journal, vol 22, no. 2, pp.215-225.

Kenney, T.C., Lau, D (1986). *Internal Stability of Granular Filters: Reply*, Canadian Geotechnical Journal, Vol. 22 No. 2, pp. 215-225.

Kjellberg, R., Norstedt, U., Fagerström, H. (1985) *Leakage in and reconstruction of the Juktan earth and rock fill dams*, 15th ICOLD Congress, Lausanne, Q59, R35, pp. 553-573.

Lafleur, J., Mlynarek, J., Rollin. A.L. (1989) *Filtration of Broadly Graded Cohesionless Soils*, J. of Geotech. Engrg, 115 (12), pp. 1747-1768.

Milligan, V. (2003) *Some Uncertainties in Embankment Dam Engineering*, J. of Geotech. And Geoenvironmental Engrg, sept. 2003, pp. 785-797.

Nilsson, Å. (1995) *Filter inventory and function analysis*, VASO report 13, English summary.

Nilsson, Å., Ekström, I., Söder, C. (1999) *Sinkholes in Swedish embankment dams*, Elforsk Report 99:34, English summary.

Nilsson, Å., Rönnqvist, H. (2004) *Measures to strengthening Embankment Dams in order to stop or control a possible through-flow process*, Intl. Seminar on Stability and Breaching of Embankment Dams, Norway.

Norstedt, U., Nilsson, Å. (1997) *Internal Erosion and Ageing in some of the Swedish Earth and Rockfill Dams*, 19th ICOLD Congress, Florence, Vol II, pp. 307-319.

Ripley, C., F. (1986) *Internal Stability of Granular filters: Discussion*, Can. Geotech. J., vol. 23, pp. 255-258.

Rönnqvist, H. (2006) *Predicting Internal Erosion in Glacial Moraine Core Embankment Dams*, Proc. of HydroVision 2006, Portland OR, USA.

Rönnqvist, H. (2007) *Assessing Internal Erosion*, International Waterpower and Dam Construction IWP&DC, vol 50, no 11, pp. 30-34, nov. 2007.

Sherard, J.L. (1979) *Sinkholes in Dams of Coarse, Broadly Graded Soils*, 13th ICOLD Congress, India, Vol. II, pp. 25-35.

Sherard, J.L., Dunnigan, L.P., Talbot, J.R. (1984a) *Basic Properties of Sand and Gravel filters*, J. of Geotech. Engrg, ASCE, 110 (6), pp. 684-700.

Sherard, J.L., Dunnigan, L.P., Talbot, J.R. (1984b) *Filters for silts and clays*, J. of Geotech. Engrg, 110 (6), pp. 701-718.

Sherard, J.L. and Dunnigan, L.P. (1989) *Critical filters for impervious soils*, J. of Geotech. Engrg, ASCE, 115 (7), pp. 927-947.

Talbot, J. R., (1991) *Discussion of "Improved filter criterion for cohesionless soils" by Honjo and Veneziano*, J. of Geotech. Engrg, 117 (10), pp. 1632-1634.

Vaughan, P., R., Soares, H., F. (1982) *Design of Filters for Clay Cores of Dams*, ASCE Journal of Geotech. Engrg. Division, vol. 108, pp. 17-31.

Vestad, H. (1976) *Viddalsvatn dam a history of leakages and investigations*, 12th ICOLD Congress, Q45, R22, pp. 369-390.

2. Dam break analysis and risk assessment

Application of the Interim Guide to Quantitative Risk Assessment across multiple dam owners by multiple Jacobs offices

A. J. BROWN, Jacobs Leatherhead
G. YARWOOD, Jacobs Exeter
S. KING, Jacobs Glasgow
J. D. GOSDEN, Jacobs Croydon

SYNOPSIS. This paper describes a number of different quantitative risk assessments (QRA) analyses carried out independently by different Jacobs offices, for different clients, using the Interim Guide to QRA. It compares the results and suggests improvements to the Interim Guide which should be incorporated when it is finalised. It also comments on the insights from, and the strategic issues relating to the use of, quantitative risk assessment in dam safety management.

INTRODUCTION

The Interim Guide to QRA (Brown and Gosden, 2004) was published in 2004 for extended trialling, prior to finalisation about five years later. Jacobs has applied the workbook in the Guide to dams owned by a number of major dam owners and also by many small owners, the application being carried out independently by a variety of teams in different offices. This paper describes and comments on the results of these analyses as a means of assessing the value and repeatability of the Interim Guide and obtaining feedback on its use.

ANALYSES

The number of dams assessed represents about 5% of UK dams; the analyses being carried out under the direction of six separate Panel AR Engineers. Groups A to D are those owned by major dam owners, whilst Group E are a group of "small dams" owned by private owners, charities and local authorities. The range of dam height, reservoir volume and age are summarised in Table 1.

The reason for the analysis varied, from being an enhancement of a Section 10 Inspection by providing an auditable justification for any works in the interests of safety, through to more general risk management practices. Similarly the process for carrying out the QRA varied, from completion by a junior engineer with review and signoff by an All Reservoirs Panel Engineer, through to completion of the workbooks by a Panel AR Engineer. In general, involving the Supervising Engineer for the dam under consideration in the process was considered to improve the accuracy of the process, in terms of understanding and scoring appropriately indicators of current condition. Moreover the Supervising Engineer considered that this assisted in focusing surveillance.

For Group E use of the Interim Guide was limited to Consequence assessment only, to provide a more repeatable assessment of population at risk and likely loss of life (LLOL), and thus categorisation of both Consequence Class (as defined in the Interim Guide) and Flood Hazard Category (as defined in Table 1 of Floods and Reservoir Safety, ICE, 1996). These are a subset of the group of dams described in Brown and Gosden (2000).

Table 1: Range of height and age of dams, and size of reservoirs

	Groups A to D			Group E		
	10%	Median	90%	10%	Median	90%
Dam height (m)	6	12	40	3	5	8
Reservoir volume ($1000m^3$)	70	600	8,000	28	50	230
Date of construction	1810	1910	1960	1550	1790	1930
% of dams > 15m height (ICOLD "Large dams")		65%			0%	

Note Dimensions generally rounded to one significant figure, and dates rounded to nearest 10 years so data does not relate to specific dams

METHODOLOGY USED FOR QRA

The methodology used was the Excel workbooks accompanying the Interim Guide to QRA as published in 2004, in some cases using the following refinements

a) Supplement No 1 to the Interim Guide was published in June 2006 on the Defra website along with the draft Guide to Emergency Planning, providing extended guidance on the estimation of the consequences of failure
b) Extended guidance on scoring internal threats, as indicated in Brown and Peters (2007)
c) Extension of the Excel worksheets to include
    - automatic calculation of seepage and settlement index,
    - the effect of infrastructure embankments on rapid dam break.
    - Current Condition Score amended to be quoted to one decimal point

- Sheets 9 and 10 were also merged, so the same property data was used for both risk to life and cost of damage.

In some groups detailed dam break maps were available, but no consequence assessment, so the existing maps were used to denote the extent of flooding with the methodology in the QRA workbook used to estimate the population at risk, likely loss of life and property damage.

The time taken to complete the workbooks was assessed as typically 3 to 4 days where the user was unfamiliar with the system, but this typically reduced by over half once the user was familiar with the methodology.
Feedback from the users included

a) "Move summary sheet for each section at the end. Anything that would help make the method less intimidating would help. Once you get into the spreadsheets they are ok.
b) The spreadsheets are a lot simpler than expected; the most complex programming is an "If statement"!
c) Generally I think the process is understandable. If time is taken to source the information required, and understand the methodology, then on an overall basis not many more changes might be required to improve it
d) The event train spreadsheets did not add to the process
e) If there is limited information for a reservoir, and no adverse indicators I believe the studies may not provide significantly different results for different reservoirs. Especially if they are done or reviewed by the same person for consistency"

In regard to item 'd', Section 2.2 of the Interim Guide notes that, although the event trains are not essential for a quantitative estimate, the event trains prompt the user to consider failure modes (BSI, 1991), and confirm and document which are to be quantified. Item 'e' is an important issue. Where there is a lack of data on dam construction and its foundation this raises the question of whether Section 10 Inspections are similarly constrained, and thus the adequacy of Section 10 Inspections. In practice a certain amount can be inferred by experienced engineers from the date of construction and properties of geological materials in the vicinity of the dam, likely to have been used in the construction and present in its foundation, in the same way as for a statutory Section 10 Inspection.

RESULTS - SUMMARY

The results are summarised as shown in Table 2 and plotted on Figures 1 to 6. Identification of data is limited to maintain anonymity.

The estimates of consequences and annual probability of failure are discussed in the following sections, including a commentary on the absolute values, differences between groups, to what extent this is real, and suggested improvements to the Interim Guide that arise from experience in its use. Comment is also made on insights in relation to reservoir safety management, provided by the use of QRA.

The discussion of improvements commences with Consequence assessment as this influences the proportionate level of detail and effort in the assessment of annual probability of failure. This is followed by a section discussing the strategic issues that arise in relation to the use of quantitative assessment of risk

Table 2: Summary of summary plots of results of analysis

<table>
<tr><th>Fig.</th><th>Axes</th><th>Data fields</th></tr>
<tr><td>1</td><td>Consequences of failure (Likely loss of life (LLOL) vs. third party flood damage</td><td rowspan="2">Median for each group, and envelope of range of all results</td></tr>
<tr><td>2</td><td>Risk as FN chart of LLOL (with and without warning) vs. annual probability of failure (APF) of dam (excluding any upstream dams)</td></tr>
<tr><td>3</td><td>Cumulative distribution of LLOL</td><td rowspan="2">All dams in Groups A to D (undifferentiated by group)</td></tr>
<tr><td>4</td><td>Cumulative distribution of APF</td></tr>
<tr><td>5</td><td>Cumulative distribution of breach discharge at dam; and "depth" and "velocity x depth" at 1km downstream</td><td rowspan="2">Group E only</td></tr>
<tr><td>6</td><td>Cumulative distribution of two alternative methods of estimating length of dambreak impact</td></tr>
</table>

In relation to the estimation of risk, Figure 2 shows both the existing median risk for each group of dams, plotted on a FN chart, and the reduction in risk where an effective warning is available. The figure also includes the zones for Acceptable, Tolerable if ALARP and Unacceptable, as defined in Clause 9 of the Annex to SPC/permissioning/09 (HSE, 2001). These lines relate to risk levels of LLOL of 1 in 100 per year (upper line) and 1 in 10,000 (lower line). It can be seen that the median values are all within the "Tolerable if ALARP" zone, which indicates that there should be a formal ALARP analysis on file to show why further risk reduction measures are disproportionate to cost. It can be seen that LLOL reduces by a factor of 5 to 10 where 60 minutes effective warning is available, the reduction being greater for higher consequence dams.

The practical outcomes from the analyses varied depending on the reason for carrying out the analyses. For Group E use of QRA provided a more realistic estimate of consequence class than that provided previously which was based on visual assessment of the dam and downstream valley.

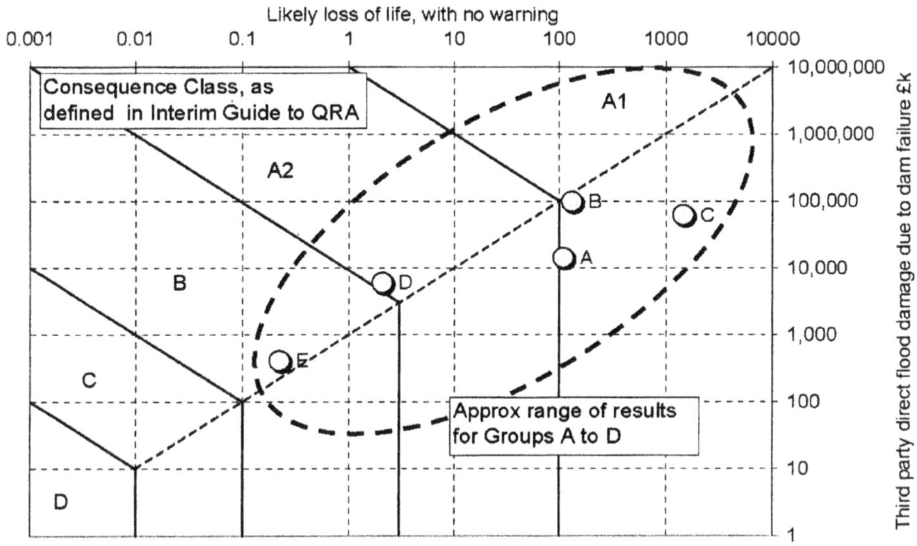

Figure 1: Consequences of failure (median for each group)

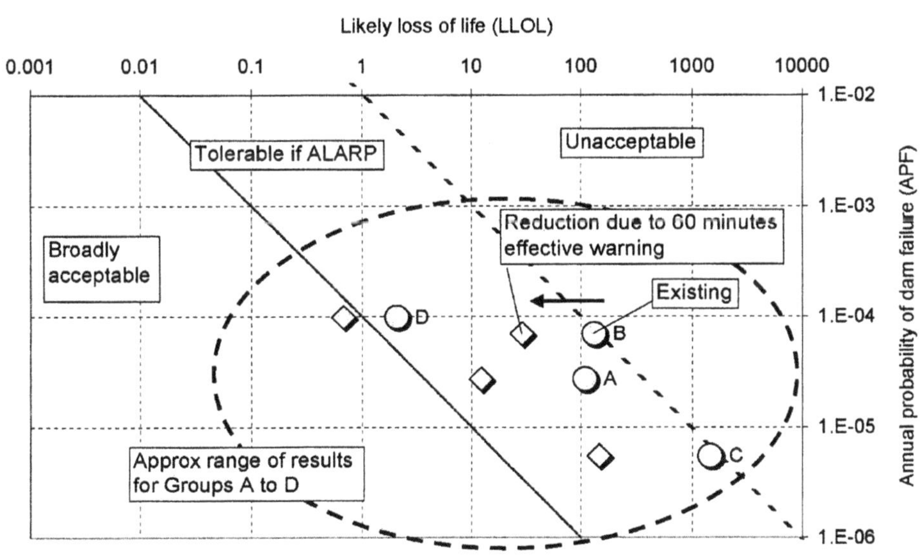

Figure 2: Risk from dams, with and without an effective warning (median for each group)

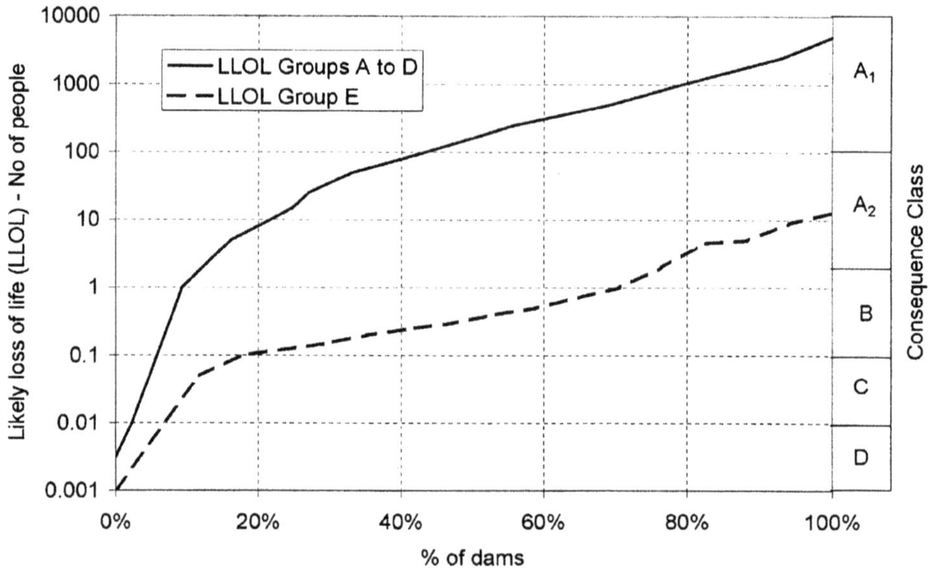

Figure 3: Cumulative distribution of likely loss of life

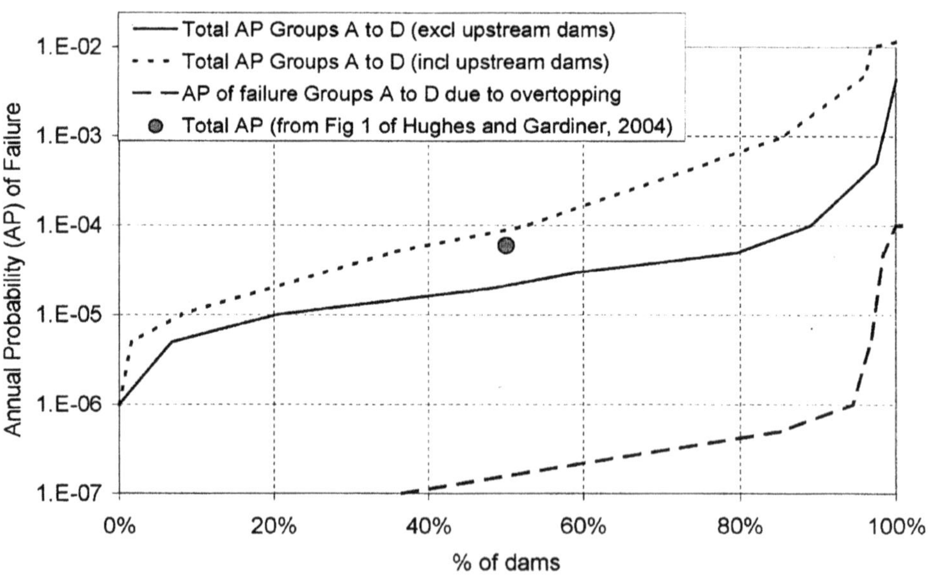

Figure 4: Cumulative distribution of annual probability of failure

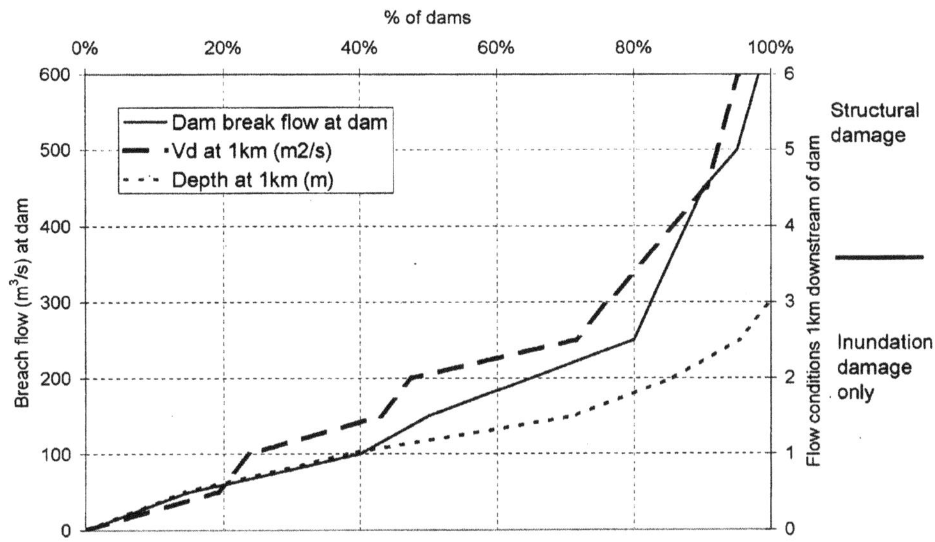

Figure 5: Group E: Distribution of dam break parameters

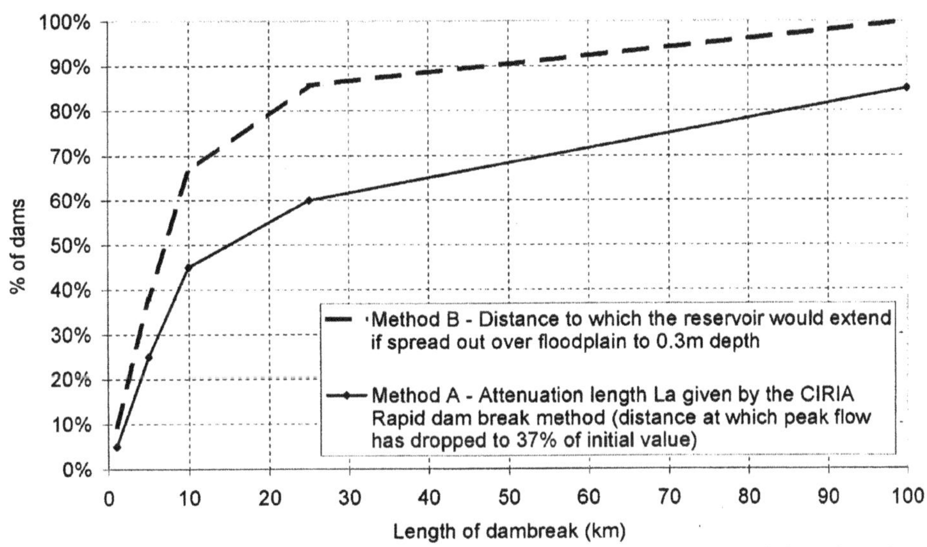

Figure 6: Group E: Alternative methods of estimating length of dam break impact

The expectations for the uses of the QRA included:

a) improved transparency and consistency in assessing the consequences of failure
b) assisting the Inspecting Engineer in deciding whether an issue was crucial to dam safety (could lead to failure), or whether it was an

operational issue which may lead to damage but unlikely to lead to release of the reservoir

c) identification of "quick fix" operation or capital investment that would materially reduce the AP failure
d) a tool which allowed an evaluation of whether the cost of candidate upgrading works were proportional to the reduction in risk achieved
e) assist in evaluating surveillance frequency, and effect of maintenance regime (e.g. grass cutting) on the probability of failure
f) prioritising works between different dams

The Interim Guide did not always fully meet these expectations as the terms of reference for its development were to compare the existing risk from floods with internal erosion, and did not include these other uses. For one group the scoring given to surveillance frequency and quality of monitoring was increased as the scoring in the Guide was considered too low. Nevertheless for all groups the analysis was felt to be of value in focusing thoughts on modes of failure, and "what really mattered".

REVIEW OF ESTIMATES OF CONSEQUENCES

General

Figure 1 shows that the Consequence Class of the median of the various groups varies significantly, as might be expected. Third party property damage is typically about £1M per likely loss of life (LLOL). Groups A and C are large dams where the fatality rate is high, leading to a higher LLOL relative to property damage. Group D dams are located in more rural areas, remote from large urban development and thus pose a reduced risk to third parties in terms of both LLOL and property damage. The differences between groups are considered a reasonable reflection of reality.

Figure 3 shows the variation in LLOL. The wide range of the consequences of failure are noted, with LLOL varying by a factor of five orders of magnitude. A risk based approach to safety management would suggest that the resources to ensure dam safety would similarly vary by five orders of magnitude.

Rapid Dambreak assessment

A risk based approach requires a methodology for rapid dam break to screen out the low consequence dams, where the consequences of dam failure would be small, and thus where significant cost in risk management measures would be disproportionate. A particular problematic area is in differentiating between Class B and C dams. For these dams there are typically up to a few tens of houses downstream. However, the depth of flooding above the flood plain is typically just over one metre, so deciding which houses have a threshold level which would be inundated by a dam break flood depth can therefore only be indicative. In reality the main risk is to those exposed in the open, as those in two storey buildings can readily rise above flood level by going upstairs, provided the building is not upstream of a transportation embankment where water would pond to a greater depth.

A Rapid method is given in CIRIA Report C542 (2000); this being reproduced in the Interim Guide (2004) with some modest refinements. This method is now used routinely to estimate dam break characteristics for the Group E dams. Figure 5 shows that the median dam break flow is 150m^3/s, whilst by 1km downstream of the dam

- 80% of the dams have a median product of velocity and depth of less than 3.5m^2/s, where damage is anticipated as inundation damage only.
- The 50% and 90% depth of flooding above the floodplain are 1.1m and 2.2 m respectively

This is likely to be an overestimate, as in some cases the breach flows from these small reservoirs are overestimated by the Rapid method, since it neglects the effect that having a wide crest width in relation to dam height would have on the breach flow.

A separate issue relates to the distance downstream which would be affected by dambreak. The problems with the CIRIA (2000) C542 method in underestimating attenuation of the dambreak wave have been identified in Supplement Number 1 to the Interim Guide (Defra, 2006). Figure 6 shows two alternative simplified approaches to estimating the downstream extent of dambreak flooding:

A – Attenuation length L_a given by the CIRIA rapid dambreak method (distance at which peak flow has dropped to 37% of initial value; noting that it drops to 14% and 5% of initial flow at $2L_a$ and $3L_a$ respectively)

B - Distance to which the reservoir volume would extend if spread out over the floodplain downstream, to a depth of say 0.3m, as representing a residual flood depth due to ponding behind roads, hedges, debris, trees etc.

It can be seen that Method B suggests attenuation would occur faster than the CIRIA rapid dambreak. Although detailed dambreak would be required to confirm the validity of this, it does provide a useful indicative check for preliminary dam break assessment.

<u>Impact assessment</u>

The various analyses used Supplement No 1, which gave extended guidance on estimating the occupancy of buildings and cost of property damage. One of the main uncertainties is whether the occupancy rates quoted in non-residential property (average $40m^2$/ person, occupancy of 25% of time) are reasonable, or whether they overestimate the population likely to be present.

REVIEW OF ESTIMATES OF ANNUAL PROBABILITY OF FAILURE

General

The vertical axis on Figure 2 shows the median annual probability of failure (APF), which varies between 3 and 10 x 10^{-5}/ year, apart from the high consequence group where APF is 0.7 x 10^{-5}. The relatively modest differentiation between groups is probably a reflection that the data points are the median for each group, and that there is a greater differentiation within groups than between groups.

Figure 4 shows the distribution of APF across the dams considered in this paper. 90% of the dams considered have a probability of failure of less than 1 x 10^{-4}/ year, whilst the median probability of failure is 3 x 10^{-5}/annum. The latter is of the same order as the overall probability of UK dams of less than 2 x 10^{-5}, inferred from the lack of failures of dams in service under the Reservoirs Act since 1975 (KBR, 2002), and slightly less than the median shown on Figure 1 of Hughes and Gardiner (2004)

The small number of dams with an APF higher than 10^{-4}/annum is noted. Although some represent lower consequence dams where safety requirements have been relaxed following a risk based approach, there are also some where QRA has identified risks which had been overlooked by the last Inspecting Engineer.

Extreme rainfall

Figure 4 includes the contribution of the probability of failure from overtopping, which is less than 1 x 10^{-6} for the great majority of dams. The exceptions include lower Consequence Class dams, where the design standard is to pass a 1 in 150, or 1 in 1000 chance per year flood, and some dams where the QRA identified inadequacies in the existing arrangements, most commonly due to some combination of a constriction downstream of the weir, blockage of the weir by trees or other debris, or structural instability of the crest wall under wave/ water loading.

Although Sheet 2.3 of the Interim Guide requires the User to assess the criticality of failure due to erosion along the sides of the chute, or structural failure of the chute, it does not provide means of quantifying the APF due to these failure modes. Both of these failure modes have been highlighted by incidents at Boltby and Ulley and worksheets should be developed to quantify these, at the screening level of accuracy of the Interim Guide.

Upstream reservoirs

For the reservoirs owned by major dam owners 37% of reservoirs have reservoirs upstream of them retained by embankment dams, which when included in the overall APF typically increases the APF by a factor of about five. Difficulties were experienced in estimating the APF where the upstream dam was in third party ownership. This analysis neglects upstream reservoirs retained by concrete dams, as currently there is no methodology for estimating the APF of concrete dams. This confirms the increased risk of failure where dams are in cascade, due to the potential for the domino effect.

Internal stability (embankment and appurtenant works)

This element of the Interim Guide was the main novel section, and as such has attracted the greatest comment. The research project preceding the Interim Guide carried out a review of systems in use worldwide, identifying 12 different systems (Table 5.3 of KBR, 2002; Brown and Gosden, 2002). Evaluation of this international practice led to the Interim Guide using historical data from the BRE database (Brown & Tedd, 2003) to define "anchor points" for poor condition dams. It is implicit in this approach that the surveillance and intervention regime effective over the period over which the data was collected continues at the same level, so incidents continue to be detected at an early stage and appropriate action taken. It was also assumed that good condition dams have a probability of failure of 10^{-7} and a scoring system was devised to interpolate in-between these anchor points.

One of the early comments that emerged with use of the Guide was that the guidance on scoring allowed different interpretation by different engineers. In addition although Table 4.3 of the Guide noted that where the construction, or condition, of a particular element was unknown, a proportion of the total score should be assigned, some users failed to notice this and assigned zero, leading to the conclusion that their dams were very safe. This feedback led to issue of extended guidance on scoring (Brown & Peters, 2007), including suggested ways to include uncertainty and suggested scores related to the age of the dam, based on typical design details at the date the dam was built. Other issues relating to assessment of internal threats are summarised in Table 3, with comment on how the system could be refined.

The process of considering behaviour of the dam as part of scoring its Intrinsic and Current Condition led to a number of insights related to surveillance. One was that where the downstream toe of a dam (or downstream outlet from a culvert through the dam) is submerged by the reservoir of a downstream dam, there is significant uncertainty over its condition, which is reflected in a high Current Condition score. Another was where instrumentation readings were not systematically evaluated within a week, again reflected in an increased Current Condition Score. Although these were in retrospect obvious, the process of quantifying their impact on the APF was valuable in highlighting their importance to scoring the current condition of the dam.

Other threats

Other threats have been identified and considered in application of the Guide, including reservoir rim landslides and coal mining in the vicinity of the dam, but none was considered sufficiently high to materially affect the overall probability of failure.

DISCUSSION – STRATEGIC ISSUES

Experience with use of the Interim Guide has showed that a risk based approach is still a new technique to many dam engineers. There is often a lack of clarity over consequences if there were no intervention, in terms of whether it is damage which does not affect the structural adequacy of a dam, or whether it affects the ability of the dam to retain water. For this reason alone a QRA approach is considered valuable, as a means to encourage clarity of thinking in terms of modes of failure and consequences.

A second major issue is that the Reservoirs Act 1975 is indiscriminate in terms of the minimum reservoir safety regime required by Undertakers, and does not differentiate between small reservoirs where the probability of loss of life in the event of dam failure is less than 10%, and large reservoirs where the LLOL exceeds 1000. Application of QRA is of significant benefit in providing a transparent and robust methodology to differentiate the level of consequences in the event of dam failure, and thus inform a risk based regime for management of dam safety.

A third area where QRA provides benefits over qualitative assessment is in assessing when the cost of candidate structural works is disproportionate to the reduction in risk that would be achieved, and providing an auditable record of that assessment. This is of benefits to owners in providing a tool to prioritise works and to demonstrate that they have reduced the risk as low as reasonably practicable (ALARP) (www.hse.gov.uk/risk/theory/alarp.htm), and to panel engineers in providing a consistent methodology. As in any analysis, understanding the uncertainties in the analysis is important. As a screening level analysis the Interim Guide provides two sets of parallel calculations for internal stability, which allows sensitivity study. More explicit inclusion of uncertainties would significantly increase the complexity of the workbook and is probably inappropriate for a screening level assessment. Such addition would, however, be a useful additional tool where the screening analysis showed that further assessment was necessary.

Table 3: Possible refinements to methodology for estimation of APF due to internal threats

Assumption	Adopted in Interim Guide	Possible refinement
What score should be assigned where the construction is unknown, or there is no measurement of settlement or seepage?	Table 4.3. Assign a % of total score depending on what is likely, in the experience of the assessor	Extend guidance on scoring, to reduce variability between assessors, as described in Brown & Peters, 2007
What is the lowest APF that can meaningfully be estimated?	10^{-7}/ year (Table 4.6)	For extremely high consequence dams with LLOL of 1000, the overall APF must be below 10^{-7} for risk to be in the broadly acceptable zone. It is suggested that for a screening system such as the Interim Guide it is not meaningful to estimate APF lower than this
What relative importance should be given to Intrinsic and Current Condition, which relate to the form of construction and actual performance of the dam respectively?	Figure 4.1. Only a small weighting is given to Intrinsic Condition.	The weighting could be increased, for example the adjustment of Anchor Point 0 for Intrinsic Condition could be increased to a factor of 100, or 1000. However, the system would then need to be recalibrated, to ensure that it provided reasonable results for a range of dam conditions.
What is the probability of failure of a dam where emergency drawdown is considered necessary to avert failure?	Table 4.2 1.4 x 10^{-2} (1 in 71) for Condition Score 10; as inferred from BRE data	The estimate of the annual probability of a serious incident could be separated from the probability of failure given an incident. This would have the advantage of focusing attention on the adequacy of arrangements in an emergency, but would complicate the system.
How is the system calibrated?	Section 4.6.2 Poor condition dams based on historic data on frequency of incidents, calibrated by trial of prototype system	The Interim Guide allows the User to enter alternative values of probability for the anchor points; guidance on this could be extended. An alternative approach would be to provide an alternative method, such as event trees (e.g. Cyganiewicz, 2007). This would need significant additional guidance to achieve consistency.

One of the issues highlighted by use of QRA is the lack of information on construction and foundation of many dams, common to all the groups considered by this paper. It is suggested that as a minimum owners could carry out portfolio wide desk study summarising the information that is available, including construction drawings (although recognising that these may be unreliable), site investigation and published geotechnical data on geological deposits in the area which are likely to have been used in dam construction. For some very high consequence dams it may then be appropriate to carry out site investigation to establish, or confirm, dam construction details, including filtering ability of the downstream shoulder and erodibility of the core.

CONCLUSIONS

It is concluded that the Interim Guide is a good first step in introducing a screening level QRA as a tool to assist in dam safety management in UK. It is now in routine use to provide an improved assessment of the consequences of failure for dams which are not necessarily Category A1 or A2. The principles are also in use to apply the ALARP test, and to inform decisions regarding dam safety management.

Nevertheless, as with any new technique, a number of areas for improvement of the accuracy of the quantitative estimates have been identified, including

a) in relation to consequences of failure
    - the method for estimating the dam break flow should take explicit consideration of crest width in relation to dam height, and crest surfacing materials in relation to breach due to overtopping
    - further research is required into the rate of attenuation of the dam break flood, particularly for reservoirs between 3 and 10m in height, and reservoir volumes between 25,000 and 100,000m^3 (which form 50% of the reservoirs registered under the Reservoirs Act 1975)
    - research and updated guidance on average occupancy rates in non-residential property
    - research and guidance on the effect of transportation embankments on extent of flooding following dam failure

b) in relation to annual probability of failure
    - improved guidance on uncertainty, particularly how to score where the construction of the dam is unknown, or its condition is obscured (e.g. by water in a downstream reservoir, thick vegetation, or permeable foundation strata are obscured by superficial soil deposits)
    - separate out the annual probability of a Level 2 Incident (emergency drawdown) from the annual probability of failure

- add an event tree method to be carried out as well as the "historic data method", to make the uncertainties clearer and promote consideration of how to manage those uncertainties
- further research into methods of reliably estimating the annual probability of less than 10^{-6}, which is necessary if very high consequence dams are to fall into the ALARP or broadly acceptable zones

The quantitative estimation of risk is considered to be a valuable process, for the discipline it imposes in terms of clarity of thinking, for the insights it provides into "what matters" and also in providing a documented audit trail. It is therefore considered that QRA should be an essential part of the toolbox of those managing the safety of UK reservoirs, and that the Interim Guide should be updated to provide this tool.

REFERENCES

Brown, A.J. and Gosden, J.D., 2000, *Safety issues at small reservoirs.* BDS conference

Brown, A.J. and Gosden, J.D., 2002, *A review of systems used to assess dam safety.* BDS conference

Brown, A.J. and Gosden, J.D., 2004, *Interim Guide to quantitative risk assessment for UK reservoirs.* Thomas Telford.

Brown and Peters A, 2007, *Update on the use of the Interim Guide to Quantitative Risk Assessment.* Dams and Reservoirs. 17(1). Pp19-24

Brown and Tedd, 2003, *The annual probability of a dam safety incident at an embankment dam, based on historical data.* Dams and Hydropower.

BSI (1991) *Reliability of systems, equipment and components.* BS 5760 Part 5. Guide to failure modes, effects and criticality analysis (FMEA and FMECA). 38pp British Standards Institution.

CIRIA, 2000, *Risk Management for UK Reservoirs* Report No.C542.

Cyganiewicz J M, Engemoen WO, Redlinger CG, 2007, Bureau of Reclamation experience with evaluating internal erosion of embankment dams. In *Internal Erosion of dams and their foundations.* Pubs Taylor & Francis.

Defra, 2006, *Supplement No 1 to Interim Guide.* Published on Defra website with informal consultation on draft Guide to Emergency Planning

HSE (2001) *Principles and guidelines to assist HSE in its judgments that duty holders have reduced risk as low as reasonably practicable.* On HSE website.

ICE, 1996, *Floods and reservoir safety, 3rd Edition.* Thomas Telford

KBR, 2002, *Floods and reservoir safety integration.* Defra research contract. Available at www.defra.gov.uk/environment/water/rs/index.htm

The influence of infrastructure embankments on the consequences of dam failure

A. J. BROWN, Jacobs. UK
A. CHAPMAN, Jacobs, UK
J. D. GOSDEN, Jacobs, UK
F. J. B. SMITH, Jacobs, UK

SYNOPSIS. Dam break inundation mapping is the first part of the assessment of the consequences of failure of a dam. Although there has been significant expenditure on research on the breach hydrograph, there has been little attention given to the influence of infrastructure embankments across the inundated area on flow paths, depth of inundation and consequences. This paper discusses the general issues and provides two case histories where the consequences of dam failure are significantly influenced by major infrastructure embankments across the valley downstream of the dam.

INTRODUCTION

Estimation of the consequences of failure of a dam is necessary for two reasons

a) to estimate the extent of flooding and population at risk (PAR) for use in planning evacuation in the event of a failure, and
b) where a risk based approach to dam safety is used, to provide the likely loss of life (LLOL) and property damage to allow the dam owner to determine whether the risk has been reduced to as low as reasonably practicable (www.hse.gov.uk/risk/theory/alarp.htm)

The first step in a consequence assessment is dam break inundation analysis and mapping. This is an evolving process with initial Defra sponsored research into the methodology being published by Binnie and Partners (1986, 1991) and a summary of early analysis in UK being given in Tarrant et al (1994). Subsequent research into extreme flood processes has largely focused on breach discharge and propagation of the flood wave down the valley (Morris & Hassan, 2004, www.floodsite.net).

One area that has received little attention is the influence of infrastructure embankments across the downstream valley on the consequences of failure. To simplify the analyses, many existing dambreak analysis have assumed that downstream transportation embankments are breached under the dambreak flood, and have therefore modelled the embankments as having pre-existing notches. Although this is conservative in terms of maximizing the flow passing downstream, it neglects additional backing up of water upstream of the transportation embankment should it not breach immediately. This paper discusses the general issues and presents two case histories of full inundation analysis where major infrastructure embankments across the valley significantly influence the consequences of failure.

GENERAL ISSUES

As with any analysis the quality of the output depends on the quality of data used, method of analysis and reasonableness of assumptions made. The important assumptions will vary depending on what is considered to be the critical output, namely the

a) areal extent of flooding,
b) likelihood of structural damage to property, or
c) risk to people who are in the open in the inundated area.

The first of these is considered to be the primary measure of potential consequences of failure, as it defines the area to be evacuated and number of properties in which people are likely to be at some risk. It has been found that the width of flooding near the dam is relatively insensitive to peak flow: typically the base of the valley is completely inundated and the valley sides are much steeper than the cross fall on the valley floor. Once the dambreak flow has travelled downstream into wide valleys where the flow is less deep then the width of flooding becomes more sensitive to peak flow, but here the risk to people may be reduced.

Table 1 summarises some of the key issues that will affect the extent of flooding due to a dam failure. Issue 1 is well covered elsewhere. The other issues are equally important but have received less attention, with this paper focussing on Issue 2a. Table 2 summarises some of the main factors affecting the impact on people.

Table 1: Key issues affecting extent of dam break inundation

Issue	Key factors affecting estimate
1. Breach formation and discharge hydrograph	Mode of initiation and progression of breach Zoning and properties of embankment fills, Crest width and surfacing, Dam height and reservoir capacity, Magnitude of inflows into reservoir
2. Attenuation of flood wave down valley	*a) Infrastructure embankments across valley:* Height of embankment; Size of openings in embankment, vulnerability to blockage; Vulnerability of openings to piping/scour failure leading to breach of embankment when water impounded.
	b) Degree of damage to buildings: Totally demolished or remain structurally intact; types of wall construction.
	c) Other factors: Hydraulic roughness; Erosion/deposition of sediment and debris during flood; Natural inflows Resolution and accuracy of topographic survey
3. Distance downstream at which dambreak analysis may be terminated	Extent to which valley widens out and catchment increases; Volume of reservoir and time to failure; Longitudinal gradient of valley, and thus whether celerity effects mean the flood travels as a catastrophic wave

Table 2: Key issues affecting impact of dam break flood on people

Issue	Comments
1. Integrity of buildings, and thus risk to occupants above the flood level	Dependent on form of building construction. Binnie and Partners (1986) suggest total destruction where vd > $7m^2/s$, and inundation only where vd< $3m^2/s$ with partial damage between these values
2. Number of people on transportation routes, or in open	This will vary with time. There is little data or guidance on which estimates may be based
3. Fatality rate (percentage of those in the inundated area that are likely to be killed)	This will vary across the flood plain, with warning time and time of day. A simplified approach using a single fatality rate for a reach of the valley is given in the Interim Guide to Quantitative Risk Assessment (Brown & Gosden, 2004)

TRANSPORTATION EMBANKMENTS

Transportation embankments across the flood plain may act as secondary dams. Where these structures are small in relation to the dam height, or are likely to be breached early in the flood, this effect may be insignificant. However there are situations in which they could significantly influence the extent of dambreak inundation and the risk to people.

The actual behaviour of transportation embankments while the dam break flood wave is passing will be complex. It will vary with time and may potentially include failure by overtopping, piping along the structure interface or scour at the downstream side of the structure. Depending on the timing of any such failure relative to the passage of the dam break flood, consequential breach of an infrastructure embankment could produce a further flood wave. The impact of such secondary events on emergency services and survivors from the initial flood wave could be particularly distressing.

Table 3 summarises the possible scenarios for the effect of transportation embankments on dam break flow. Where an embankment breaches this may lead to higher flows downstream, which in turn makes it more likely that transportation embankments further downstream would themselves fail. Thus overall there are a significant number of possible scenarios, depending both on the behaviour of individual embankments and the likelihood that the failure of any one embankment influences the probability of failure of any other.

Assessment of the risk to those in the potential inundation area is further complicated by the location of the population at risk:

a) upstream of an embankment, the number of people at risk is maximised if the embankment remains intact, as this maximises the flood depth, though the fatality rate could be lower because of reduced velocity;
b) downstream of an embankment, the risk to life is maximised if the embankment breaches at an early stage in the dam break flood because attenuation of the flood wave is minimised, leading to greater depths and velocities.

Table 3: Possible scenarios for the effect of transportation embankments on the extent of inundation

Response of transportation embankment to dam break flood	
Intact	Breached by dam break flood
Effect upstream of embankment	
Increased inundation upstream (water level backed up as determined by the size of culvert opening, the degree and timing of blockage and the height of embankment) but reduced velocity	Two stages of behaviour: a) Increasing inundation, limited to the overtopping depth necessary to breach the embankment b) After the breach, the upstream water levels will be governed by the breach dimensions; where the breach is wide, the control may change to the valley cross-section and slope
Effect downstream of embankment	
Peak flows attenuated by storage upstream of transportation embankment	Several phases a) Flow attenuated to capacity of openings b) Dam break wave from breach of the embankment c) Breach dimensions govern flow through breach, and thus downstream depths and velocities, when fully developed may be governed by the valley cross-section and bed slope

CASE HISTORY A : FLOOD BREACHES INFRASTRUCTURE

Case History A was a dambreak analysis recently carried out by Jacobs of a 24m high dam retaining a 1.3Mm^3 capacity reservoir. The analysis was run in an InfoWorks RS model built for the dam break analysis, with two scenarios analysed as shown in Table 4. There were no scenarios modelled with alternative combinations of breached and non-breached embankments. Breach discharge of the dam was estimated as 1800m^3/s, with a time to peak of around 20 minutes and an overall duration of about 40 minutes. The dam break flood was estimated to reduce to the natural 1% annual probability fluvial flood about 19km downstream, where the dam break model was terminated.

The sequence of modelling was an initial rapid impact assessment using the Interim Guide to QRA to determine screening level parameters, followed by a reconnaissance visit to the river valley prior to the detailed modelling. Although a detailed hydraulic model of the river channel was available for the immediate area downstream of the reservoir, it was only suitable for fluvial flooding as it did not include details of all the structures on the valley floor which would influence dam break floods. Crest elevations for infrastructure embankments were obtained directly from ground model data, whilst dimensions of openings not on the line of the main watercourse were measured during the site visit.

Table 4: Scenarios considered in breach analysis for Case History A

Scenario	Assumption	Comment
Intact	All infrastructure embankments intact	Provides an indication of the likelihood of failure of each structure from overtopping together with a maximum impounded water level behind each structure
Breached	All structures that are assessed as likely to breach fail instantaneously at the beginning of the simulation.	Provides an evaluation of the maximum effect of peak breach flow down the valley

Address Point data was used to give direct counts of the number of properties at risk from the breach inundation extent. The cost of damage to property was also estimated for the dam breach event and required the floor area of non-residential properties, which were manually measured using GIS. This proved to be time consuming and it is recommended that data on commercial property values and areas, as included in the National Property Database (developed for the Environment Agency), is obtained to streamline the process whenever possible.

The paper focuses on a railway embankment located 300m downstream of the reservoir, just downstream of the confluence of the impounded watercourse with a major river, and the effect of this on flood flows in the first 6km downstream. The railway is the first major hydraulic obstruction affecting the attenuation of the flood wave down the valley, and is assumed to breach in the "breached" scenario.

The railway embankment crest is approximately 8m above the surrounding floodplain. The crest width is 10m and the base width is 40m. There are four openings through the embankment, details of which are provided in Table 5. Figure 1 shows one of these openings, where a road is spanned by a cast iron bridge.

The likelihood of openings through the embankment being blocked was assessed in accordance with Table 3.4 of the draft Guide to Emergency Planning (Jacobs, 2006). This considers the minimum opening dimension and provides estimates of blockage at constricting structures in a dam break scenario. The effective open area is a product of the blockage assessment.

Table 5: Openings in the railway embankment

	Bottom width (m)	Height of crown (m)	Open area (m^2)	% blockage	Effective open area (m^2)
Culvert	10.0	1.7	11	50%	5.5
Road 1	4.5	5.0	15	80%	3.0
Road 2	12.7	7.9	70	50%	35.0
Road 3	7.6	7.1	30	80%	6.0

Likely breach dimensions for transportation embankments were determined using Froehlich (1995), and were assessed for this railway embankment as 30m base width and 45m width at the embankment crest and assumed to be additional to the existing openings. This would require removal of about 8000m^3 of fill. Comment on the need for research into breach dimensions in infrastructure embankments is given in section on debatable issues.

It was assessed as likely to breach on the basis of the fill being relatively loose as it was originally placed without modern compaction equipment and a 0.5m depth of overtopping, together with non-uniformity of crest levels leading to concentrated flow. The 10m crest width, and the good condition of the structures at the various openings, will however tend to prolong the period before full breach occurs.

Figure 2 presents a plan view and the inundation extent upstream of the embankment for the two scenarios in Table 4 (embankment intact and breached) whilst Figure 3 illustrates the peak flood levels down the river valley for the two scenarios. Table 6 summarises the change in flood depths and flow after failure of the embankment.
Key conclusions from this analysis regarding the effect of the transportation embankment are that if it does not breach:

a) Flooding backs up into the adjacent valley
b) Upstream water depth increases by 50%
c) Potential reduction in water level and fatality rate downstream, due to attenuation.

Table 6: Effect of transportation embankment on dam break inundation

Scenario (as Table 4)	Upstream of embankment			Downstream of embankment		
	Depth just upstream (m)	Flooded width[1] (m)	Properties flooded	Velocity. (m/s)	Discharge/ width (m^2/s)	Fatality rate (%)
Breached	9.0	180	255	1.7	7.9	3.0
Intact	13.5	195	327	1.5	7.3	2.7
Difference %	50%	8%	28%	-12%	-8%	-10%

Note: Width across valley at distance u/s of 2 x flooded width

Figure 1: Case History A - Bridge opening through railway embankment

CASE HISTORY B: FLOOD RETAINED BY INFRASTRUCTURE

Case History B is a situation where the transportation embankment is so high that it can retain the complete dam breach discharge without failing, thereby significantly reducing dam break flood flows downstream but increasing water depth upstream of the embankment. The dam break analysis was performed using DAMBRK.UK in the 1980s' and recently reviewed.

The dam which is considered in the analysis is 28m high and retains a 0.8Mm^3 capacity reservoir. The breach discharge is estimated as 1000m^3/s with a time to peak of just under 20 minutes. About 2.5km downstream of the dam a modern road passes across the valley on a 10m high embankment, as shown in Figure 4. The only openings are two 2.7m diameter corrugated steel culverts. This embankment could retain 1.5Mm^3 without overtopping and was considered as a potential Environment Agency flood storage site. The plan extent of dam break flooding with an intact and breached embankment is shown on Figure 5. There is a railway crossing just upstream, and although it would complicate flow conditions, it would be drowned out by water backing up from the much larger road embankment.

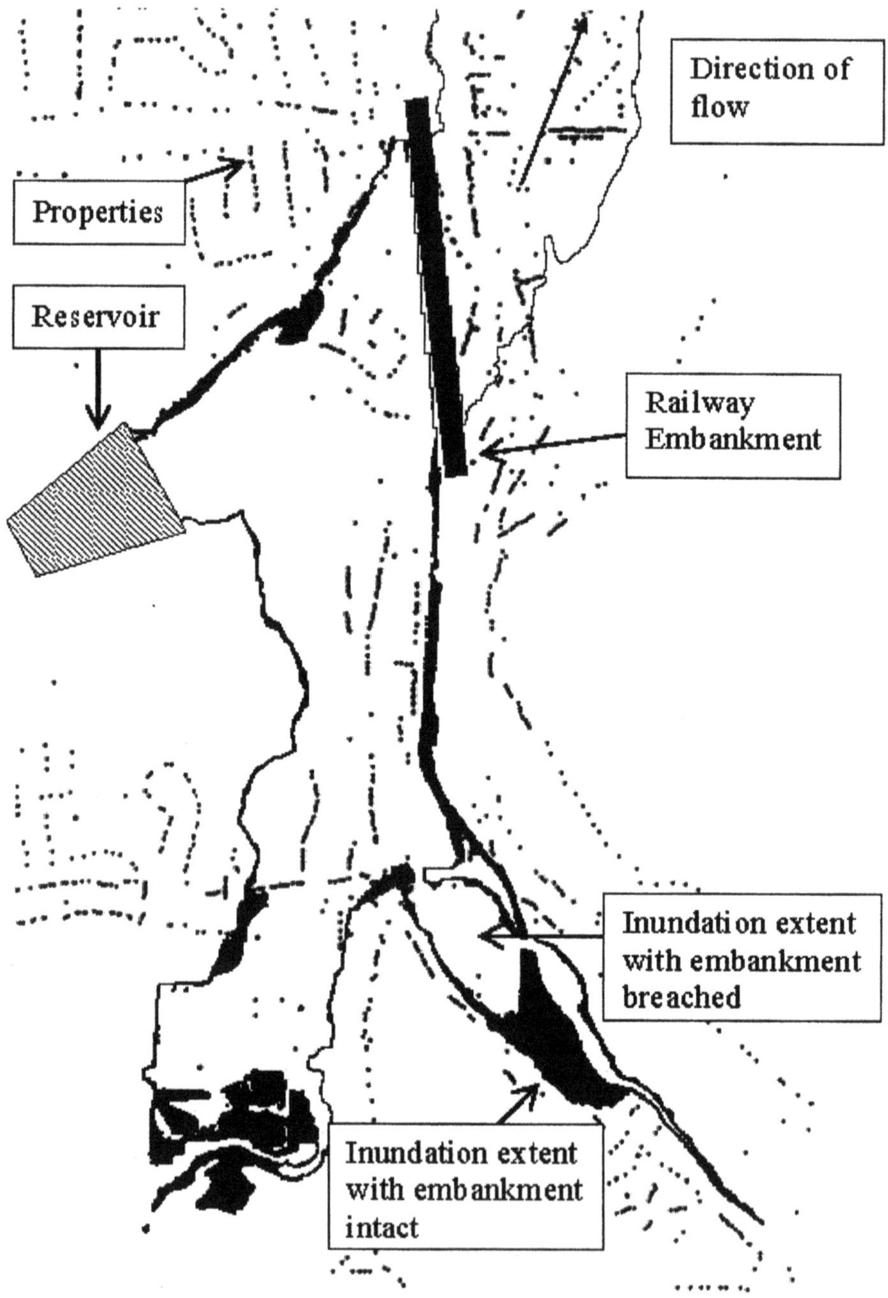

Figure 2: Case History A - Extent of dam breach inundation

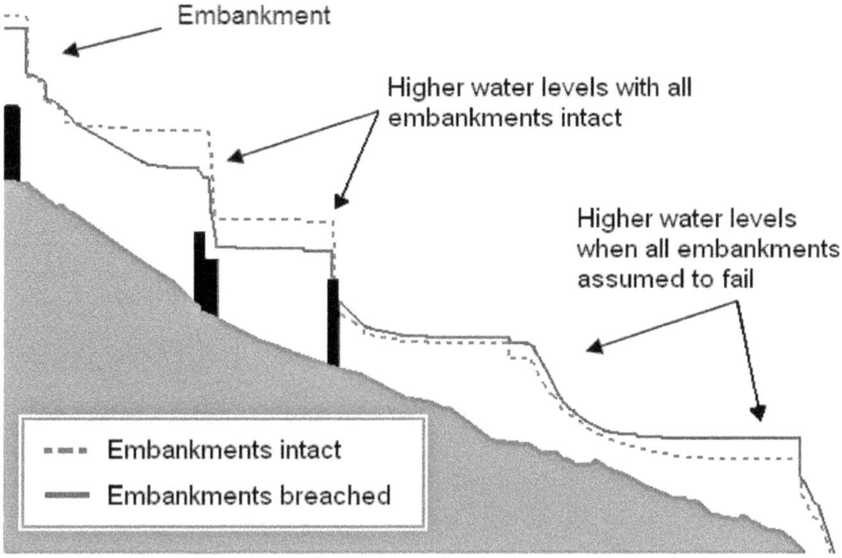

Figure 3: Case History A - Long section down river valley

There are small towns just upstream and downstream of the road embankment, with populations at risk in the breach scenario of 200 and 1600, respectively. The river is culverted through part of the town downstream. When the effect of the transportation embankment is taken into account the number of people at risk upstream will more than double, whilst that downstream would effectively reduce to zero.

The significant findings from considering the road embankment are

a) The entire storage in the reservoir could be retained upstream of the road embankment so that in a "sunny day" failure scenario there would be no major flooding in the town downstream
b) In a rainy day scenario, the embankment could be overtopped but breaching by overtopping would require prolonged overtopping to cut through the substantial crest width and tarmac road. Failure may alternatively occur at lower flows by piping or scour along the road culvert
c) In both cases the peak flood depth upstream of the road embankment would be up to 10m, compared to the 2.5m depth shown on the current maps. This would approximately double the number of houses flooded in the village upstream of the dam and significantly increase the extent of the area to be evacuated
d) Floodwater would back up a side valley parallel to the road, which includes isolated houses that are unlikely to be aware of their risk

Table 7: Effect of transportation embankment on dam break inundation

Scenario (as Table 4)	Upstream of embankment		Peak flow downstream of embankment (m^3/s)	Total population at risk (u/s and d/s)
	Depth just upstream (m)	Flooded width[1] (m)		
Breached	2.5	70	900	1800
Intact	10	240	90	450
Difference %	400%	340%	-90%	-75%

Note: Width across valley at distance u/s of 2 x flooded width (breached)

Figure 4 Crest of road embankment (Case History B) from left abutment

DEBATABLE ISSUES

What allowance should be made for infrastructure embankments in dam break analysis?

It is clear that transportation embankments can have a significant effect on the extent of flooding in the event of dam break, and thus on the location and magnitude of the population at risk and likely loss of life. It is also clear that the actual behaviour in the event of dam failure is likely to be complex, and thus difficult to predict accurately, particularly where there are several embankments in the downstream valley. It is suggested that the approach set out in Table 4 provides a good initial approach to bracket the potential range of behaviour. In some cases, for example where there is only one major centre of population downstream, it may then be appropriate to carry out a further analysis with the likely worst case for that population. This could be done using the assumption that all upstream transportation embankments are breached, but that those just downstream are not breached.

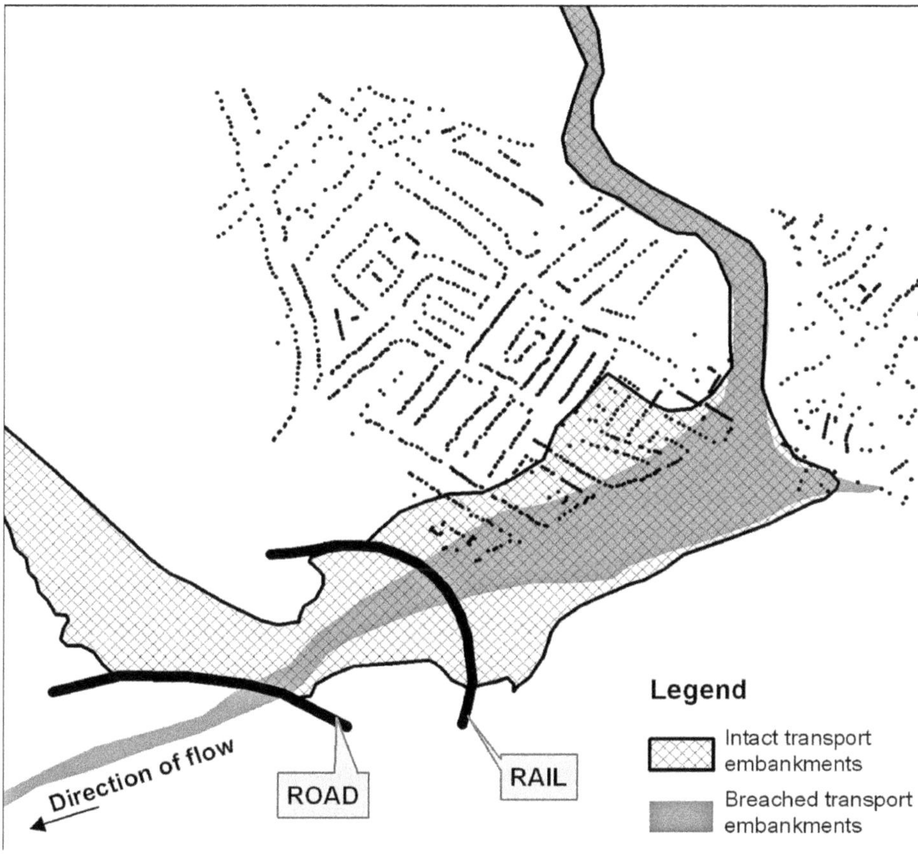

Figure 5: Case History B - Dam breach inundation extents

How should breach of transportation embankments be modelled?
The main issue is whether a transportation embankment would breach, with a secondary being how it would breach; whether by overtopping, or at the edges of existing openings through the embankments (e.g. Figure 1 and Table 5). It is also likely that existing methods for predicting initiation, rate of breach formation and final dimensions are not suitable for transportation embankments. It is suggested that further research is required into methods for prediction of the conditions of when breach of transportation embankments is likely to occur under dam break loading.

Where multiple scenarios of flow condition are analysed, which scenario should be used to estimate likely loss of life?
Techniques currently available for estimating the likely loss of life arising from dam failure include

a) Figure 9.1 of Interim Guide to QRA (2004), which is relatively simplistic, using averaged fatality rates inferred from previous incidents

b) Defra R&D project report FD2321 (2006) which requires knowledge of flood depth and velocity varying spatially with distance from the river, and also the percentages of very old and disabled or infirm in each sub-area

To obtain estimates of velocity varying spatially normal to the direction of flow requires 2D flood modelling which is not always available, and requires higher quality ground data to produce reliable results. The former has the significant benefit of simplicity, but introduces uncertainties where there are several dam break scenarios as to which set of flow parameters are used to estimate the fatality rate. Case History A conservatively used the velocities with breached embankments in combination with maximum depth. Initial fatality rates may not be the sole driver for emergency services: large areas of potential inundation affecting many people could be seen as a higher priority than small numbers of lives at risk in isolated areas since deaths from the secondary effects of flooding – exposure, infection – could be just as big a concern.

PRACTICAL IMPLICATIONS FOR DAM BREAK IMPACT ANALYSIS

The case histories presented depend on good quality data to make meaningful estimates of the consequences of failure. This includes

a) topography of the valley downstream from the dam
b) for transportation embankments and other structures crossing the valley,
    - the crest level (to allow an assessment of the depth and location of overtopping) – from Lidar data
    - the size and construction of openings through the structures
    - other information necessary to asses the likelihood of being breached
c) data on properties at risk should include floor area in non-residential property (e.g. as given in National Property Database)

Items 'a', the Lidar data in 'b' and 'c' are, subject to licensing issues, available from the Environment Agency. They would be expensive if they had to be purchased in the open market. Item 'c' is not normally readily available, and is most easily obtained by on-site survey which can be carried out to sufficient accuracy with hand-held laser "tapes". This is not possible where there is no public access but approximate assessments can be made from aerial photographs and mapping. The other alternative is asking the owner of the infrastructure for the dimensions of clear openings in his infrastructure, but this is likely to be time consuming. There could be benefit in a national database of data on infrastructure embankments relevant to dam breach analyses.

CONCLUSIONS

The prediction of the extent of dam break flooding, and its consequential impact on people, is inevitably subject to significant uncertainties including:

a) the speed of failure of the dam, and thus the magnitude of the peak breach outflow
b) infrastructure embankments across the valley downstream which can act as secondary dams attenuating the peak dambreak flow but causing additional flooding upstream, whilst subsequently failing causing secondary flood waves
c) the fatality rate, which will vary with time of day, vulnerability of the population and effectiveness of emergency planning, including effective warning
d) the quality of data used in making the above assessments

This means that any dambreak should only be considered as indicative, with effort put into defining the upper and a lower bound extent of flooding, rather than any single "correct" answer.

It is noted that in many situations the width of flooding near the dam is relatively insensitive to breach discharge and the presence of transportation embankments. This is because where the base of the valley is completely inundated, the sides of the valley are much steeper than the cross fall on the valley floor. However, changes in breach discharge and transportation embankments will have a greater impact on the risk to people and fatality rate, which are a function of depth and velocity.

This paper has examined the effect of transportation embankments downstream of dams and shown how the assumptions made have a significant affect on the consequences of dam failure. In the extreme they may store the dambreak flood, and effectively prevent flooding downstream. On the other hand they may be overtopped and fail, causing a secondary flood wave with major implications for emergency services and recovery operations.

It is recommended that all dam break analysis should catalogue and give site specific screening level consideration to infrastructure embankments in the valley downstream. The rapid dambreak method given in the Interim Guide to QRA has been extended to facilitate this. The screening assessment should then include the potential effect of downstream embankments on the quantitative estimate of the consequences of failure, of which the population at risk and likely loss of life are the key outputs. The outputs from the screening assessment may be used to determine how the transportation embankment should be included in a hydraulic model. It is suggested that in most situations the approach set out in Table 4 would be appropriate.

REFERENCES

Binnie and Partners (1986) *Modes of dam failure and flooding and flood damage following dam failure.* Final Contract Report to DoE Contract No PECD/7/7/184.

Binnie and Partners (1991) *Estimation of flood damage following potential dam failure: guidelines.* 1989 Report for DOE Contract no 7/7/259. Published by Foundation for Water Research FR/D 0003 March 1991

Brown, A. J., Gosden, J. D. (2004), *Interim Guide to Quantitative Risk Assessment for UK Reservoirs.* Thomas Telford.

Defra (2006) *Flood risks to people.* R&D project FD2321/TR2. www.defra.gov.uk/environ/fcd/research

Froehlich DC (1995) *Embankment dam breach parameters revisited.* Int. Water Res Eng Conf. Vol 1. pp887-891.

Jacobs (2006), *Draft Engineering Guide to Emergency Planning for UK Reservoirs.* Revision A04. www.defra.gov.uk/environment/water/rs/

Morris MW and Hassan M (2004) *European research on dambreak and extreme flood processes.* Proc BDS Conf. pp315-326

Tarrant FR, Hopkins LA, Bartlett JM (1994) *Inundation mapping for dam failure – lessons from UK experience.* Proc BDS Conf. pp282-291

Reservoir Flood Plans – Towards Implementation

IM HOPE, Technical Manager Reservoir Safety, Environment Agency
AK HUGHES, Director of Dams & Water Resources, Atkins Ltd

SYNOPSIS. The Water Act 2003 established a new role for the Environment Agency as the enforcement authority for the Reservoirs Act 1975 in England and Wales. Further change is heralded with the introduction of reservoir flood plans (or emergency action plans).

The Department for the Environment Food and Rural Affairs (Defra) is producing guidance on technical standards for reservoir flood plans entitled an *"Engineering Guide to Emergency Planning for UK Reservoirs"* ('the Guide'). Reservoir flood plans, which will be used in emergency planning by Local Resilience Forums, will increase awareness of the extent of potential inundation areas from reservoirs and the challenges faced by all those involved in managing and communicating flood risk.

The current programme for introducing reservoir flood plans is to carry out a public consultation and impact assessment on the proposals, led by Defra, in autumn 2008, followed by the Ministerial direction and the introduction of the formal requirement for reservoir flood plans in spring 2009.

This paper provides a review of the current developments made towards completing the Guide and the major changes which will result from the requirement to produce reservoir flood plans. It also highlights the issues which arose during the earlier informal consultation process.

INTRODUCTION

Following the Water Act 2003, responsibility for reservoir safety in England and Wales was transferred to the Environment Agency. As the enforcement authority, the Environment Agency is responsible for assuring the safety of the nation's 2,100 reservoirs by enforcing the Reservoirs Act 1975. The Environment Agency aims to work alongside reservoir undertakers (operators, users and owners), supporting them in meeting the requirements of the Reservoirs Act 1975. This role is carried out by the Reservoir Safety team, based in Exeter.

The Environment Agency is not directly responsible for the safety of reservoirs. Reservoir undertakers are responsible for ensuring safety, compliance with the law and assessing the flood risk posed by their reservoirs. As described in Hope (2006) the Environment Agency is responsible for enforcing the Act by making sure that undertakers fully comply, warning and ultimately prosecuting those that don't.

The Environment Agency has compiled a register, which can be viewed at its Area offices, of reservoirs in England and Wales to which the Reservoirs Act 1975 applies (those capable of holding at least 25,000m^3 of water above lowest natural ground level). Known as 'large raised reservoirs', these are owned by some 772 businesses and individuals, and include 185 flood storage reservoirs, which the Environment Agency itself owns and operates. An investigation has recently been concluded into a further 320 existing reservoirs, 119 of which are now subject to the Act (Goff & Hope 2008).

RESERVOIR FLOOD PLANS

Flooding from reservoirs can be caused by an uncontrolled breach of the dam, perhaps caused by over-topping during a severe rainfall event. The consequences of this kind of flood could be catastrophic. However, the chance of a dam failing is considered to be 'low'. The average age of dams in Great Britain is over 110 years and there are between four and six emergency drawdowns of reservoirs each year (this action is seen as a last resort to prevent dams failing).

Great Britain is arguably behind the rest of the developed world in producing reservoir flood plans (or emergency action plans) for reservoirs. Under the Water Act 2003 undertakers must produce flood plans for their reservoirs where directed by the Secretary of State.

Defra is the government department responsible for reservoir safety. Defra is currently funding the production of an "Engineering Guide to Emergency Planning for UK Reservoirs". This work, carried out by Atkins, will set out the requirement for flood plans in detail and how it will be applied to the different categories of reservoirs. A consultation on the government direction is expected to be published by Defra during autumn 2008 and the direction will be given to undertakers in spring 2009. It is expected that undertakers will have five years to prepare and submit compliant reservoir flood plans.

The draft proposals, developed in 2006, envisaged that the reservoir flood plan would include some or all of the following, depending on the consequences of the reservoir failing:

- on-site emergency plan. To set out what the undertaker would do in an emergency to try to prevent the dam failing.
- full dam breach and inundation analysis. To provide a plan of the area inundated and information on velocities and depths of flow.
- plan for liaising with external organisations. To define and test channels of communication between the undertaker and the Local Resilience Forum.

On-site emergency plan

An on-site plan should make sure the undertaker is prepared for an emergency at the dam. The aim of preparing and maintaining an on-site plan is to:

- identify ways of preventing the dam failing in an emergency or incident and to make sure that methods are in place, that they are workable, and staff understand what to do;
- show what to do to delay the dam failing (if it cannot be prevented) and reduce the amount of water released so that work can be done off-site to reduce the loss of life and damage caused

Undertakers will be able to use the on-site plan to:

- brief staff and subcontractors who are unfamiliar with the dam, so that they can help to prevent the dam failing in an emergency;
- set out what steps to follow, and any other information needed to manage an emergency.

The on-site plan must be kept up-to-date. Staff and external organisations, who have a role to play in the emergency, also need to be trained and to practice carrying out the plan.

Last year the Environment Agency produced a guide to help undertakers prepare and maintain an on-site plan which can be downloaded from www.environment-agency.gov.uk/reservoirsafety. Undertakers are advised to produce on-site plans immediately, if they haven't already, rather than waiting until it becomes a regulatory requirement.

Inundation analysis

The inundation analysis will model the effect of a dam breach and the flood water passing downstream to establish the consequences of a dam failing. It will identify property and critical infrastructure that could be affected. It will also provide information on depth and velocity of flow which will help in emergency planning and in assessing casualties.

Most water companies and a few other reservoir undertakers have carried out inundation mapping to identify the extent and consequences of a dam failure leading to release of impounded water. Generally, this information is paper-based and varies in standard. Although this information has, in many cases, been available for some years, Defra/CPNI (Centre for the Protection of National Infrastructure) previously advised that it must be kept confidential.

It has now been agreed, however, that the existing inundation maps should be released to Category 1 responders under the Civil Contingencies Act 2004, but that the full information should only be used for emergency planning at this stage. This early release of information does not affect the plans in progress to introduce the formal legal requirement (under the Water Act 2003) to produce reservoir flood plans, but is a useful pilot process.

Following a request from Defra in March 2008, it is now proposed that simplified inundation mapping will be carried out nationally by the Environment Agency to provide a baseline assessment of all reservoirs falling within the Act. This will ensure consistency, confirm dam categories (ICE 1996), allow emergency planners to identify effects on critical infrastructure, and identify the area of influence for future town and country planning. Under the proposal undertakers will not have to provide maps for all dams, but they may have to provide more detailed information and analysis for the highest risk dams.

The Environment Agency awarded a contract to Mott MacDonald and JBA Consulting for a trial inundation mapping project in May 2008. The trial is being carried out in North West England, supported by Government Office North West, to define the specification for this work. This work is being carried out under the Environment Agency Strategic Flood Risk Mapping Framework

The incident at Ulley reservoir in 2007 brought a number of issues to the fore. Although Ulley was a category C reservoir, there was significant infrastructure downstream, including two high pressure gas mains, electricity pylons, a sewage treatment works, a substation, an 'A' road and the M1 motorway. During the incident three communities were evacuated as a precaution. The reservoir has now been classified category A. Local and Regional Resilience Forums hold details of the type and location of critical infrastructure. Whilst some of this information may be available to the public (through OS maps, Google Earth etc.), it will not always be obvious.

Critical infrastructure will affect the categorisation of the reservoir. The flow chart in Figure 1 shows the process for populating the maps and how the issues that arise might be resolved. This proposed method will ensure a consistent approach to categorising dams. It will inevitably lead to a number of reservoirs being re-categorised. However, the project still has to resolve how accurate the maps should be (which will impact on costs) and how/if undertakers can appeal against re-categorisation of their asset by a third party based on information that by its nature, must remain confidential.

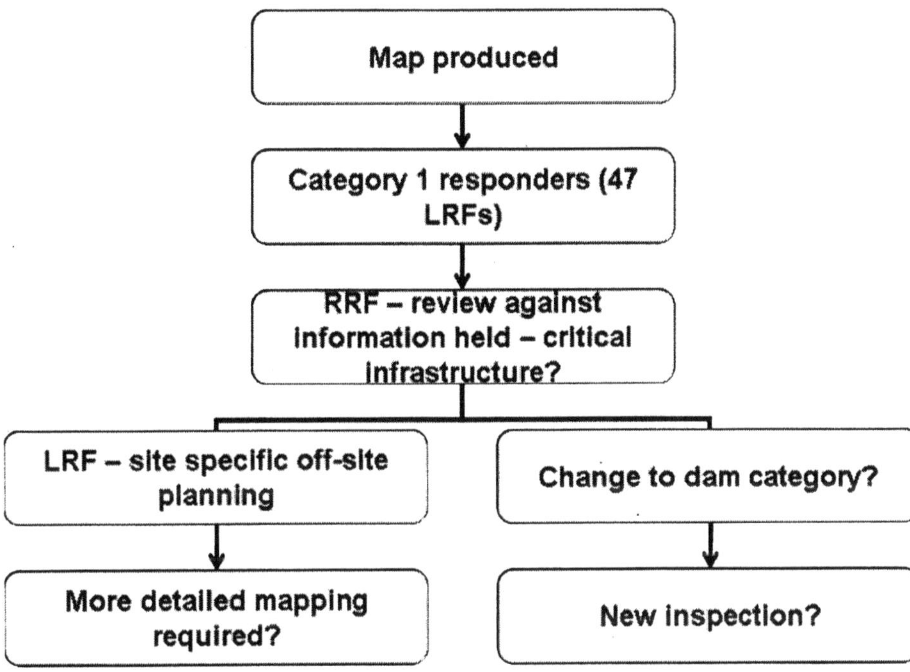

Figure 1: Process for population of inundation maps with critical infrastructure.

Developing inundation maps will allow emergency planners to provide a site-specific response which is more useful than the general advice already provided in the National Risk Assessment for major dam failure (H44). This information will be used to identify whether further, more detailed mapping is needed.

To gain the greatest benefit from the proposed national mapping exercise, discussions have been held with emergency planning representatives to identify the level of information needed for emergency planning purposes.

Dambreak analysis

The accuracy of current dambreak inundation mapping is variable – depending greatly upon assumptions made, models used and topographic and modelling data resolution. In particular, assumptions for dam failure, including breach modelling, can contribute greatly to the overall uncertainty. For example, the IMPACT project (Morris & Hassan 2005) demonstrated that uncertainty in modelling dam breach (for the Tous case study used) was responsible for up to 50 per cent of the uncertainty in extreme flood level prediction further along the valley.

More recent work has been carried out under Task 6 of the FLOODsite project, addressing the modelling of breach initiation and growth. FLOODsite report T06-06-03 *"Breaching processes: a state of the art review"* gives a summary of current state of the art methods, along with details of key developing areas (for breach prediction) and current 'best practice' models.

As part of the inundation mapping trial, Atkins with HR Wallingford will be reviewing dam break modelling as a whole, and as used in the UK, in order to advise on the most appropriate modelling techniques to use. This will be incorporated into the Engineering Guide and the specification for Environment Agency national inundation mapping.

Off-site emergency planning

Off-site emergency planning is being carried out under the Civil Contingencies Act 2004. Part 1 of the Act provides a statutory framework for civil protection at the local level, which applies across the UK. It sets out clear expectations and responsibilities for front line responders to make sure they are prepared to deal effectively with emergencies. It divides local responders into two categories:

a) Category 1 responders - central to most emergencies (for example emergency services police, fire, ambulance, local authorities, NHS bodies, Environment Agency).
b) Category 2 responders (for example Health and Safety Executive, transport and utility companies) - required to share information and co-operate with Category 1 responders as part of emergency planning.

Category 1 and 2 responders form Local Resilience Forums (LRFs), which help co-ordinate emergency planning, training and exercises locally. Category 1 responders have a duty to carry out risk assessments and produce a Community Risk Register covering their area. LRF members will have to assess the need for, and carry out, detailed off-site emergency planning for those reservoirs posing a significant risk. It is anticipated that this role will be led by upper tier local authorities (County or Unitary Councils).

The risk of a dam failing is a combination of the likelihood of failure and the consequences of it happening. Assessing the likelihood of any dam failing is difficult. The dam failing due to overtopping would be as a result of a flood greater than the spillway capacity (unless it is blocked). This is a relatively low level of risk compared to other hazards. In practice, an earthfill dam is more likely to fail as a result of leakage or piping, rather than overtopping or perhaps due to a spillway failure (as at Ulley). No dams have failed causing deaths in England and Wales since the Reservoirs (Safety Provisions) Act 1930 came into force, but recent evidence given to the Pitt Review (Hinks & Mason 2007) suggests the risk of dam failure resulting in loss of life in the UK could be as low as 1 in 45 years.

Whilst there are a number of incidents each year, where reservoirs have to be drawn down to prevent the dam failing, we can make limited conclusions about the chance of dams failing. Following instruction from Defra, the Environment Agency has recently introduced a post-incident reporting procedure to share lessons learned from reservoir incidents. This process was jointly developed with Halcrow Group Ltd and is starting to provide much needed information. However, reporting is voluntary and the database is still in its infancy. A copy of the first annual report can be found on the Environment Agency's website.

Whilst undertakers have information on the condition of dams, this has not generally been ranked in terms of severity of risk. Risk ranking has certainly not been applied between undertakers. This contrasts with the situation in other developed countries, for example in Australia where Stewart et al (2007) describes how Portfolio risk assessment informs prioritisation and funding of dam safety upgrading projects. Portfolio risk assessment is also applied to the regulation of dams. See www.damsafety.nsw.gov.au.

An Interim Guide to Quantitative Risk Assessment (QRA) was produced by Defra to address this problem. The Environment Agency is sponsoring an overhaul of the interim QRA Guide as part of the Reservoir Safety Research and Development Programme. Once developed, this will allow a more accurate assessment of the risk to our dams.

RESPONSE BY LOCAL RESILIENCE FORUMS

LRFs have prioritised their response to dam failure against the other risks in their area based on generic advice. Currently LRFs are not informed about the probability of individual dams failing.

The extent to which Category 1 responders will prepare detailed off-site emergency plans for dam failure will depend on how high they believe the risk to be, compared to other emergencies they may have to deal with. Therefore, LRFs across the country may respond differently to potential dam failures. Their response can range from:

- detailed plans setting out their response for individual dams to
- a standard emergency plan for all dams setting out roles and responsibilities of everyone involved.

If a detailed plan is produced, it will fit within the group of emergency plans set out in the Civil Contingencies Secretariat document *Developing a Multi-Agency Flood Plan* (CCS 2008).

The police are normally responsible for co-ordinating major emergencies and deciding whether to evacuate the local area. This control is exercised through the Gold, Silver and Bronze Command structure. The Defra project is seeking to address how this is reflected in the specification for reservoir flood plans.

The need to have and share information was reinforced in the "lessons learnt" report following an exercise on a dam failure in a major northern city in December 2006. One of the key findings from silver control was that emergency plans "need to be shared between responding organisations".

PLANNING GUIDANCE
Inundation maps will help to make decisions about land use planning policy. Although the likelihood of inundation is low, the following important elements need to be considered:

- safety of people within buildings
- safety of buildings
- safe entry and exit from buildings
- emergency services being able to evacuate or rescue people from buildings
- location of critical infrastructure.

Inundation maps will also provide a vital link between developers, planners and the reservoir undertaker, so that the undertaker is more closely involved in the planning process.

This is important because the category of the dam can change as a result of development. If this happens, the undertaker can face significantly higher costs (for example constructing a larger spillway) after the next inspection unless these costs are identified and perhaps borne by the developer.

As a result of lobbying, flooding from reservoirs has now been identified as a potential flood risk in the recently published Planning Policy Statement. Appropriate guidance principles are currently being developed.

RESPONSE TO FLASH FLOODING

Flooding from a dam failure is similar in some ways to flash flooding from small catchments as a result of intense rainfall.

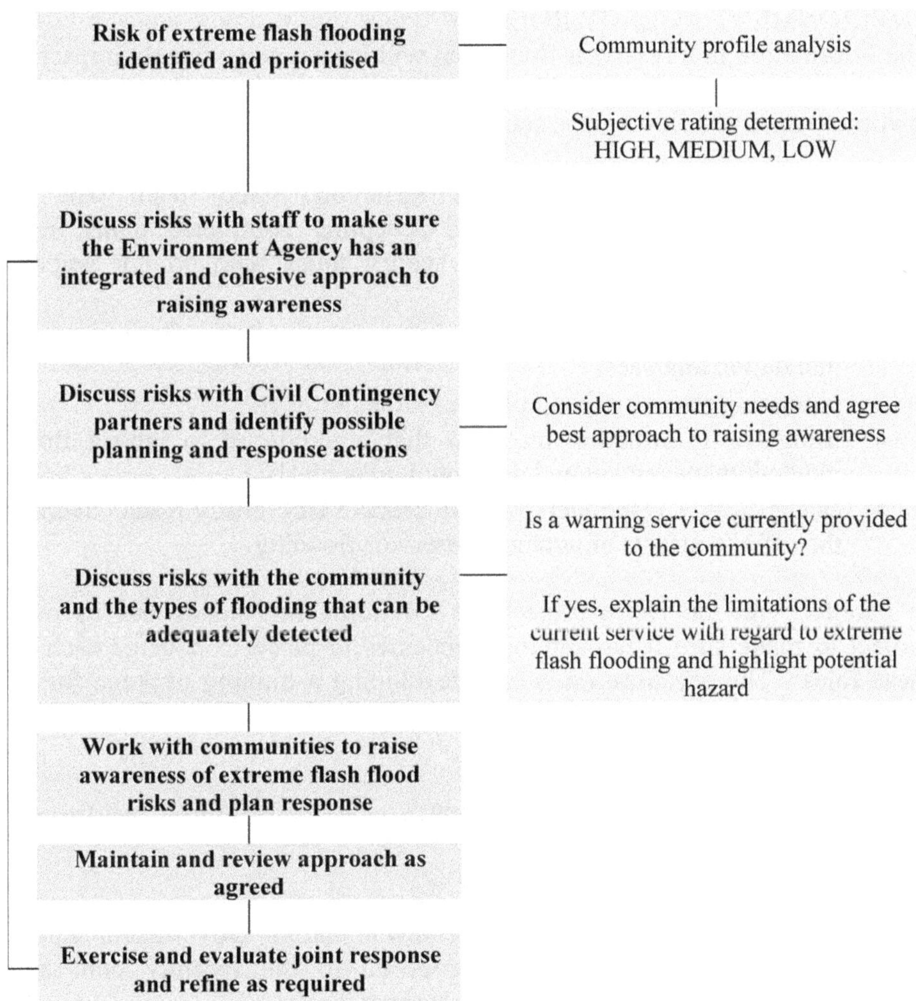

Figure 2: Procedure for flash flooding

The Environment Agency has developed a procedure for identifying rapid response catchments and involving communities at risk to raise awareness and recommend suitable mitigation measures. A similar process may be appropriate for catchments below dams.

The Environment Agency's policy is to make sure that all those living, working, and on holiday in areas at risk from extreme flash flooding are made aware of the hazard and know what actions to take if flooding occurs. Work in communities situated in catchments that could suffer from extreme flash flooding is set out in Figure 2 and will be mainly focussed on planning and raising awareness.

ENVIRONMENT AGENCY ROLE

The information in a reservoir flood plan will have a considerable impact in Environment Agency Area offices. The Environment Agency's role in relation to reservoir flood plans can be summarised as follows:

- as enforcement authority the Reservoir Safety team will be responsible for examining and accepting completed plans from undertakers and taking action against those who do not prepare them;
- providing information to undertakers to help them prepare inundation analyses;
- producing reservoir flood plans as an undertaker;
- using the information on areas that could flood to inform flood incident management and development control;
- as Category 1 responders, to help prepare emergency plans covering the off-site effects of potential reservoir flooding.

In parallel with the Defra project the Environment Agency has its own project to make sure it has internal processes in place to manage each of these roles. The organization is also developing a training package for its staff, panel engineers and undertakers.

THE PITT REVIEW

The Pitt Review interim report (Pitt 2008) contains two interim conclusions that directly relate to reservoir safety.

IC62 "The interim conclusion of the Review is that the Government should implement the legislative changes proposed in the recently published Environment Agency biennial report on reservoir safety."

The key changes proposed are outlined below:

- **Better risk-based definition of a reservoir within the Act** - At present the definition is purely volumetric, based on a minimum capacity of 25,000 m^3 of water. A definition of reservoirs falling within the Act that takes into account the consequences of a dam breach is required.

- **Funded powers to act at reservoirs with no owner** - At present there is a gap in the law, so that reservoirs that are situated on land that is disclaimed following business failures have no legal owner and, therefore, no legally responsible reservoir undertaker, unless the Crown chooses to exercise its right as 'keeper of last resort'. The Crown does not always choose to take ownership in these circumstances.
- **Mandatory post-incident reporting** - Defra has charged the Environment Agency with keeping a database of incidents, so that lessons can be learned and disseminated to the reservoir industry, but at present the system is voluntary. Undertakers may not always choose to inform the Environment Agency about emergency incidents at their reservoirs, for commercial or other reasons. It is believed that it would be in the public interest for it to be made mandatory for reservoir undertakers to report their emergency incidents to the Environment Agency.
- **More flexible enforcement powers** - At present, enforcement options are prosecution, formal caution or warning letter. The Government's Macrory Review is bringing in a variety of alternative intermediate enforcement penalties, such as administrative fines and reputational sanctions. The Environment Agency believes that applying these options to reservoir safety will make enforcement more efficient and more effective.
- **Better quality inspection reports** – The Environment Agency believes that there should be a quality review process for inspection reports, managed by a professional and technical body such as the Institution of Civil Engineers.
- **Better enforcement powers for reservoir flood plans** - There is currently no statutory requirement for a panel engineer to sign off a reservoir flood plan. Neither is there a power for the enforcement authority to serve notice on a reservoir undertaker to prepare a reservoir flood plan, nor to exercise reserve powers to prepare one on behalf of an undertaker and recover the costs of doing so.

IC63 "The interim conclusion of the Review is that all reservoir undertakers should be required by Defra to prepare inundation maps and share them with Local Resilience Forums to improve Community Risk Registers and emergency planning."

In his speech to RUSI Critical National Infrastructure Conference on 16 April 2008 Sir Michael Pitt called for critical infrastructure operators and security organisations to be more open about the risks that exist. He observed that a masonry dam had inundation maps, which showed that a school was in the path of a flood should the dam fail, but the head teacher is unaware of the risk and no escape routes have been prepared because the maps are not available to the public. He concluded that we need to move to a situation where we are making more effort to communicate risk accurately and debate risk in a more public way.

COMMUNICATION STRATEGY

The many strands of this project need to be supported by a comprehensive communication strategy. This will range from the formal impact assessment before the new regulatory requirement is introduced, to making sure that public access to inundation maps does not cause unnecessary alarm. This is seen to be fundamental to the success of the system.

This communication strategy will cover both internal and external needs and extend to specifying and implementing training courses for reservoir panel engineers, undertakers and relevant Environment Agency staff.

The strategy will build on the experience of other countries. For example, in Switzerland inundation maps have been in the public domain for a number of years and they have recently been introduced in Spain. De Cea Azanedo et al (2007) highlight that communicating information to the public at risk was the most complex part of implementing emergency action plans for dams in Spain.

WAY FORWARD

This paper has outlined most of the issues being developed. By forging a strong working relationship with those who will be using the reservoir flood plans, a practical and workable specification and process can be developed.

Whilst there will always be a minor risk from both extreme floods and dam failure, this project will make sure that emergency action plans are in place to further improve the safety of our reservoirs and those that live and work near them.

ACKNOWLEDGEMENTS

The views expressed in this paper are the personal views of the authors and not necessarily those of the Environment Agency or Atkins.

REFERENCES

Civil Contingencies Act 2004, HMSO London

Civil Contingencies Secretariat 2008. *Developing a Multi-Agency Flood Plan (MAFP)* – Guidance for Local Resilience Forums and Emergency Planners

De Cea Azañedo, J.C., Yagüe Córdova, J. and Fruns Montoya, A. 2007. *Emergency action plans for dams: A new method for public safety in Spain.* ICOLD 75th Annual Meeting, Saint Petersburg, Russia.

Environment Agency, 2005. *A Better Place to Live – Working Together for the Safety of Our Reservoirs*

Environment Agency, 2006. *Creating a Better Place: corporate strategy 2006-11.* Bristol: Environment Agency

Exercise Loxley Final Report 2006.

Goff, C.A. and Hope, I.M. 2008. Investigation of potential reservoirs. *Proceedings of the 15th British Dam Society Conference.* Thomas Telford, London.

Hinks, J.L. and Mason, P.J. 2007. *Reservoir Safety in the United Kingdom.* Paper to the Pitt Review

Institution of Civil Engineers 1996. *Floods and Reservoir Safety Third Edition*

Hope, I.M. 2006. *Reservoir Safety in England and Wales – A Time of Change.* ANCOLD Conference 2006.

Morris, M.W. 2008. *Breaching processes: a state of the art review.* FLOODsite Report T06-06-03

Morris, M.W. and Hassan, M. 2005 *Conclusions & Recommendations from the IMPACT Project WP5: Combined Risk & Uncertainty*

Pitt Review – Interim Report March 2008

Planning Policy Statement 25, HMSO, London

Reservoirs Act 1975, HMSO, London

Stewart. D, McGrath. S and Nabbs. D, *Prioritisation of Dam Safety Management in a Large Water Business.* ICOLD 2007.

Water Act 2003, HMSO, London

Assessing Dam Breach Hazard in Cyprus

T. D. BLYTH, Mott Macdonald Ltd, Cambridge, UK
Dr K. KYROU, Water Development Department, Nicosia, Cyprus
S. PATSALI, Water Development Department, Nicosia, Cyprus

SYNOPSIS. Accession to the European Union has driven dam safety reform in Cyprus. The Water Development Department (WDD) a sub-directorate of the Ministry of Agriculture, Natural Resources and Environment is responsible for ensuring Cyprus meets the requirements of the European "Directive 2007/60/EC on the Assessment and Management of Flood Risks". The directive requires member states to undertake flood risk assessments for all river basins and produce flood hazard risk maps and management plans where a significant risk is found to exist. The WDD is now in the process of developing Reservoir Flood Plans for its 56 International Commission on Large Dams (ICOLD) registered dams. The WDD engaged Mott Macdonald to provide advisory support and technical training in Hazard Ranking, Dam Break Hazard Analysis (DBHA), Flood Mapping and Emergency Planning.

Initial Hazard Ranking, based on dam height, reservoir capacity, downstream reach, and community characteristics, provided a, quick, semi-rigorous and rational basis for prioritizing WDD's many reservoirs for future hazard analysis and planning. Current UK guidelines and methodology - modified to account for local conditions – were then applied to assess hazard posed by individual reservoirs and to develop contingency plans for those reservoirs. Rapid and Standard methodology outlined in the Draft Engineering Guide to Emergency Planning for UK Reservoirs (2007) (herein The Draft Guide) were used to assess flood risk for a number of dam breach scenarios, at each reservoir. Risk assessments based on the Interim Guide to Quantitative Risk Assessment for UK Reservoirs (Brown & Gosden, 2004)) was also undertaken.

A number of observations and conclusions were drawn from the work which may prove valuable for others preparing Reservoir Flood Plans and on-site Emergency Plans for Reservoirs in the UK and overseas.

BACKGROUND TO STUDY

Key Drivers

In an era inexorably linked to Climate Change, community perception of the risk posed to society by engineered structures such as large dams, is at an all time high. The Ulley Dam incident in late 2007 helped raise awareness in the UK whilst widespread damage and loss of life in New Orleans following Hurricane Katrina in 2005 – a result of failed flood levees- has been a catalyst for flood management policy review and dam safety reforms globally.

The European Floods Directive

The European Directive on the Assessment and Management of Flood Risks[1] (herein *The Floods Directive*) came into force on 26 November 2007 after EU environment ministers agreed there was a need for greater European co-ordination on flood risk management. The Floods Directive is designed to help Member States prevent and limit floods and their damaging effects on human health, the environment, cultural heritage, infrastructure and property. Whilst The Floods Directive does not specifically refer to flooding resulting from dam breaks it does require flood hazard maps to be produced for floods with a low probability, implying consideration of extreme event scenarios, such as dam break. The timetable for implementation of The Floods Directive is clearly set out. Member States have until 2009 to transpose the Directive into domestic law. In addition, Member States have to undertake preliminary flood risk assessments of their river basins and associated coastal zones by 2011, to identify areas where potential significant flood risk exists. By 2013, where real risks of flood damage exist, they must complete flood hazard and risk mapping for these areas. By 2015, flood risk management plans which document measures to reduce the probability of flooding and potential consequences must be developed for these zones. The management plans will address all phases of flood risk management but focus particularly on prevention, protection and preparedness.

UK Legislation

In the UK, Safety legislation for reservoirs was first introduced in 1930 after several reservoir disasters resulted in loss of life. This Act was superseded by the Reservoirs Act 1975, which today provides the legal framework to ensure the safety of large raised dams with reservoir capacities above ground level, greater than 25,000 cubic metres. The Water Act 2003, supplemented the Reservoirs Act and gave the Secretary of State powers to

---

[1] EC Directive 2007/60/EC http://eur-lex.europa.eu/LexUriServ/LexUriServ.do?uri=CELEX:32007L0060:EN:NOT

direct reservoir undertakers (operators, users and owners) to prepare Reservoir Flood Plans, setting out the action they would take in order to control or mitigate the effects of flooding likely to result from any escape of water from the reservoir. The Department for Environment, Food and Rural Affairs (Defra) has funded the production of an *Engineering Guide to Emergency Planning for UK Reservoirs* to assist panel engineers and reservoir owners comply with the requirements of both Acts. The guide is currently available in draft form and will go out to formal consultation in Summer 2008. It will be published in 2009 when the Secretary of State will issue a direction to undertakers, making Reservoir Flood Plans a legal requirement. This extends the guidance on assessment of the likely consequences of dam failure developed in Brown & Gosden (2004).

In parallel with the aforementioned reforms, Defra is coordinating the transposition of The Floods Directive into UK law. Since August 2007, a UK Floods Directive liaison group involving both Defra and Environment Agency (EA) staff have been investigating options for capturing the requirements of The Floods Directive. The EA anticipate that many of the requirements will be covered by the existing Acts, with those outstanding to be potentially incorporated into the proposed 2009 Flood Bill.

A Water Scarce Island

Located 33° east of Greenwich and 35° north of the Equator Cyprus, the Mediterranean's third largest island, is water scarce. An arid climate of long, hot and dry summers and mild winters, gives rise to a mean annual precipitation (including snowfall) of just $460mm^2$. This equates to 2.670 million cubic metres (MCM) of rainfall over the total surface of the Government controlled southern region. The area's modest rainfall is irregular and unevenly distributed. Most of the precipitation falls in the Trodos Range (1,952 metres) in the centre of the island with minimum precipitation observed in the eastern plain and the coastal areas. Water scarcity is further compounded by Cyprus's abundant sunshine (11.5 hours per day on average in summer and 5.5 hours in winter) and winds, which together result in evapotranspiration of around 86%. This leaves just 370 Mm^3 of surface water available for development. None of Cyprus's 14 major rivers have perennial flow. In recent times, consecutive dry years and depleting groundwater reserves - a result of seawater intrusion into coastal aquifers - has lead to increased strains on water resources.

Since the neolithic period Cypriots have been drilling deep wells to combat water scarcity. Hardly surprisingly, today Cyprus's Water Policy is governed by the philosophy "not a drop of water to the sea". Cyprus's

[2] WDD Website http://www.cyprus.gov.cy

population of 900,000 rely heavily on groundwater and reservoir storage for reliable, year- round water supply. Following independence in 1960 a number of large projects were mounted to increase reservoir storage capacity, which today stands at about 330 million cubic metres. Providing this storage are 56 ICOLD registered[3] dams, ranking Cyprus as the most densely dammed country in Europe.

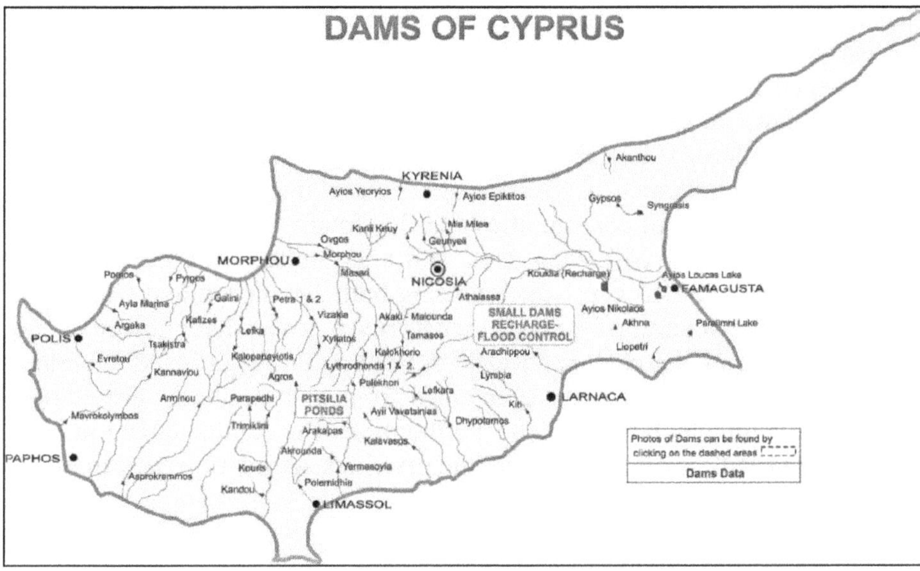

Fig 1. Location map of large dams in Cyprus *(Courtesy Water Development Department, Nicosia, Cyprus)*

Cyprus's major population centres such as Limassol and the capital Nicosia, lie in the floodplain downstream of large dams. Cyprus's 2004 accession to the EC means that it is now committed to the Flood's Directive. The WDD's traditional roles of implementing Cyprus's Water Policy and securing its water resources have been expanded to include implementation

---

[3] For inclusion on the ICOLD Register of Dams, a large dam is defined as any dam above 15 metres in height (measured from the lowest point of foundation to top of dam) or any dam between 10 and 15 metres in height which meets at least one of the following conditions: a) the crest length is not less than 500 metres ; b) the capacity of the reservoir formed by the dam is not less than one million cubic metres ; c) the maximum flood discharge dealt with by the dam is not less than 2 000 cubic metres per second ; d) the dam had specially difficult foundation problems ; e) the dam is of unusual design.
ICOLD : Technical Dictionary of Dams 1994

of the Floods Directive. Mott Macdonald was commissioned by WDD to provide advisory support and internal capacity building via technical training in Hazard Ranking, DBHA, flood mapping and emergency planning.

METHODOLOGY

Hazard Ranking

Faced with a large number of dams to analyze, WDD required a tool to focus investigations, identify priority dams and plan for future analyses. Hazard Ranking provided a quick, semi-rigorous and rational assessment of the Hazard potential of dams. In this paper 'Hazard' is defined as the consequence of a failure occurring whilst 'Risk' is the probability of that failure. Brown & Gosden (2004) recommends two approaches to Hazard Ranking. Cullen (1989) and Thompson (1994) both use a number of physical parameters such as dam height; reservoir capacity, the size and distance downstream of communities and floodplain topography to quantify the hazard associated with each dam. An analysis was carried out for all large dams under WDD's jurisdiction using both Thompson and Cullen. In practice, the application of both methods proved somewhat subjective (i.e. valley shape parameter can vary widely depending on the point of reference). Nonetheless, similar results were obtained from both methods, providing a rational basis on which to proceed. Furthermore, the results agreed in large with the expectations and confirmed the priorities set by experienced WDD Engineers.

Dam Break Hazard Analysis

DBHA provides information on the consequences of possible dam break scenarios for use in emergency action planning and risk analysis (downstream risks to population, property and environment). Depending on the level of risk involved, dam-break analyses may vary considerably in scale and detail. As a result, the *Draft Engineering Guide to Emergency Planning for UK Reservoirs* proposes two different approaches to DBHA: The Standard Method and the Rapid Method. The choice of method to apply to a particular situation should be based on sound engineering judgment and consideration of the complexity of the downstream floodplain and the population and infrastructure at risk.

In this study DBHA's were undertaken on four pilot reservoirs identified as having relatively high hazard potential during initial Hazard Ranking. Both Standard and Rapid Methodology were employed with mixed success.

Standard Method

The Draft Guide does not specifically define the process to be followed when undertaking a DBHA, but rather stipulates the scenarios to be analysed and the content of outputs. The Standard Method requires computational hydraulic modeling of the flood progression downstream of the dam to achieve the detailed outputs that are specified. Outputs required from each modeled scenario include: detailed (<1:10,000 scale) flood inundation maps; flood risk maps showing the extent of flooding and properties flooded where velocity is less than or greater than 2m/s; damage parameters of both velocity and depth at key points such as transportation infrastructure; flood hydrographs at specific internal (flood zone) boundaries; peak flow variation down the valley; and longitudinal sections down the valley showing thalweg and peak water levels.

The Standard Method was applied at two pilot dams for which Mott Macdonald provided technical advice during design and construction: Tamasos Dam, a 30m high earthfill dam sited on the Pediaios River, 23 kilometers upstream of the capital Nicosia (population 309,000); and the 38m high Akaki Malounda Dam on the Akaki River upstream of both Akaki and Malounda townships. Inundation maps and flood plans were produced for both these reservoirs. In urban Nicosia where the flood plain was both complex and extensive, a standard 1D modelling approach was found to be inadequate. Psuedo 2D modelling was undertaken as a stop-gap, but full 2D modelling is necessary to enable refinement of the flood mapping in future. Insufficient baseline data and constraints of the commission meant that it was inappropriate to undertake 2D modelling at this stage of the project.

Rapid Method

As its name suggests the Rapid Method is a quick and simplified approach to undertaking a DBHA. The Rapid Method can be applied to low consequence dams where the cost of undertaking the Standard Method is disproportionate to the increased accuracy and resolution of outputs achieved. It can also be applied as a screen to determine whether a Standard Analysis is required. The detailed outputs of the Standard Method are not required of the Rapid Method and the approach is thus less rigorous, less time consuming and requires less base data. Hydraulic parameters such as maximum average velocity and maximum depth are required only at certain cross section locations (flood zone boundaries).

The methodology employed in this study was based on excel spreadsheets developed by Brown & Gosden (2004). These were used to obtain breach outflow hydrographs for each scenario, route the flows downstream, and assess the hazard potential based on likely loss of life and damage. The Rapid Method was trialed for DBHA's at two pilot reservoirs: Yermasoyia; and Polemidhia.

Whilst the method requires relatively little input data and can be quickly undertaken (the requisite outputs and even crude flood outlines were developed in a day or two) the methodology can not be used to reliably predict flood outlines on wide or complex floodplains or where significant attenuation occurs in the flood plain or upstream of road embankments. This is because gross assumptions must be made regarding attenuation in upstream reaches. In such reaches, application of the Standard Method utilsing 1D or 2D hydraulic modeling has been recommended.

Software

Numerical modeling of a dam break is a computationally demanding routine due to the rapidly varying flow regime. The choice of numerical model and modeling platform for a Standard Method analysis is therefore an important aspect for consideration at project inception. Given the large number of reservoirs that WDD needed to analyse and the unavailability of reliable digital topographic datasets, a 1D hydraulic modeling platform was deemed the most economical and applicable solution. WDD sought an inexpensive and user friendly platform on which to develop their in-house modeling capability. At project inception, Mott Macdonald's in-house HYDRO-1D software and the United States Army Corp of Engineers HEC-RAS software were considered for the task. Research had previously shown that HEC-RAS might be unsuitable for dam break analysis given its inability to specify a computational time step less than 1 second.[4] In contrast, additional testing had shown HYDRO-1D to be robust under the same conditions.[5] HYDRO-1D has an additional advantage over HEC-RAS in that it can be seamlessly linked to a 2D model via HYDRO-2D in reaches where a 1D modeling approach might be unsuitable. To test model suitability both HYDRO-1D and HEC-RAS were used to model dam break in parallel for the first pilot reservoir. HYDRO-1D performed well and by using sound modeling techniques HEC-RAS also was able to successfully simulate the dam break. Given that HEC-RAS is free software the WDD opted to continue to use HEC-RAS for future analyses.

Treatment of transportation embankments

By virtue of their ability to impound large volumes of water, the behavior of transport embankments within the flood plain can drastically affect flood impacts. The net effect can be both positive and negative. They can increase inundation upstream whilst lessening downstream impacts by attenuating peak flows. Alternatively if the embankment breaches early in

---

[4] Crowder, R.A., et al., (2004), Benchmarking and Scoping of 1D Hydraulic River Models, Environment Agency Research Technical Report W5-105/TR2L – Test L (Contraction & Expansion), pg 15

[5] Wardlaw, R (2006) Benchmarking the HYDRO-1D Hydraulic River Modeling Package, The University of Edinburgh, pg 109

the dam break flood, then this may add to the overall peak and compound existing flooding. A subjective analysis must be made of the treatment of each in turn. Consideration must also be given to the extent of blockage of culverts or openings in the embankment by debris. Advice is provided in the Draft Guide.

Derivation of Breach Outflows
The Draft Guide specifies Standard Scenarios to be analysed such that Reservoir Flood Plan users can comprehend the wide range of possible scenarios and the severity and assumptions in the "standard scenario" and so there are repeatable, conservative estimates of the extent and impact of dam failure. The Standard Analysis Scenario outlined in The Draft Guidelines proposes the following two scenarios as a minimum:

Rainy day "dam failure" – 10 000yr inflow to reservoir causing dam failure. Provides a likely maximum dam break flood from a full reservoir;
Rainy day "No dam failure" – 10 000yr inflow to reservoir. Required as a baseline for determining the incremental loss of life and damage in the event of dam failure.

In addition a third "optional" scenario, was analysed in this study:

Sunny day "dam failure" – No reservoir inflow, reservoir water level at spillway crest. This can be useful for emergency planning where the peak dam breach flow is significantly less than the rainy day "dam failure" scenario.
The Draft Guide recommends estimating peak breach discharges using methodology outlined in Brown & Gosden (2004). For embankment dams the peak discharge equation is proposed by Froehlich (1995):

$$Q_P=0.607\ V^{0.295}H^{1.24}$$

where:
Q_p = Peak breach discharge (m^3/s);
V = Reservoir capacity (m3);
H = Height of peak reservoir level above dam base (m);

Brown & Gosden (2004) use methodology outlined in CIRIA Report C542 (2000) to derive a simplified triangular breach hydrograph with total volume equal to the reservoir volume, subject to reducing the Time to Peak (Tp) (initially taken as 120 times dam height) and Time of Event (Te) to ensure that the hydrograph volume remains the same as the reservoir volume.

It should be noted that modern predictive breach models offer a much more dam specific approach to developing breach discharge hydrographs since they consider the physical development of the breach through the core of the dam. Given that these models are still in early stages of development and the number of reservoirs to be analysed by the WDD, these models were not considered for this study. Future studies should consider the use of these models.

Consequence Assessment

The final part of the DBHA is the consequence assessment. This is intended to identify areas likely to be inundated allowing those responsible for emergency planning and response to make informed decisions when planning for possible future events, and when dealing with actual events. It also provides data which can be used by others in assessing the risk to infrastructure.

Again, the Draft Guide recommends methodology specified in Brown & Gosden (2004). The methodology is used to broadly estimate the:

a) Number of buildings in the inundation area, and the area and type;
of non-residential property, and the degree of damage;
b) Number of people (population) at risk;
c) Likely Loss of Life (the base case is with no warning);
d) Third party property damage.

The Consequence class of the dam is a function of the Likely Loss of Life and Third Party direct flood damage.

Whilst the data required for such an analysis is generally readily available in the UK, it is anticipated that in many overseas situations this data may not be available or may require considerable effort to source. In this study, base line data such as average house prices and average number of inhabitants had to be estimated adding further uncertainty to the analysis.

PROBABILITY OF FAILURE ASSESSMENT

Probability of failure assessment is carried out in order to determine the risk of failure of a dam. Methodology provided in Brown & Gosden (2004) was again used as the basis for the assessment. Standard spreadsheets that estimates the risk (or probability) of dam failure were modified to account for Cypriot conditions. For instance, one would expect the risk of failure from a seismic event to be considerably greater in Cyprus than in the UK and hence the changes were made to reflect this. Similarly many of Cyprus's dams remain dry year round, filling only with the infrequent rainfall. Such dams are susceptible to maintenance issues not normally encountered in the UK.

CONCLUSIONS

A number of observations and conclusions were drawn from this work which may prove valuable to others undertaking similar work in the UK and overseas:

-Hazard Ranking is a quick, semi-rigorous and rational tool for undertaking an initial assessment of the relative hazard posed by dams and where there are a number of dams to consider can assist in the prioritization of emergency planning efforts.
-DBHA's should focus strongly on the requirements of the end user. There is an overriding need for a consistent approach to undertaking DBHA and reporting results;
-In undertaking DBHA's it is important not to lose sight of the uncertainty inherent in both dam breach modeling and flood mapping. The approach should always err on the side of caution but the resources attributed to a particular component should reflect the uncertainty and in turn the flood impact associated with the uncertainty. i.e. there may be little point analyzing in detail debris blockage at a transportation embankment if there is uncertainty regarding the Time to Peak of the breach which generally has a considerable impact on the peak breach outflow. However, it is important to consider the hazard envelope i.e. the overall flood outline from all scenarios considered;
-Assumptions regarding DBHA's should be clearly identified on published outputs. Modern flood mapping and Geographic Information Systems (GIS) techniques produce detailed outputs which don't necessarily reflect the quality of the underlying data or uncertainties in the methodology or model used. They can therefore be easily misrepresented. Assumptions made in developing the flood models (there are lots of them) are not readily apparent from viewing a flood map and these can influence the results considerably;
-Generally speaking, current UK DBHA and Reservoir Emergency Planning guidance can be successfully adopted for use overseas provided local idiosyncrasies and differences are recognized and the methodology is modified accordingly. Overseas dams can be subject to different failure mechanisms and have different failure probabilities to their UK counterparts. Care is needed with emergency planning as both the scale and nature of the hazard and the emergency response can be vary considerably;
-The application of UK guidance overseas can be difficult where certain base data (e.g topographic survey, socioeconomic, utility and services, dam monitoring records, etc) are not available, or not to the standards typically encountered in the UK.

REFERENCES

Brown & Gosden (2004) *Interim Guide to Quantitative Risk Assessment for UK Reservoirs*, DEFRA, London

CIRIA (2000) *Risk Management of UK Reservoirs*, London.

Cullen, N (1989) *Risk assessment of earth dam reservoirs*, DOE Report 0002-SW/3.

Crowder, R.A., et al., (2004), *Benchmarking and Scoping of 1D Hydraulic River Models, Environment Agency Research Technical Report W5-105/TR2L – Test L (Contraction & Expansion).*

Defra (2006) *Engineering guide to emergency planning for UK reservoirs*

Thompson, G and Clark, P B (1994) *Rapid hazard ranking for large dams* Reservoir Safety and the Environment Proceedings of the 8th British Dam Society Conference.

Wardlaw, R (2006) *Benchmarking the HYDRO-1D Hydraulic River Modelling Package,* The University of Edinburgh.

Dambreak and Emergency Planning: meeting end user needs

MW MORRIS, HR Wallingford Ltd, Wallingford. UK.
AK HUGHES, Atkins Ltd, Epsom. UK.
F BUIJS, HR Wallingford Ltd, Wallingford. UK.

SYNOPSIS. With changes to UK legislation and the European Floods Directive on the point of implementation, dam owners will face more stringent requirements to consider flood risk from reservoirs and to take preparatory steps in terms of dambreak flood risk assessments and emergency planning. Guidance on emergency planning in the UK is under review and a number of significant research projects are underway to advance our understanding and ability to analyse and manage flood risk. With the continued rapid pace in development of computing power, modelling and mapping technology, it is a good time to review current practice and capabilities, and to identify what should be considered as reasonable practice for dambreak modelling and emergency planning.

In reviewing practice, it is essential to consider the range of end user applications that the results of any study might be used for, and to ensure that the uncertainty and resolution of any predictions are suited to such applications. A failure to recognise the significance and magnitude of uncertainty within these predictions, and applications, undermines the value of undertaking and using such studies.

This paper presents a brief history of dambreak and emergency planning development in the UK leading towards the current convergence of a number of projects and drivers. This includes how modelling and data management tools have developed over the last decade, particularly within the field of flood risk management in general, to now provide us with powerful tools for flood risk assessment and management planning. Different approaches for dambreak analysis and emergency planning are considered, leading to a recommended type of approach that steps towards greater integration with wider Environment Agency flood risk management principles and is driven by the practicalities of end user needs and a consistent approach to dealing with uncertainties.

Ensuring reservoir safety into the future 2008, Thomas Telford, London

INTRODUCTION

The UK is going through a very interesting period of change and potential for change at this time. The Reservoirs Act 1975 has served us well for the last 20 years but as with the review of the Reservoirs (Safety Provision) Act, 1930, it is appropriate to review and undertake change. The Water Act 2003 gives the Secretary of State power to direct owners to prepare a flood plan. This direction will specify the matters to be included in the plan to a specification to be advised. These plans will include dam break analysis and inundation mapping and will be used for emergency planning but with revisions planned for the Reservoirs Act 1975 provide an opportunity to develop a methodology which might be appropriate for a number of uses including legislative change, spatial planning, emergency planning and asset management.

DEVELOPMENT OF DAMBREAK AND EMERGENCY PLANNING IN THE UK

The failure of Dale Dyke Reservoir in March 1864 inundated the city of Sheffield after travelling 11 km at an average speed of nearly 30 km/hr. Parts of the city were flooded to a depth of 3 metres and large areas were covered with a thick layer of wood, mud, sand and stones. More than 250 people were killed, 798 houses were destroyed and more than 4000 seriously flooded. Dozens of mills, factories, shops and workshops were totally or partially destroyed (Smith, 1972). At the time the coroner stated that 'by one Act of Parliament dams should be subject to frequent, sufficient and regular inspections'. Unfortunately no legislation was enacted until there had been 3 failures in 1925.

In the 1980's a simple dam break modelling programme – DAMBRK was developed in the United States and this was subsequently modified and adapted to enable it to be used in the UK; indeed the information on time of travel from the Dale Dyke tragedy was used to 'calibrate' the model.

Over the last 20 years there has been a gradual and in recent years impressive improvement in modelling techniques. However, in the UK most of the analysis carried out to date has been fairly crude and rudimentary and have often been carried out without a clear reason for doing the work because the information has not been shared with others. In contrast many other countries around the world have adopted a far more rigorous assessment process often working with 2D modelling, public consultation and emergency planning/rehearsal and publication of maps.

The Water Act of 2003 states that

- 'The secretary of State, by written notice served on the undertakers in relation to a large raised reservoir, direct them to prepare a plan (a "flood plan") setting out the action they would take in order to control or mitigate the effects of flooding likely to result from any escape of water from the reservoir'.

The EU Floods Directive seeks to reduce and manage the risks that floods pose to human health, the environment, infrastructure and property. Member States of the EU will have to reduce flood risk where the risk is deemed to be significant. This is done by first determining the extent of flood risk (through hazard mapping and flood risk mapping. Whilst reservoirs are not specifically mentioned, the directive requires that all flood risk be considered and thus it is likely that the flood risk arising from reservoirs will be included.

AN OPPORTUNITY

It is clear that the UK is entering a period of change where opportunities exist to review and update our approaches and specifications and take the UK into the 21st Century. A review of the Reservoirs Act 1975, the Water Act 2003 and the EU Floods Directive all provide reasons and opportunities for improvements.

In reviewing the actions of the past it is clear that many of the studies undertaken in the 1990's lacked consideration of how the data was going to be used, and indeed the suitability of the modelling for the application. It might be that, in the event of failure, such plans could result in unnecessary losses and / or actions.

In moving forward it is essential to consider the various users and end applications of any modelling and mapping that is undertaken. It is essential to ensure that the results are meshed with an overall system supporting a range of applications which might include not only emergency planning but also legislative change, spatial planning and asset management. The data provided and the methods used must be appropriate for those applications and support more detailed analysis, perhaps where the consequences of failure are high, where it is required.

POTENTIAL APPLICATIONS AND END USERS

Historically, dambreak analyses have been undertaken purely for emergency planning purposes. Even then, the results of many studies were locked firmly away, only to be reviewed at the time of an emergency. However, with growing recognition of the range of actions that can be taken to contribute towards overall flood risk management it is becoming more widely recognised that the results of dambreak analyses may be used in a variety of applications, both at a national or local scale. For example, these may include:

- National risk assessment (including risk to Critical National Infrastructure)
- Local (reservoir specific) risk assessment
- National emergency planning
- Local emergency planning (including evacuation planning)
- Asset and business management
- Spatial planning

A *National Risk Assessment* would be undertaken to meet specific government needs and may also facilitate local risk assessments and enforcement of dam safety regulations. Flooding in 2007 highlighted the real risk that infrastructure can pose; identifying the real risk from national infrastructure is a key driver for undertaking such a study. Results from such an assessment would also support any national emergency planning measures (for example, high level provision of large capacity pumps).

Current plans to undertake broad (national) scale modelling of flood risk from reservoirs offers a number of opportunities, but must be undertaken to an appropriate resolution and accuracy in order to underpin the wider range of potential applications and hence be of greater industry value other than providing a 'one off' snapshot of broad scale flood risk.

Reservoir specific risk assessments may be undertaken to meet the requirements of UK and European legislation. Results from such studies will also support more detailed risk management measures, such as *emergency planning*, including evacuation planning, and asset and business management measures. Prioritisation of works to mitigate and manage flood risk (liability) is a key business activity.

Spatial planning uses Environment Agency (EA) guidance regarding flood risk and acceptability for land use. The significance of flood risk in the planning process is recognised and weighting regarding development decisions is likely to grow. Consideration of flood risk from reservoirs needs to be integrated into this process in a logical, transparent and consistent manner.

With such a range of different potential end uses for dambreak analyses it is important to ensure that the accuracy and resolution of an analysis is fit for a particular application, or group of applications. Current modelling and GIS technologies allow for flood plans to be produced overlaying satellite images or ordnance survey plans, but typically fail to reveal the true accuracy of the plans. This can be misleading once the plans pass from the originator on to a variety of users who are remote from the original studies.

Uncertainty is the Key Issue

Uncertainty within the base data, models used and end user application or management processes are key to ensuring that maximum value is obtained from any future dambreak analyses, whether undertaken at a national scale or as more detailed individual reservoir studies.

A series of questions should be addressed in order to ensure that a suitable approach is taken. These include:

1. What do we need (from dambreak modelling)?
2. What have we got (in terms of data and methods)?
3. What could we get (in terms of data and methods)?
4. Can we meet user needs with the available tools, techniques and data?
5. If yes, we can match methods and data as appropriate; if no, we should not proceed with modelling that is not fit for the required application. Instead, we should focus upon refining tools, techniques and data to match our needs. If this cannot be achieved, the results should not be routinely used for that particular application.

Answering these questions properly requires careful consideration of the dambreak analysis processes. The following section provides a suggestion as to what issues may arise. This is intended only as an example and requires more detailed validation before wider use. Values given here as examples should not be used directly to justify existing works or approach.

CONSISTENCY WITH EXISTING ENVIRONMENT AGENCY APPROACH TO FLOOD RISK MANAGEMENT

Defra and the EA are implementing a risk based approach to flood risk management for risks arising from flood and costal erosion). The Flood Risk Management Research Consortium projects (FRMRCI / FRMRCII; www.floodrisk.org) and FLOODsite projects (www.floodsite.net, Morris & Samuels, 2006) support and integrate with this programme. Concepts developed for flood risk management can, and should, be used as a framework for considering flood risk from reservoirs.

The Source – Pathway – Receptor model (Figure 1) is recommended by the EA for considering flood risk. By adopting a risk based approach for each stage (source, pathway, receptor) a true assessment of flood risk may be made. Within this model it is possible to identify the contributions that each component (stage or process) makes to the overall flood risk and to systematically identify uncertainties within the flood risk analysis process.

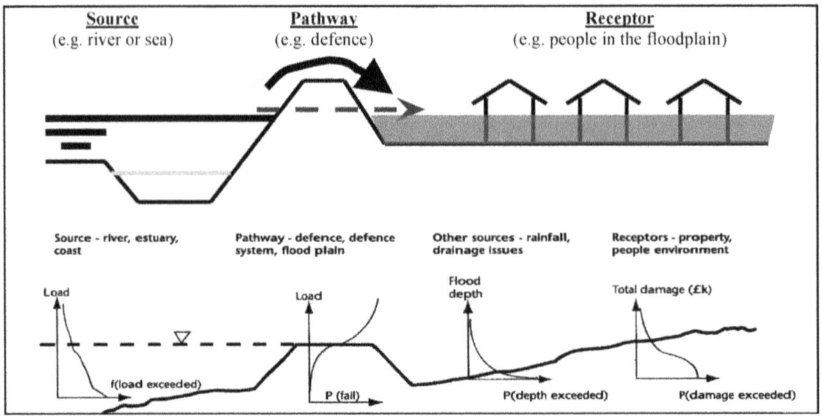

Figure 1: The source-pathway-receptor-consequences model for flood risk (Defra, 2002).

The process of flood risk assessment from dambreak may be considered in a similar way to that shown in Figure 1. The source is formed by the hydraulic loading of the dam, the pathway is represented by performance (failure or not) of the dam, and the receptor is represented by assets and people in the inundation affected areas. The consequences are driven by the speed and extent of flood spreading. The dam break flood risk is formed by the likelihood of the hydraulic loading conditions, the likelihood of the dam break scenarios and the consequences due to inundation.

Historically, dambreak analyses have not dealt rigorously with risk. Assumptions are typically made regarding load conditions (PMF, sunny day failure etc.) and how the dam fails (partial failure, catastrophic etc.). Recent developments in quantitative risk assessment (Brown & Gosden, 2004) consider risk in more detail, but do not offer the complete 'model' as outlined in Figure 1. A particular difference here lies in the representation of pathway (dam) performance – i.e. how it fails under different load conditions. The S-P-R approach represents pathway performance through a fragility curve (HR Wallingford, 2005). This relates the probability of failure to a given load condition. Work is required here to determine to what detail dam performance can be related to load conditions. For example, how far beyond a simple fail – no fail load condition can we reliably predict behaviour and how important is this detail for establishing dambreak flood risk?

The EA adopts a hierarchical approach to flood risk assessment, with high, intermediate and detailed levels of assessment. These broadly relate to national, regional strategic and site specific levels of study. Table 1 below (Sayers et al, 2002, following from Defra, 2002) provides a summary of these approaches in relation to end user ('decision to inform'), data sources and methods for analysis. Similar consideration may be given to national or site specific dambreak studies, and hence appropriate sources of data and methodologies to use in relation to end users (or 'decisions to inform'). This is outlined in a comparable table (Table 3) later in this paper.

Table 1 shows that risk assessments inform different policy levels, e.g. a national, regional and site specific level. The required detail of the risk assessment models to inform the decisions differs for each policy level. Table 1 illustrates that as the detail of the decision-making increases, the data requirement and the complexity of the applied methodologies increases. It is advantageous, however, for some data sets to be common to all levels of hierarchy. For example, working from a common topographic data set, but using different densities of data, will aid modelling speed without compromising consistency between mapping of flood risk at different hierarchical levels.

Table 1: Tiered hierarchy in objective setting and data supply for fluvial and coastal flood risk assessment (Sayers et al., 2002).

Level	Decisions to inform	Data sources	Methodologies
High	National assessment of economic risk, risk to life or environmental risk Prioritisation of expenditure	Defence type Condition grades Standard of Service Indicative flood plain maps Socio-economic data Land use mapping	Generic prob. of defence failure based on condition assessment and crest freeboard Assumed dependency betwn. defence sections Empirical methods to det. likely flood extent
Intermediate	*Above plus:* Flood defence strategy planning Regulation of development Prioritisation of maintenance Planning of flood warning	*Above plus:* Defence crest level and other dimensions where available Joint probability load distributions Flood plain topography Det. socio-economic data	Prob. of defence failure from reliability analysis Systems reliability analysis using joint loading conditions Modelling of limited number of inundation scenarios
Detailed	*Above plus:* Scheme appraisal and optimisation	*Above plus:* All parameters required desc. defence strength Synthetic time series of loading conditions	Simulation based reliab. analysis of systemSimulation modelling of inundation

Detailed emergency planning, land use planning and asset management requires a detailed representation of hydraulic boundary conditions, dam failure and deterioration, breach formation and inundation scenarios. The source-pathway-receptor consequences and tiered decision-making approach enables modelling multiple failure and inundation scenarios in an integrated way. In flood defence management the possibilities to implement performance based asset management have been explored (Defra / Environment Agency, 2004a, b & c). Such an approach ideally integrates asset management, emergency planning, land use planning and other user needs.

POTENTIAL APPLICATIONS, REQUIREMENTS AND ACCEPTABLE UNCERTAINTY FOR DAMBREAK ANALYSES

The following section details potential end uses of dambreak analyses and indicative measures of uncertainty (in terms of predicted water level) that might be appropriate for such applications. It is recognised that predicted water level is perhaps not the best measure of uncertainty here, given its dependency on site specific factors, however it does provide a practical measure that is widely recognised by all end users. Table 2 provides a summary of indicative levels of accuracy that might be required for different end uses. More detailed consideration is required to refine these values, whilst this initial summary presents the concept.

Table 2: Indicative required accuracy for flood level prediction in relation to potential different end uses

Indicative acceptable accuracy of flood level prediction for different end applications						
National risk assessment	±0.1m	**±0.5m**	±1.0m	**±2.0m**	**±5m**	±10m
Critical national infrastr.	±0.1m	**±0.5m**	±1.0m	±2.0m	±5m	±10m
Emergency planning	±0.1m	**±0.5m**	**±1.0m**	±2.0m	±5m	±10m
Spatial planning	**±0.1m**	**±0.5m**	±1.0m	±2.0m	±5m	±10m
Asset management	±0.1m	**±0.5m**	±1.0m	±2.0m	±5m	±10m

National Risk Assessment at a broad national scale would provide an indication of overall national risk arising from reservoirs through indicative inundation plans to an accuracy of perhaps ±2-5m. However, the risk assessment may also be used to categorise risk in relation to individual reservoirs (i.e. reservoir risk categorisation A, B, C, D etc) and to provide base data for more detailed studies of high risk dams, in which case a greater degree of certainty is required (perhaps ±0.5m). In this case the modelling would need to adequately reflect the resolution between reservoir risk categories which currently differentiates between a threat to a community of more than 10 people, as compared to 0-10 people. Such resolution requires a far more accurate prediction of flood timing, velocity and depth and hence level accuracy of ±0.5m is suggested instead of ±2-5m.

Risk to Critical National Infrastructure can be assessed at both national and detailed individual levels. The cost implications of risk to critical infrastructure are large. When infrastructure is found to be at risk, logical further steps are to establish what measures might be needed to protect the infrastructure. Hence, accuracy of mapping for both initial identification and protection options needs to be better than initial broad national scale modelling, and is suggested as ±0.5m. Assessing the physical threat to infrastructure (which may be above or below ground) can require detailed consideration of flow velocity, damage potential and timing.

Emergency Planning can be implemented in a number of ways, ranging from the broad to the complex. The objective here is to provide predictions of scenarios that will provide maximum value to emergency services during an event. Information that is of value includes inundation extent, timing and damage zones. Damage zones implies prediction of flow depth and velocity to a reasonable resolution. The way in which flow modelling is undertaken affects the accuracy of all of these parameters. For example, if flood wave attenuation is not correctly modelled, not only will flows and water levels be incorrectly predicted, but so too will the timing of the flood wave.

Emergency services are likely to evacuate areas based upon geographical features or zones (i.e. whole streets) rather than the predicted line of inundation. However, the uncertainty in the inundation plan is a function of uncertainty in flow prediction combined with topography. Uncertainty in flat areas relating to ±0.5m can cover large areas in comparison to uncertainty in inundation of populated areas within a valley.

Spatial Planning could make use of reservoir plans to limit risk in the event of a dam failure. The current position is 'grey' with some developers, planners, owners and the Enforcement Agency recognising the issues whilst others ignore it. As a dam owner, any construction within the potential inundation zone of your reservoir may (currently) change the risk category of the dam and hence actions needed to maintain the dam. With threat to life being a key issue, required accuracy is likely to be high.

It is the author's experience that development costs typically dwarf the costs of undertaking reasonably detailed flood risk studies, hence the likely appropriate (cost beneficial) accuracy for such applications is quite fine.

If this suggestion is correct, it raises an interesting moral question in that society would appear to place greater value on establishing true risk for a specific development / planning issue than generally for the emergency planning / evacuation planning of the wider population!

Asset Management requires a risk based approach in order to determine how best to spend limited resources (Defra / Environment Agency, 2004a, b & c). The objective here would be to use dambreak analyses, including full impact assessments, to identify potential risk (liability) and hence to identify what risk reduction or management measures might be appropriate. Portfolio Risk Assessment procedures help to ensure maximum value is obtained in working through various risk management measures. The degree of accuracy appropriate for asset management will depend upon the number and type of dams being considered, and the issues that arise. It might be appropriate to first review indicative risk at a level similar to national broad scale modelling. Subsequently, higher risk dams might be considered in more detail, probably at a level similar to that adopted to assess risk to critical infrastructure.

In conclusion, we can see from this indicative overview of potential accuracy attached to different end user applications, that prediction of flood levels ranging from ±2-5m down to ±0.1-0.5m might be needed. We also see that where results from one application, such as national mapping, are used to inform a variety of decisions, then the required accuracy of the application is dictated by the decision requiring the most stringent conditions.

MATCHING CURRENT DATA AND MODELS TO END USER NEEDS

Having identified different end users and uses for dambreak modelling results the challenge is to now match available data sets and modelling capabilities appropriately.

Topographic Data

Topographic data can pose a significant (cost) barrier to undertaking dambreak analyses. A number of different data sets exist:

- National scale SAR data (resolution 5m grid; accuracy ~±0.5m at best
- LIDAR data (resolution ~2m grid; accuracy ~±0.1m) – not available nationally, but considerable coastline and river valleys covered
- Composite LIDRA/SAR data (resolution 5m grid; accuracy ~±0.1-0.5m depending upon source
- Site specific survey or LIDAR data collected for dambreak studies (resolution as required; accuracy ±0.01-0.1m depending upon method)

If national scale modelling is to be used to support reservoir risk categorisation and more detailed site specific studies (for example, for high risk category reservoirs) then it is likely that national scale modelling will require LIDAR resolution and accuracy data.

Modelling Capabilities

There are many stages to a dambreak assessment, and the potential to use a range of different models at each stage. The importance of various model contributions will vary between dam type and location; for a national assessment however, a single most stringent approach should be adopted to ensure applicability to all.

A good example of analysing and understanding uncertainties within a dambreak analysis is given within Work Package 5 (WP5) of the European IMPACT project (Morris & Hassan, 2005). Here, the Tous Dam failure in Spain was used as a case study and combined breach and flood routing modelling undertaken using a number of models. A practical approach to modelling uncertainty gave a range of predictions for flood levels at a downstream town; a back analysis of the source of uncertainty contribution to flood level was also made, identifying sources such as breach modelling, topographic data, choice of flow model, roughness assumptions etc.

Existing Modelling Capabilities - Hydrology

Research is ongoing here to resolve differences in extreme event prediction between FSR and FEH predictions. Whilst there may currently be uncertainty in both method and prediction, this should not encourage the use of unduly simplified methods elsewhere within the overall dambreak analysis. Instead, the accuracy of methods should be refined to meet specific end user needs.

Existing Modelling Capabilities – Structure failure

Modelling capabilities here vary between different structure types. The need for a consistent approach across structures is recognised and identified within the recent Reservoir Safety R&D Strategy project (Atkins, 2008). Prediction of breach is an area where research has advanced and methods typically fall to predictive modelling versus the use of simple assumptions built around peak discharge equations. The uncertainty attached to peak discharge equations is potentially huge; the accuracy of peak discharge from a predictive model is now in the region of ±20-30%.

Existing Modelling Capabilities – Routing models

There are now a range of flow models available that offer 1D, 2D or even 3D flow simulation. Links to GIS tools for mapping are useful, but can hide the true resolution and accuracy of a simulation. Developments in 2D modelling mean that it is now as easy to produce a 2D model as a 1D; as always, use of appropriate topographic data dictates output, but this is not a reason to differentiate between choice of 1D or 2D approaches. More important is the basis of the model code and suitability for simulation of dambreak flood conditions. Some 2D codes offer a fast but simplified representation of hydraulics. This simplification can introduce large inaccuracies for some applications, particularly where rapidly varying flow conditions occur, as with dambreak. Suitability of a model for fluvial flow modelling and mapping is not a guarantee of suitability for dambreak flood modelling.

Matching Capabilities to End User Needs

Table 3 below starts to match needs to applications. However, the table does not detail conditions other than the two extremes of simple national risk assessment and individual detailed studies. It is anticipated that detailed interim measures should be identified as part of the national broad scale modelling project commissioned by the Environment Agency this year.

Table 3: Indicative tiered hierarchy in objective setting and data supply for reservoir flood risk assessment

Level	Decisions to inform	Data sources	Methodologies
High	National risk	National LIDAR data sets Location of reservoirs within Act Basic type, structure, dimensions etc. Indicative land use	Extreme hydrology Simple – but smart – failure mechanisms Coarse grid, main topographic features – 2D flow simulation
Intermediate	Is this appropriate for reservoir flood risk assessment?		
Detailed	Individual reservoir risk assessment and emergency planning	Local LIDAR data Reservoir bathymetry Dam construction and condition details (survey and sampling if necessary) Land use within impact zone	Detailed hydrology Realistic structure failure mechanisms (predictive modelling) Dynamic modelling of reservoir and routing downstream using integrated 1D and 2D models as appropriate.

CONCLUSIONS, OBSERVATIONS AND SUGGESTED ACTIONS

A key aim of this paper is to highlight the current opportunity (through a range of linked EA/Defra projects) to adopt end user focused approaches to dam break analysis. This will help to avoid inappropriate modelling that can provide results of limited value and application.

Adopting concepts and procedures already used for wider flood risk analysis and management within the EA will help to integrate reservoir flood risk analysis within the developing EA risk management framework. Ultimately, risk from reservoirs should be managed consistently with other flood risks. This approach will also help to ensure that requirements arising from the European Floods Directive are also addressed in a consistent manner.

Ensuring that the chosen methodology for dambreak analysis is suited to a range of different end user needs and applications requires a thorough review and understanding of the uncertainty contributions at each stage of the dambreak analysis process. It is thought likely that the real magnitude of uncertainty inherent in different approaches is far larger than many users appreciate. Also, that the resolution and accuracy required for applications other than a high level indicative risk assessment are more stringent than many users appreciate. However, recent trends in 2D and predictive model development, making more accurate predictions of flood conditions easier and faster, may help to offset this mismatch.

Where a mismatch occurs between user needs and available data and modelling methodologies, it is important to firstly recognise this in any use of existing data and methods, and to secondly focus R&D and national projects towards providing suitable data and methods. At all costs, we should avoid blindly modelling dambreak scenarios and using predictions without a clear understanding of their basis and inherent uncertainty.

REFERENCES

Atkins (2008). *Defra reservoir safety research strategy.* Draft final report, 2008.

Brown, A.J. and Gosden, J.D. (2004). *Interim guide to quantitative risk assessment for UK reservoirs.* Thomas Telford, 2004. ISBN 0 7277 3267 6.

Defra / Environment Agency (2002), *Risk, Performance and Uncertainty in Flood and Coastal Defence – A Review,* SR587, Environment Agency.

Defra / Environment Agency (2004a), *PAMS – Phase 1 Scoping Study, Task 1 interim report: Logical framework for O&M activity*, Environment Agency.

Defra / Environment Agency (2004b), *PAMS – Phase 1 Scoping Study, Task 2 interim report: Review of existing methodologies,* Environment Agency.

Defra / Environment Agency (2004c), *PAMS – Phase 1 Scoping Study, Task 3 interim report: options for performance based asset management systems,* Environment Agency.

HR Wallingford (2005), *Performance and Reliability of Flood and Coastal Defences,* FD2318/TR1, Wallingford.

Morris, M.W. and Hassan M.A.A.M. (2005). *Conclusions and Recommendations from the IMPACT Project WP5: Combined Risk and Uncertainty.* Available via www.impact-project.net. EC Research Project EVG1-CT2001-00037.

Morris, M.W. and Samuels, P.G. (2006). *FLOODsite (Integrated Flood Risk Analysis Management Methodologies) – Research Relevant to the Dams Industry?* British Dam Society 14th Biennial Conference, University of Durham, 6-9th September, 2006.

Sayers, P.B., Hall, J.W., Meadowcroft, I.C., (2002), *Towards risk-based flood hazard management in the UK, Proc. ICE Civil Engineering*, v150, n1, special issue: Floods – a new approach, pp. 36-42.

Smith, N (1972). *A history of dams*. Citadel Press.

A step change in reservoir safety management: Quantitative Risk Assessment and its strategic implications

A. J. BROWN, Jacobs. UK
J.R. CLAYDON, Independent Consultant, Yorkshire, UK
J. D. GOSDEN , Jacobs, UK

SYNOPSIS. Quantitative risk assessment (QRA) techniques now provide the ability to make meaningful estimates of the probability of failure, its consequences and thus the risk (probability x consequences) of dam failure. This has the potential to provide a major, step, improvement in reservoir safety management in UK. This paper describes the areas where QRA may be applied, the issues on which a strategy for use of QRA on dams needs to be developed and agreed within the UK reservoir engineering community, together with the requirements of the tools for the QRA.

INTRODUCTION

There are now several published alternative tools to assist in carrying out Quantitative Risk Assessment (QRA) of dam safety, including ICOLD Bulletin 130 (2005), ANCOLD Guidelines (2005), Canadian work (Hartford et al, 2004) and the UK Interim Guide (Brown and Gosden, 2004). Related approaches to asset management for flood and coastal risk management are also being developed in the UK using the concept of fragility curves, the variation of probability of failure with loading conditions (Defra Project FD2318, 2007).

Adoption of QRA as a tool for dam safety management provides significant benefits, but also significant challenges in that it may be in conflict with traditional "deterministic" standards and challenge preconceptions as to traditional priorities for reservoir safety.

This paper explores areas where QRA offers opportunities to improve management of reservoir safety, discussing the benefits and disbenefits of each, and the consequential strategic issues and associated requirements for QRA tools.

WHAT IS THE PURPOSE OF QUANTITATIVE RISK ASSESSMENT?

Quantitative Risk Assessment for reservoirs has different objectives, depending on the perspective of the stakeholder. Communication between stakeholders requires a common understanding of processes and terminology. The current situation is that there is no universally agreed methodology or suite of tools for QRA, which can lead to misunderstanding and lack of acceptance of results. In order to come to agreement on the method to be used it is first useful to consider the needs of the different stakeholders.

The Regulator

The Regulator needs to understand the level of risk across the range of dams and reservoir which are present in the UK, and how these compare to other risk to the public, so that appropriate risk control measures are put in place. The Reservoirs (Safety Provisions) Act 1930 and the Reservoirs Act 1975 are both based on the volume of water stored. Although this is, in principle, simple to measure it does not measure the risk to the public, i.e. the product of the probability of failure and the consequences if a dam did fail.

The UK government prefers a risk based approach to management of residual risk, where risk is defined as the product of probability of an event, and its consequences. Evidence of this includes the paper on flood risk assessment (Defra, 2004) accompanying the consultation on the new government strategy "Making space for water" and the Health and Safety Executive (HSE) approach to regulating health and safety, as set out in "Reducing risk, protecting people" (2001). Similarly in the last few years Defra have renamed their "Flood and Coastal Defence" function as "Flood Risk Management".

The Interim Pitt report into the summer 2007 flooding includes a number of important recommendations which are likely to affect reservoir safety, including

> *IC33 – flooding legislation should be updated and streamlined under a single unifying Act that addresses all sources of flooding and facilitate flood risk management*
>
> *IC62 – the Government should implement the legislative changes proposed in the recently published Environment Agency biennial report on dam and reservoir safety*

It is clear from the report that a risk based approach is preferred by Government, and this highlights the need for the dam engineering industry to provide the appropriate technical tools.

The Owner

The owner has a wide range of needs with which QRA can assist. These include

a) To understand the magnitude of the consequences of dam failure. This may be obvious in a qualitative way to most owners – the consequences would be so severe they could go out of business. However there will be some owners who are unaware of the potential hazards and many who will have no idea of the scale of the economic damage that could be caused.
b) To have information on the likely extent of flooding and the number of people that would need to be evacuated, in the event of a dam failure
c) For a single reservoir to:
    - Understand the potential failure modes of the reservoir.
    - Consider the effect of changes in inspection and monitoring.
    - Assess options and alternatives to reduce the risk
    - Optimize the level of risk during any remediation process
    - Define the post remediation residual risk and identify appropriate residual risk control strategies
d) For a portfolio of reservoirs to:
    - Compare reservoirs for prioritized attention.
    - Provide reassurance to others especially those at risk of flooding.
    - Provide information for corporate risk management and insurance.
    - Demonstrate due diligence in the management of the asset

The Panel Engineer

The Inspecting Engineer is responsible for reviewing the adequacy of the safety of a dam. As such Engineering Guides have been developed to assist in this process, the earliest on Floods and Reservoirs being issued in 1976. QRA provides improved tools for use in this assessment, by providing

- A systematic means of reviewing and quantifying modes of failure.
- An analysis which can be updated.
- A means of showing the effect of improvements.

Summary

The various stakeholders in dam safety have a range of different objectives, which would benefit from a commonly agreed approach and toolbox of methods for quantitative risk assessment.

WHAT IS A TOLERABLE LEVEL OF RESIDUAL RISK?

Adoption of a risk based approach is significant in that it recognises that residual risk can never be zero, but is reduced to a tolerable level. The quantification of risk provided by QRA allows comparison of the flood risk from dams with the risk from other forms of flooding and thus a rational basis for deciding what level of risk is tolerable.

The loss of life in some historic dam failures, and fluvial and coastal flood events, in the UK is shown in Table 1. The data from the historic fluvial and coastal events are plotted on a FN chart in Figure 1, together with the estimated fatalities in the event of failure of major UK dams from a number of QRA studies for major dam owners.

It can be seen that the risk from fluvial floods is similar to the risk of failure from the highest risk dams, in that although a failure of a higher risk dam has the potential to kill say 100 more people than a major fluvial flood, the annual probability is estimated to be 100 times less, such that the risk is similar.

In relation to coastal flooding, it is noted that coastal defences in the UK are generally designed only to retain a 1 in 200 chance per year (0.5% annual probability) flood event and historically have not been specifically designed for overtopping and are therefore likely to fail in say a 1 in 1000 chance (0.1% annual probability) flood. Major flood defences protecting London are designed to retain a 1 in 1000 chance per year flood, so the annual probability of failure would be correspondingly lower. In contrast dam embankments are designed to pass the Probable Maximum Flood (a probability of the order of a 1 in a million annual chance per year) safely. When other threats are considered the overall annual probability of failure of dams is as shown on Figure 1, but still generally less than the probability of failure of coastal defences.

This comparison of the consequences and probability of loss of life from different types of flooding shows how QRA can be a powerful tool to compare different risks to life.

The strategic issue that then arises is what level of residual risk to flooding is tolerable, and where national resources are limited which type of flood risk should be prioritised for risk reduction. A similar approach can be adopted to compare the risk from dam failure with other high hazard industries.

The requirement for tools for QRA of dam safety is therefore to provide quantitative output which is on the same basis as, and thus comparable with, quantification of other risks to society.

Table 1 : Loss of life in some major floods in the United Kingdom

Year	Source of flooding	Location	Annual chance	Number of properties flooded	Loss of life
2007	Fluvial	Widespread	Commonly 4 times average monthly rainfall	48,000	13
2005	Fluvial	Carlisle	1 in 185	1,800	3
2000	Fluvial	Widespread	Commonly 1 in 15	9,000	0
1998	Fluvial	Widespread at Easter	1 in 150 to 1 in 50	Not avail.	5
1953	Coastal	East coast	1 in 500	24,000	307
1952	Fluvial	Lynmouth	1 in 750	165	34
1925	Dam failure	Skelmorlie	Not avail.	Not avail.	5
1925	Dam failure	Eigau/ Coedty	Not avail.	Not avail.	16
1912	Fluvial	Norwich	1 in 800	1200	4
1864	Dam failure	Dale Dyke	Not avail.	Not avail.	250
1852	Dam failure	Bilberry	Not avail.	Not avail.	81

Figure 1 : Comparison of risk from various forms of flooding (UK data)

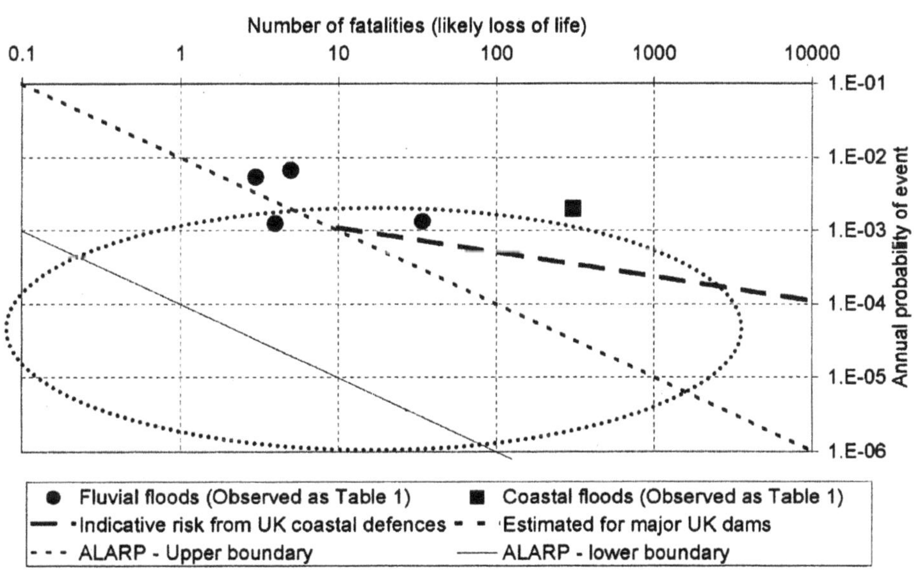

HOW SHOULD ENGINEERS DECIDE WHEN A DAM IS SAFE ENOUGH?

QRA provides a means to quantify how a dam may fail, including explicit consideration of modes of failure, and quantification of each of the steps in that process. An example of this is the methodology used by US Bureau of Reclamation to assess the safety of their dams against internal erosion, where the dam safety review team answers each of the questions shown in Table 3 (Cryganiewicz et al, 2007). Although panel engineers may consider modes of failure in a generalized way when assessing the adequacy of a dam, it is often not carried out systemically and it is rare for it to be written down.

It is suggested that more systematic use, and documentation, of event trains to capture likely modes of failure is desirable as a means for owners to demonstrate their safety case and panel engineers to provide an audit trail of their safety review. Such event trains should be part of the QRA toolbox of techniques. The process of thinking through the most likely failure modes of a dam has also proved invaluable to Supervising and Inspecting Engineers in identifying likely indicators of adverse performance, and thus elements of the dam where surveillance should be focused.

Table 3 US BOR Event Train for failure from internal erosion

Q	Adverse step in event tree	Example for failure along bottom outlet pipe
1	What is the event that initiates internal erosion?	Pipe fractures
2	Will high flows occur sufficient to begin erosion?	Flows may be limited by pipe diameter
3	Is there an unfiltered exit?	Rock toe may limit
4	Will a roof form completely through the core?	Depends on type of core material (unknown)
5	Are flows limited by an upstream element?	Magnitude of flows determined by size of pipe, size of fracture and head on the pipe
6	Will early intervention be unsuccessful?	Use of existing drawoff
7	Will breach process initiate?	
8	Will heroic intervention be unsuccessful?	Full emergency drawdown, with additional imported pumps

The second part of a QRA analysis, of quantifying the probabilities of each of the steps which could lead to failure, requires a clear definition of failure. Traditional techniques such as Floods and Reservoir Safety (ICE, 1996) provide a recommended design standard, in terms of a specified flood with wave freeboard. This has the weakness that the calculation provides no measure of how close to failure the dam may be. QRA would normally define failure as the loss of the ability of the dam to retain the reservoir, and by defining a probability this provides a measure of the proximity to failure. This change would be similar to the change in structural design of reinforced concrete from elastic design in CP114 to plastic design in CP110 in 1975.

It is suggested that Floods and Reservoir Safety should be rewritten to be a risk based approach, where the Critical Flood (the flood that would just cause failure in the duration of a single storm) is related to a tolerable probability of failure. The techniques and tools for this analysis already exist; the main uncertainty relates to the magnitude of overtopping flow to cause failure which is the subject of ongoing research.

HOW IS THE ADEQUACY OF SAFETY EVALUATED IN OTHER INDUSTRIES?

The Control of Major Accident Hazards Regulations 1999 (COMAH), which implemented European Directive 96/82/ EC, requires that operators of top tier sites produce a "Safety Report", with the purpose and content of the report set out in Schedule 4 of the Regulations. Such a report can be in several parts, with contents including

a) Demonstrating a safety management system is in place
b) Accident hazards (failure modes) have been identified and necessary measures taken to prevent such accidents
c) Safety has been incorporated in the design, construction, operation and management
d) On-site emergency plans have been drawn up
e) Information to allow planners to make a decision relating to the siting of new development around establishments

Further detail on interpretation of the COMAH regulations, including when risk has been reduced "As low as reasonably practicable" (ALARP) and how QRA assists in this process is given in HSE publications, including Appendix 3 of Reducing Risks Protecting People (HSE, 2001a) and Guidance Notes to Inspectors (HSE, 2001b). There are similar requirements for the safety case in other high hazard industries.

The principles and administration of review of the safety case in different industries was reviewed in the HSE Discussion Document "Regulating Higher Hazard Industries" published in 2000, and the Cullen Report into the Ladbroke Grove Rail incident published in September 2001. The latter endorsed the use of safety cases, but raised important questions about what makes them effective. These led to the HSC Policy Statement (2003) on "Our approach to permissioning regimes", which sets out their approach to management of the risk from high hazard industries. This includes ten principles. The key items relevant to application to reservoir safety are contained in principle 3 and include:

- Permissioning regimes build on the fact that the legal duty to manage risks lies with the organizations that create them
- The key to receiving "permission" will normally be a description and demonstration of how duty holders manage their risks
- The process of describing and demonstrating requires duty holders to think through their actual operations, from beginning to end, identify their hazards and consider the risk and control measures or systems needed to comply with the requirements of the regime

DISCUSSION – NEED FOR A UK STRATEGY FOR RESERVOIRS

Should we move towards a safety case for reservoirs?

The question "Is the dam safe enough?" (Hatford et al, 2004) is a tricky one to answer. As Low As Reasonably Practical (Appendix 3 of HSE, 2001) may satisfy Health and Safety Executive (HSE) but not consultees in the flood plain of a proposed new reservoir. Comparison between different risks in society is difficult in terms of agreed methodologies, communication and acceptability. A way forward might be for each reservoir to have a Safety Case prepared at construction or the next inspection and periodically reviewed and updated subsequently. This would have the advantages of transparency, and ensuring that owners of reservoirs thought through and documented how they manage the safety of their reservoirs.

The Safety Case would be a documented statement of the risk to third parties posed by the dam, and how those risks were being managed to ensure they were tolerable. Such a safety case would be similar to the "safety report" defined in COMAH regulations, and could include

a) Description and drawings of dam and reservoir
b) Impact assessment of likely consequences if the dam failed
c) Data on performance of dam, including Reservoir Record, Instrumentation and Monitoring
d) Statement as to design criteria for dam, and supporting calculations such as floods, slope stability, internal erosion and failure modes (where available; if not available prepared only when considered necessary to support the safety case).
e) Evaluation of most likely failure modes and for higher hazard dams quantification of probability of failure
f) On-site emergency plan
g) For high consequence dams inundation maps for use for evacuation in the event of imminent dam failure
h) Statement as to Surveillance and dam safety management regimes

This would be building on the current provision of information to Inspecting Engineers under Section 21(5) of the Reservoirs Act 1975, but would differ in that the owner would add "h", a formal statement of how he manages the safety of his dam, and that he endorses this as reasonable. The big advantage of the safety case system is that it focuses liability where it lies. It would also have the advantage that the owner can include how the subject dam relates to the whole of his portfolio, and thus provide a basis for an owner to prioritise work across his portfolio.

It is recognised that non-technical owners will require assistance from dam engineers in preparing the safety case, but the final document must be signed off by the Undertaker as being their responsibility

A range of options for the preparation and review of a safety case regime for reservoirs are given in Table 4. The safety case would apply to all new reservoirs, alterations and existing reservoirs. The role of the "qualified civil engineer" would remain, similar to the Registered Professional Engineer in some countries, but could be as advisor to the owner, or in an independent role. The closest to the current regime would be Option C, but would mean that the safety case was prepared independent of accredited engineers. It would be more logical for the safety case to be prepared by accredited engineers, which would mean adoption of Options B or E, the difference relating to the system for review.

Related issues include that

a) The Safety Case would have to be kept as a live document by the owner, being submitted to the Regulator for consent at prescribed intervals, which could be ten years for Consequence Class A2 and B dams, and five years for Consequence Class A1
b) To maintain the independence of a review this is not just independent of the undertaker but also independent of the preparer of the safety case
c) If the cost was considered disproportionate for small, low consequence dams, then a two tier approach could be adopted, with a safety case only required for Category A dams, or dams which come under the ICOLD definition of "large dam" (which applies to only the largest 20% of UK dams)

<u>Should we move to risk based engineering assessment?</u>

Application of deterministic standards have several disadvantages, including that it can lead to overdesign, lack of thinking about the problem, lack of recognition of adverse circumstances in which the standards are inappropriate and lack of recognition of uncertainty. A risk based approach requires explicit consideration of failure modes which has the advantage that this develops a greater understanding of the problem, requires explicit consideration of uncertainty and enables meaningful/measurable use of ALARP principles to decide when additional costs are proportionate to the reduction in risk achieved. It is considered that a risk based approach ultimately leads to better decision making in terms of management of residual risk to both the public and the owner of the asset.

Table 4: Options for Safety case regime for reservoirs

Option	Preparation of safety case	Regulator	Involvement by Panel Engineers
A	Owner prepares	Technical review (Note 1)	None
B	Owner prepares, but employs Panel Engineer to oversee preparation, and sign off	Technical review (Note 1)	Employed by Owner to oversee and sign-off safety case.
C	Owner prepares	Check that independent review completed and actions being taken	Independent review employed by Owner
D	Owner prepares	As C	Independent review employed by Regulator
E	Owner prepares, but employs Panel Engineer to oversee preparation, and sign off	As C	a) As B, and b) Independent review employed by Regulator

Note 1. Where the Regulator carries out a technical review this would be by staff in the sole employ of the Regulator, to avoid commercial conflict of interests. For the 2000 dams in England and Wales this could be the equivalent of 3 or 4 full time staff.

What level of residual risk is tolerable, and how does it relate to other residual risks faced by society?

One measure of an acceptable probability of flooding from dam failure is to compare it with standards for the probability of fluvial floods, with Table 5 showing the definitions given by both Planning Policy Statement (PPS) controlling new development in UK, and the European Floods Directive. However, a fluvial flood is a natural event whereas a reservoir is a man-made structure/threat. Arguably the probability of flooding due to dam failure should be say 10 times less. A more comprehensive measure of tolerable risk is As Low as Reasonably Practicable (ALARP) analysis of risk (probability x consequences). This provides a tool to evaluate when the cost of further risk reduction measures is proportional in relation to the reduction in residual risk.

Table 5 Terminology for probability of floods

Probability of flooding	European Union Floods Directive (2007/60/EC)	Table D.1 of PPS 25 (2006)	
		Zone	Annual probability
Low	Not defined	1	Less than 1 in 1000 (<0.1%)
Medium	Likely return period ≥100 years	2	1 in 100 to 1 in 1000 (1% -0.1%)
High	Not defined	3a – High probability	1 in 100 or greater (>1%)
		3b – Functional flood plain	1in 20 (5%) or greater

CONCLUSIONS

Quantitative risk assessment (QRA) techniques now provide the ability to make meaningful estimates of the probability of failure, its consequences and thus the risk (probability x consequences) of dam failure. This provides the ability to use QRA to make more informed dam safety decisions, but also challenges the status quo. This paper has described some of the opportunities and the practical consequences.

The following issues are identified where it is recommended that current practice is modified

a) the dam owner should be made responsible for preparing a "safety case" which sets the risk to the public from that dam, and how he is managing that risk ALARP, this safety case being sent to the Enforcement Authority. This safety case would be subject to independent review under the existing periodic Section 10 Inspection process. The level of detail would depend on the level of risk or consequences, and could vary from simple qualitative assessment, through event trees to full quantitative risk assessment
b) the existing "Floods and Reservoir Safety" prepared to assist Panel Engineers and dam owners, is based on "deterministic" standards" and should be rewritten to be risk based. Similarly other existing and future engineering guides should be rewritten to be risk based.
c) the UK should have a toolbox of different techniques for QRA, suitable for a range of uses by the stakeholders involved in dams.

The differences between particular dam engineers is noted and suggests that a panel of users should be set up to develop the toolbox of QRA tools, rather than leaving development to a single contractor or user group. It is anticipated that the specification for the techniques in this toolbox would include:

- quantitative output which is on the same basis as, and thus comparable with, quantification of other risks to society
- suitable for a range of level of detail of application, from preliminary screening analysis of low risk dams, to detailed analysis of extremely high consequence dams
- be practical and useable by a knowledgeable practitioner not just theoretical and not a 'black box' written by the developers
- methods for dam owners to use in support of their safety case, including justification where the cost of further risk reduction works is disproportionate to the reduction in risk that would be achieved
- output in the format that could be annexed to both the safety case and a Section 10 Inspection Report under the Reservoirs Act 1975, as an audit trail

REFERENCES

ANCOLD, 2005, *Guidelines on risk assessment.* 156pp excl appendices

Brown, A.J. and Gosden, J.D., 2004a, *Interim Guide to quantitative risk assessment for UK reservoirs.* Thomas Telford.

Cyganiewicz J M, Engemoen WO, Redlinger CG, 2007, *Bureau of Reclamation experience with evaluating internal erosion of embankment dams. In Internal Erosion of dams and their foundations.* Pubs Taylor & Francis.

DCLG (2006) *Planning Policy Statement 25 (PPS25): Development and Flood Risk*

Defra (2004) *Flood and coastal erosion risk assessment and prioritisation.* Background paper accompanying consultation on "Making Space for water". Flood Management Division. July

Defra, 2007, *Performance and reliability of flood and coastal defences.* R&D Technical Report FD2318.

Hartford DND, Baecher GB, 2004, *Risk and uncertainty in dam safety.* CEA technologies Dam safety Interest group. Thomas Telford. 391pp

HSC (2003) Policy statement *Our approach to permissiong regimes.*

HSE (2000) *Regulating higher hazard industries: exploring the issues.* Draft for discussion. London, HSE

HSE (2001a) *Reducing risks, protecting people*

HSE (2001b) *Principles and guidelines to assist HSE in its judgments that duty holders have reduced risk as low as reasonably practicable.* On HSE website.

ICE, 1996, *Floods and reservoir safety*, 3rd Edition. Thomas Telford

ICOLD, 2005, *Risk assessment in dam safety management.* Bulletin 130

3. Design, analysis and monitoring

Spillway gate design features which can cause vibration

JACK LEWIN, Independent Consultant, Richmond-upon-Thames, UK

SYNOPSIS. Gate vibration is a serious problem which can result in structural damage or restrict operation at certain gate openings. Operators are mostly unaware that a risk of gate vibration exists. In some cases vibration of a gate will occur under specific hydraulic conditions which may become manifest only years after commissioning of the installation. Even when these conditions have been identified it may not be easy to reproduce them so that they can be investigated. Apparently steady state conditions may be subject to minor hydraulic disturbances which overcome the damping forces acting on the gate and initiate an unstable motion, giving rise to oscillations of increasing magnitude. A number of design features can cause gate vibration; the paper briefly describes these and the conditions under which their effect on the discharge flow at spillway gates is critical. Gate vibration may also arise due to maintenance and servicing problems, or to degradation over time.

INTRODUCTION

Few problems are more difficult to cope with than the vibration of hydraulic equipment (Cassidy 1996). It is not possible to provide an in depth understanding of generation of vibrations in one technical paper. Extensive work is available setting out the analytical factors of the vibrating system (Kolkman 1979, Naudascher et al 1994, ICOLD 1984); for an introduction to the subject, see Lewin (2001).

Surveys of existing installations suggest that the designers were not aware of critical hydraulic aspects which can affect the reliability of the plant. Even when the potential for vibration is present, gates may operate for a long time without incident. A change in hydraulic conditions or operating practice can result in unexpected gate vibration. When this occurs in extreme conditions or at unattended installations it is not certain that the problem will be recognised or detected in time.

The natural frequency of a gate depends on the rigidity per unit mass. When oscillation occurs the water at the gate face moves with it (the added mass).

Any force due to the excitation frequency but in opposition is a damping force. At radial gates damping is due to the friction of the side seals and is relatively low. At vertical lift gates additional forces are caused by bearing friction at the guide rollers and, to a lesser extent, roller friction. Seal friction also depends on wear.

Because the natural period of vibration of a gate depends on its total mass (structural mass plus added mass), it will vary with the gate opening and the upstream water level. These two factors will also affect damping but not necessarily in the same ratio.

Because it is difficult and often impossible to predict the conditions when gate vibration is likely to occur, it is important that design features which can cause fluid excitation are not present.

CAUSES OF GATE VIBRATION

Side Seals

Figure 1 shows different side seal arrangements. Figure 1a, a gate side seal in the shape of a musical note is a conventional solution. 'L' shaped seals, Figures 1b and 1c, are used when the upstream water head is moderate (8 m or less) because the seals are more flexible and permit a larger clearance between the skin plate and the embedded seal contact plate. At higher heads the possibility of seal breakthrough increases.

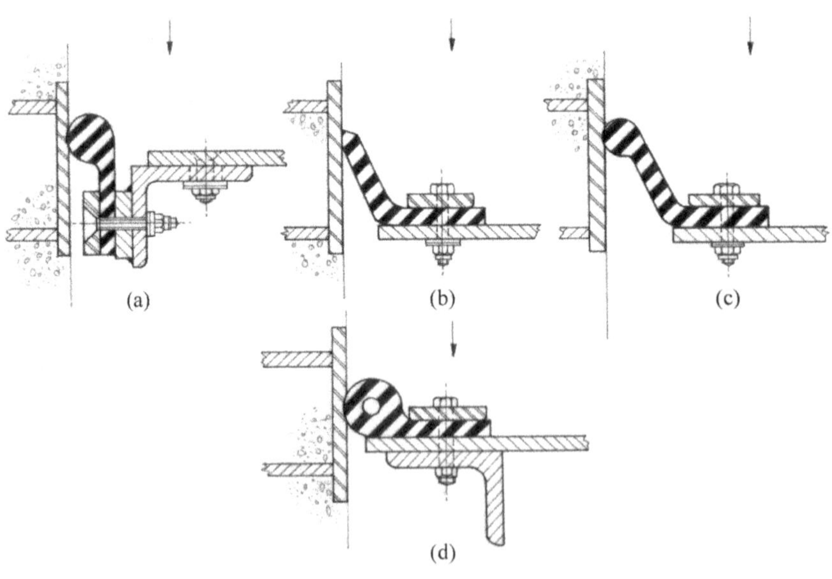

Figure 1. Side seals

The seal arrangement in Figure 1d should not be used because it only permits deflection by compression of the seal, not by flexure. This severely limits the width of the gap which the seal can close.

Fluctuating discharge through a seal is a frequent cause of gate vibration. A breakthrough of upstream water results in a reduction in the water pressure which, in turn, causes the seal to close. This initiates another leakage flow. A regular repetition of fluctuating flow causes the gate to vibrate. Intermittent leakage flow through side seals requires rapid remedial action before the amplitude of gate vibration builds up.

Many old vertical lift gates use staunching bars as side seals. Because of the rigidity of the bars leakage is continuous, but at high gates leakage jets have caused cavitation erosion at low level. Kolkman (1984) mentions that noise, vibration, hammering, wear at the edges of leaking parts and damage to wheels have been caused by the shocks induced by the collapse stage of cavitation bubbles or voids.

Sill Seals

The seal should provide a sharp separation of flow under the gate. A seal in the shape of a musical note is therefore unsuitable, and only rectangular shaped rubber or elastomeric seals should be used (Figure 2). In both cases the flow can reattach, and when this is intermittent and regular, gate vibration can occur.

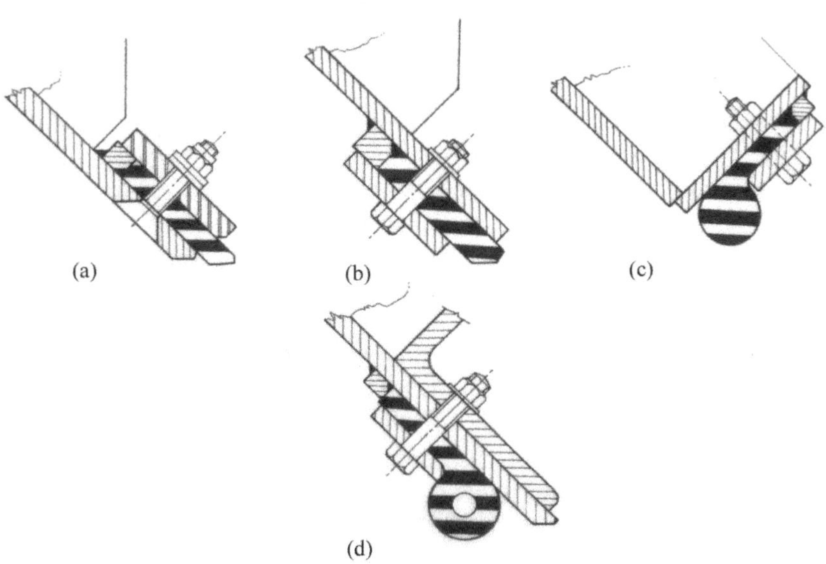

Figure 2. Sill seals

The correct method of mounting the seal is shown in Figure 2a.

Figure 2b, which shows the seal upstream of the skin plate, causes flow attachment at the upper edge of the seal clamping plate due to the downward flow close to the gate lip – also a feature of Figure 2d.

In the case of Figure 2c, the upper end of the seal clamping plate tends to become more horizontal when the gate opening increases and cuts into the expanding flow under the gate. In Figure 2d, the skin plate projects into the flow as the gate rises, causing reattachment after discharge under the seal. Figures 2c and 2d are reproduced from 'as constructed' drawings of existing gates and are not isolated cases.

Leakage at the sill seal can cause gate vibration. Differential tightening of the screws holding the seal and the retaining plate can introduce slight waving of the seal which can cause leakage. A ferrule is usually fitted around the screw section through the seal. It is slightly less thick than the seal and permits even tightening of the nuts.

<u>Unsteady Flow</u>
High velocity flow through a narrow gap is unsteady. The natural frequency of a gate and its added mass of retained reservoir water tend to regularise the leakage discharge, causing gate vibration.

Gate sills can be distorted by closure when debris is lodged on the sill. In view of the risk of gate vibration, remedial work to eliminate discharge on this account should not be delayed.

If an initial opening of a gate is insufficient the flow is unsteady. To avoid vibration, gate manufacturers specify a minimum first gate opening of between 100 mm and 150 mm. Where gates are operated automatically this is incorporated into the control system. However, there are instances of gate installations which are manually operated where staff are not aware of this requirement, and it is not made explicit in the operating instructions.

<u>Structural Considerations</u>
Gate design guidelines to avoid vibration can be formulated (Lewin 2001).

(a) No structural member upstream or downstream of the control point should protrude into a line at 45 degrees from the point of flow control, as shown in Figure 3. Two references (Kolkman 1984; Hart et al 1979) suggest that members upstream of the control point should preferably be 60 degrees clear of the control point.

(b) It is better to arrange for the vortex trail to be shed from the extreme downstream edge of a gate in order to achieve flow conditions that are as steady as possible (Vrijer 1979).

(c) A sharp cut-off point should be provided at the lip (Schmidgall 1972).

Figure 3a illustrates vertical lift gates where a reinforcing girder is placed low. Discharge under the gate causes flow to attach to the web of the girder and the flange. The attachments are intermittent and severe gate vibration can occur.

Figures 3b and 3c show construction at the bottom of vertical lift gates which comply with the criteria to avoid vibration. The shape also reduces hydraulic downpull forces which can become a significant factor at high gates.

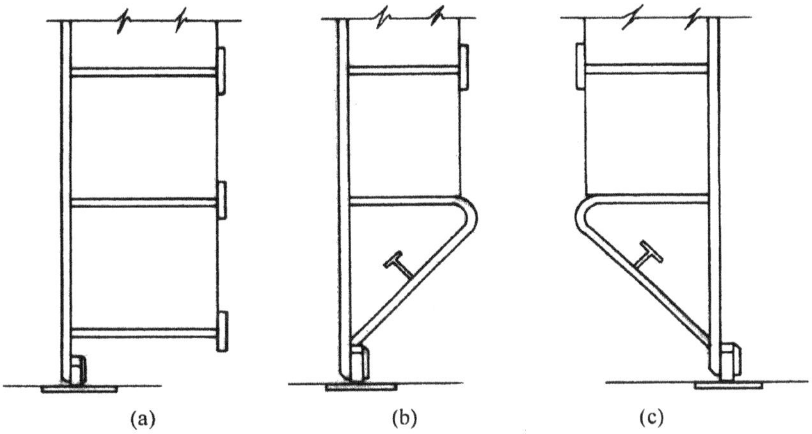

Figure 3. Structural reinforcement of gate bottoms

Other examples of a skin plate reinforcing member placed low are shown in Figure 4. An extreme case is Figure 4b, where the bottom of the gate was flat. It resulted in severe vibration. A number of gates of this design were installed at Pershore Mill (Bruce et al 1978).

The examples of gate construction illustrated in Figure 4 are not unique and a few gates of this type are still operational despite considerable age. They may have survived because they have been rarely used, and then only because they were mostly in the open position.

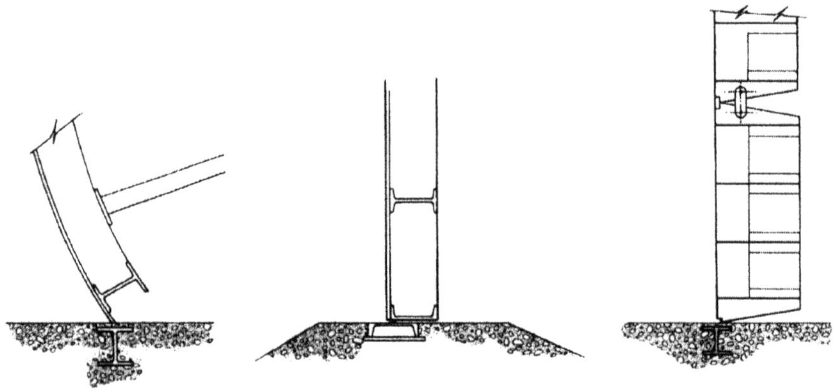

(a) (b) (c)

Figure 4. Unsuitable arrangement of structural stiffening members at the bottom section of gates

Figure 5 illustrates a radial gate where a low structural member stiffens the bottom of the skin plate. In the shut position it appears that the expanding flow under the gate will, on opening, satisfactorily clear the skin plate stiffening member. However, when the gate is raised the stiffening member rotates about the trunnion axis and the flow discharged under the gate attaches to the stiffening member. Detachment and reattachment occur and when this regularises it is likely to cause vibration.

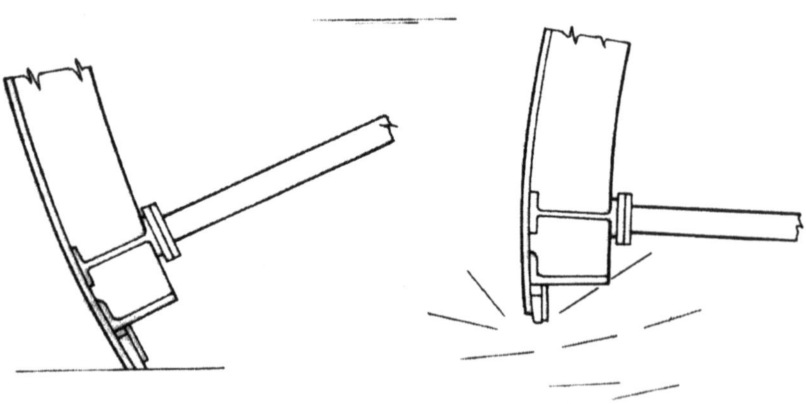

Figure 5. Structural reinforcement of the bottom of a radial gate which can cause flow reattachment when the gate is raised

Other Causes of Gate Vibration

Eddy Shedding
Regular eddy shedding can occur when there is an appreciable downstream water level, Figure 6. New radial gates in Australia suffered from persistent flow induced vibration (Hardwick et al 2000). The extended periods of large gate opening and low flow velocity were unusual. A model suggested that instability of a shear layer at the gate lip could be responsible for the excitation. The vibration was eliminated by mounting spoilers on the upstream of each prototype gate skin plate to break up the two dimensional character of the shear layer.

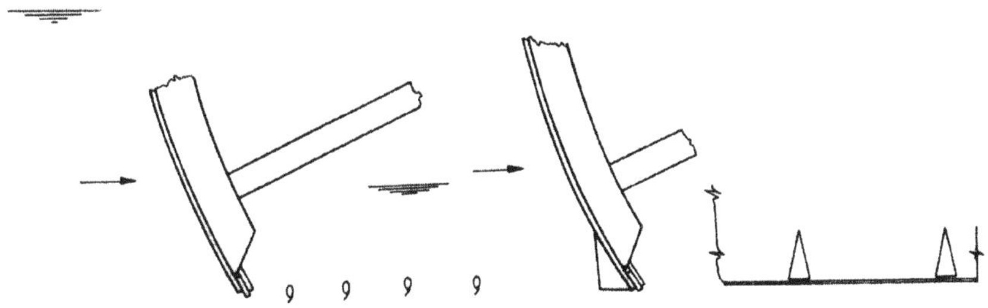

Figure 6. Regular eddy shedding at the bottom of radial gate

Off-Centre Trunnions
In 1967 a spillway gate 12 m high and 9 m wide on the Washii Dam in Japan collapsed (Ishii et al 1968). It was swept downstream, Figure 7. The cause was instability introduced by eccentricity of the trunnion bearings. The trunnions were located about 0.5 m above the true origin of the radius of the skin plate.

Eccentric trunnions were deliberately introduced on some large radial gates to reduce the hoisting forces. There is no record of violent vibration of these gates because the eccentricity was much lower than that of the Washii gate. The safe limit of eccentricity appears not to have been researched.

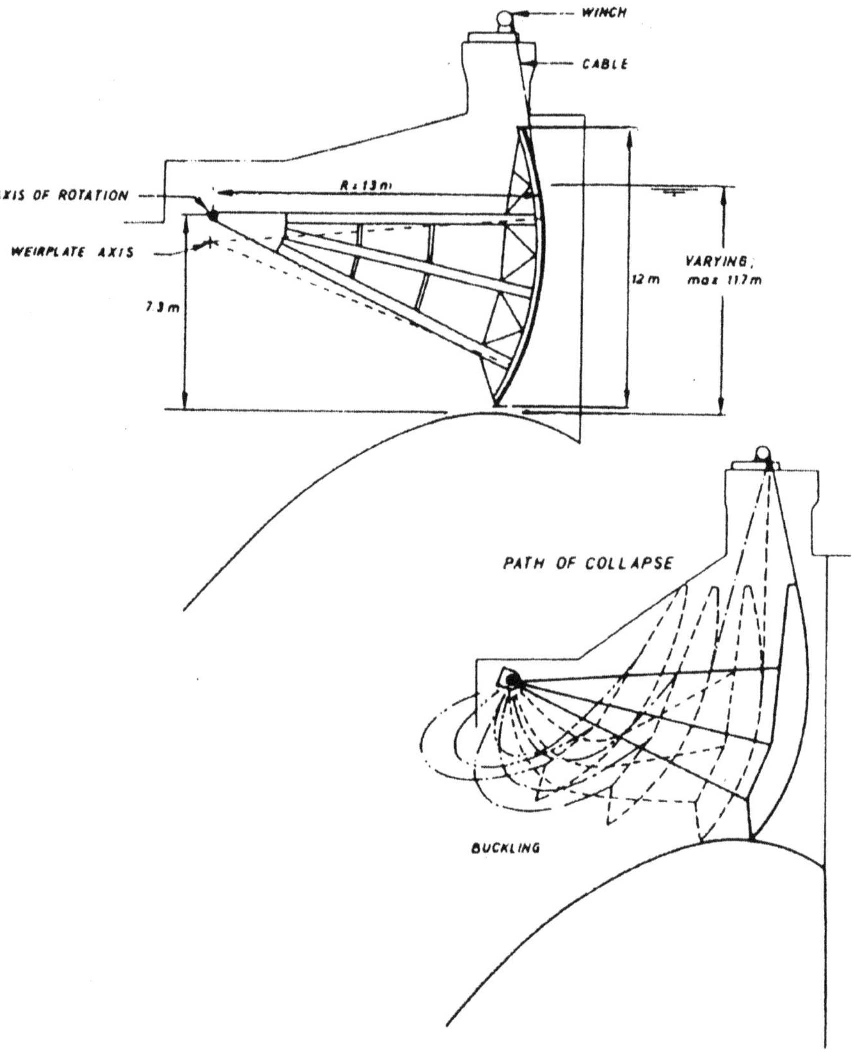

Figure 7. Spillway gate vibrations leading to the collapse of a gate (after Ishii et al)

Venting Overflow of Gates

Certain types of gates are designed to discharge by overflow, such as bottom-hinged gates, vertical lift gates and radial gates with hinged overflow sections. The objective may be to discharge floating debris or, at tidal barrages, to prevent salt water penetrating inland. At some river barrages, bottom-hinged gates are selected to improve appearance. A number of different designs are illustrated in Lewin (2001).

The volume of air enclosed beneath a nappe varies periodically and is excited by pressure fluctuations beneath the nappe. The pressure fluctuations excite the nappe which, in turn, fluctuates. The amplitude of nappe fluctuation is greatest at its lower end.

The falling nappe absorbs air, so unless air is admitted the nappe will eventually collapse, which can cause a serious shock to the system. Figure 8 shows two methods of effecting air supply. Flow dividers are one solution; a high overflow can drown them out and venting ducts in gate piers and abutments must be provided. Venting will also increase the pressure under the nappe, and the discharge and resulting velocities will decrease. This is important in controlling the occurrence of cavitation. Spacing of flow dividers and information to calculate air supply are given in Lewin (1983).

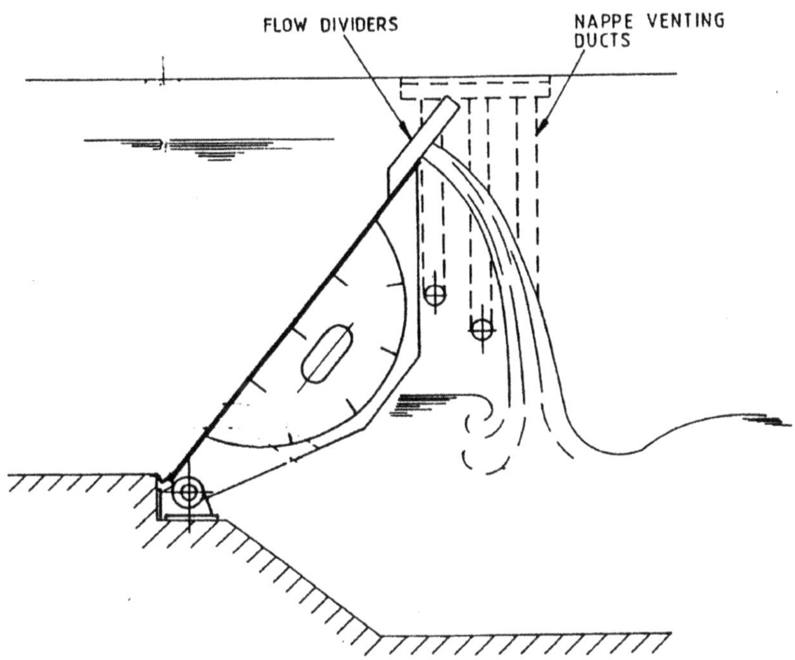

Figure 8. Venting the overflow of a gate

Structural Arrangement of Gate Arms and Stiffening Girders

Figure 9 illustrates preferred structural arrangements of the gate arms of a radial gate and the stiffening girders of wide span vertical lift gates.

There are no recorded examples where structural members close to the bottom of a gate of the type shown in Figure 9 have resulted in gate vibration, but in rivers which carry sand during flood discharge they have been seriously eroded. An example is the collapse of a gate on the Sukkur Barrage on the River Indus in Pakistan.

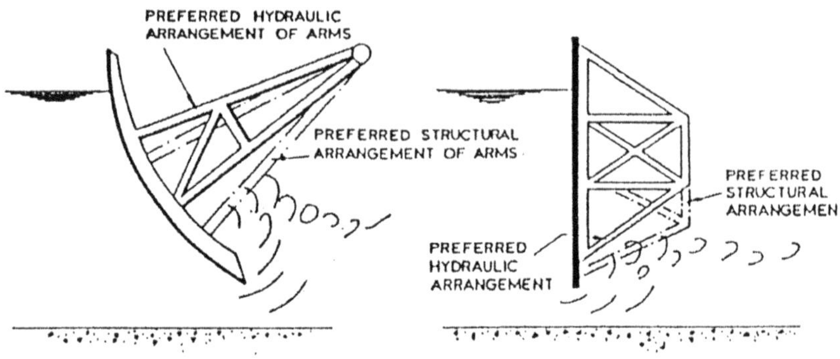

Figure 9. Preferred hydraulic structural arrangement of a gate subject to drowned discharge

<u>Slack of Vertical Lift Gates in Guide Channels</u>

Slack can be an important contributory cause of gate vibration at vertical lift gates. With Stoney Roller gates, slack is difficult to avoid because of their construction. The vibration of gates at the Toome Sluices in Northern Ireland was due to the interaction of a low webbed girder stiffening the Stoney Roller gates and slack of the roller assemblies in the guide channels.

<u>Hollow Cone Valves (Howell Bunger Valves)</u>

Causes of vibration of hollow cone valves are included here because many spillway gate installations use them to control compensating flows. Because of their simplicity and high discharge coefficient they are often the first choice of discharge valve.

The lower section of Figure 10 shows an open hollow cone valve. In this position the flow control may shift from the sleeve to the valve body (A to B) and intermittent attachment and reattachment may occur, resulting in severe vibration. Under these conditions the opening of the valve, that is, the sleeve travel, has to be limited.

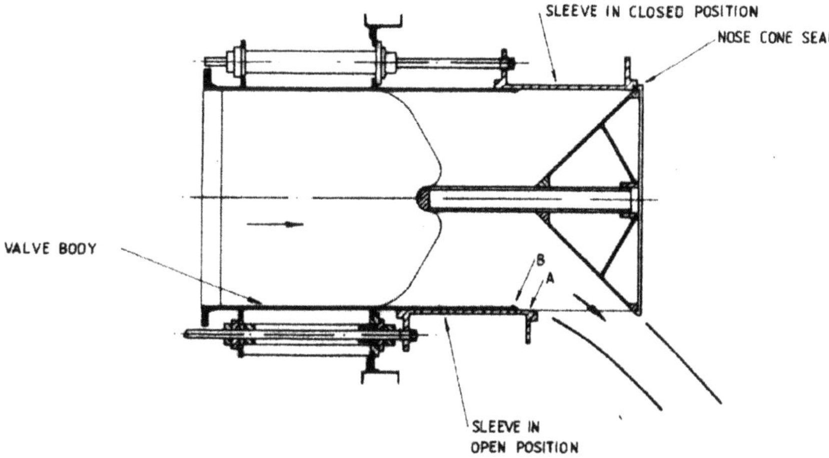

Figure 10. Hollow cone valve

Vane failures have occurred and have been attributed to a number of reasons, but the most likely one is hydroelastic instability causing vibration normal to the chords and twisting about the longitudinal axis. Destructive resonance occurs at a critical velocity at which the flow couples the two forms of vibration in such a way as to feed energy into the elastic system. Possible modes of vibration for a hollow cone valve are shown in Figure 11. Mercer (1970) has suggested equations for calculating parametric values which avoid failures.

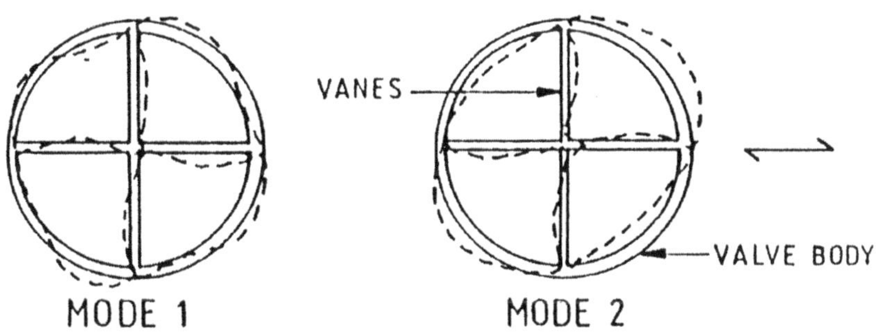

Figure 11. Possible vibration modes of hollow cone valves (after Mercer)

CONCLUSIONS

The design aspects discussed and illustrated in this paper which can cause or contribute to gate vibration are not comprehensive. For an understanding of gate vibration, references quoted in the introduction are recommended.

While it is possible to predict when gate vibration is likely to occur this is of limited practical application because causative factors, such as the vibrating mass, change throughout gate operation. However, model investigations can be valuable (Novak et al 1981; Kolkman 1976, Haszpra; 1979).

Vibration of low level gates at high head dams can be more severe as additional factors, such as demand for air supply, are present . A summary of design elements which are inimical to gate vibration would require a longer paper.

REFERENCES

Bruce, B.A. & Crow, D.A. (1978). *Hydraulic model studies of the Pershore Mill sluice gates.* B.H.R.A. report RR 1495, Jul.

Cassidy, J.J. (1995). Vibration of Hydraulic Equipment for Dams. *ICOLD Bulletin 102*, Foreword.

Hardwick, J.D., Attari, J., Lewin, J. (2000). Flow-induced Vibration of Torrumbarry Weir Gates. *Proc. 7th Int. Conference on Flow-induced vibration*, Lucerne, Switzerland, Jun, Balkema, 219–224.

Hart, E.D. & Hite, J.E. (1979). Barkley Dam Gate Vibrations. *19th I.A.H.R. Congress*, Karlsruhe, paper C15.

Haszpra, O. (1979). *Modelling hydroelastic vibrations.* Pitman Publishing.

ICOLD. (1995). Vibration of hydraulic equipment for dams. *Bulletin 102.*

Ishii, N., Imachi, K., Hirose, A. (1968). Instability of elastically suspended tainter-gate system caused by surface waves on the reservoir of a dam, *Am. Sos. Mech. Eng. Fluids Eng. Div.*, Joint applied mechanics, fluids engineering and bio-engineering conference, New Haven, Conn, Jun., paper No. 77–FE–25.

Kolkman, P.A. (1984). Gate Vibrations. Chapter 3 of *Developments in Hydraulic Engineering – 2*, ed P. Novak, Elsevier Applied Science Publishers.

Kolkman, P.A. (1976). Flow-induced gate vibrations. *Delft Hydraulics Laboratory*, publication 164.

Lewin, J. (1983). Vibration of hydraulic gates. *Journ. I.W.E.S.*, 37, 165–173.

Lewin, J. (2001). *Hydraulic Gates and Valves, 2nd Edition.* Thomas Telford.

Mercer, A.G. (1970). Vane failures of hollow cone valves. *I.A.H.R. Symposium*, Stockholm, paper G4.

Schmidgall, T. (1972). Spillway gate vibrations on Arkansas River dams. *Proc. A.S.C.E.*, Journ Hydr Div. HY1.

Vrijer, A. (1979). Stability of vertically movable gates. *19th I.A.H.R. Congress*, Karlsruhe, paper C5.

A method of mitigating tree blockage to bridges on a spillway

M.J. WEARING, CRM Rainwater Drainage Consultancy Ltd, Bolton, UK
R BAKER, CRM Rainwater Drainage Consultancy Ltd, Bolton, UK
I.C. CARTER, MWH, High Wycombe, UK

SYNOPSIS. On a wooded catchment there is a risk that fallen trees may be washed over the spill weir and trapped under bridges, which might result either in a loss of capacity and a potential overtopping of the dam or else a change in flow down the spillway and overtopping of the side walls. In the paper a trash barrier consisting of vertical poles placed upstream of the spill weir is described and results are presented from model tests showing the efficiency of the poles at trapping trees. It is concluded that the problem of tree blockage of a spillway can best be mitigated by the introduction of a double row of poles in front of the weir and that suitably placed poles would not affect the operation of the spillway.

INTRODUCTION

It is very common for access to be provided across the crest of an embankment dam with a bridge across the spillway. On many Pennine dams the engineers have built aesthetically pleasing multiple arch bridges some of which have very small waterways. On a wooded catchment these small arches present a safety hazard since a fallen tree washed under the bridge on a storm flood could result in total blockage of a waterway.

This could have two detrimental effects on the spillway performance. Firstly, the loss of capacity could result in the spillway weir drowning and rising reservoir levels causing the dam to overtop. Secondly, the change in flow pattern downstream of the bridge will alter flow depths and could accentuate cross-waves resulting in wall overtopping in different places to those expected under normal operation.

The paper describes model tests on rows of poles placed upstream of the spillway weir to trap large items of debris in the reservoir for removal after the flood has subsided.

PREVIOUS STUDIES

Previous work has been published on the use of rows of poles to trap trees upstream of hydraulic structures but most of these have concentrated on heavily wooded catchments where extreme numbers of floating trees could be expected during floods. The emphasis of those studies has thus been on the best way of trapping large quantities of debris without impeding flow.

The US Army Corps of Engineers manual (Wallerstein and Abt, 1997) on the subject contains examples of different types of structure that exist throughout USA and Europe, in particular the Treibholzfange scheme in the Bavarian Alps, where model tests showed that a row of poles in a downstream facing 'V' formation was very effective at trapping large quantities of debris. The manual also gives advice on mechanical trash racks that would self-clean during a storm event, although this would be an excessive solution for the issue of trapping trees during severe or extreme events, which was the key objective of the research discussed in this paper.

Godtland *et al* (1996) carried out model tests on trash barriers to trap trees and highlighted that branches on model trees are inherently stronger than on the prototype: a model tree will tend to trap where branches on the prototype would have broken allowing the tree to pass. Bearing this in mind the authors decided not to uses model trees in this study that had foliage or a proliferation of small branches.

Godtland *et al* also noted that the behaviour of a single tree passing through a free waterway was quite different to a tree approaching a waterway that was already partially blocked. In a free waterway the tree aligns with the flow and passes through the structure but the alignment did not happen with partial blockage and thus an addition to the blockage was more likely forming a 'log jam'. This finding was also observed in the current study.

LABORATORY TESTS

Laboratory tests were performed using scale representatives of tree trunks, with or without branches. Five different configurations were used as shown on Figure 1. The trunk of the tree was made from 15mm dowel while branches were made from 8mm dowel. The simple log had a critical width of 15mm unless turned sideways. The log with one branch (ref. 1B) had a critical width of 60mm but it could rotate around a single pole. The log with two branches presented at an angle to each other (ref. 2B-3D) had a critical width of 80mm. The critical width for logs with two branches laid flat (ref. 2B) was 120mm and for logs with three branches (ref. 3B-3D) it was 80mm in any direction.

The barrier poles were made from 6mm dowel and spaced at 50, 75 and 100mm. These model sizes could be scaled up to any size dependent upon the size of the object that was to be trapped, so for a 10m long log the model would simulating poles at 5, 7.5 and 10m spacing.

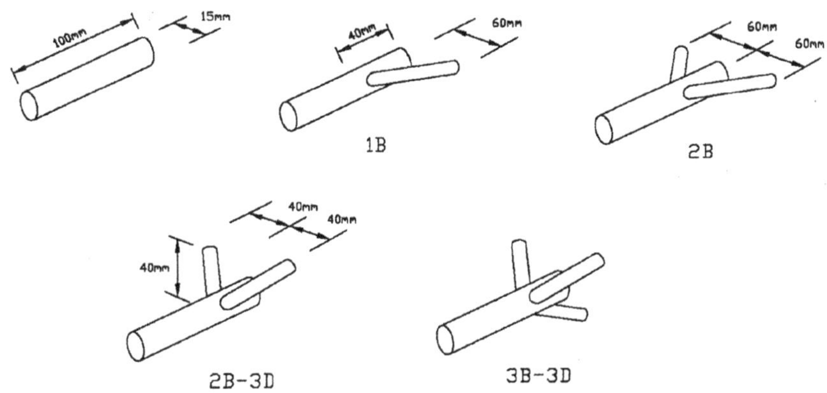

Figure 1 Configuration of the model trees

Model tests were performed on a 750mm wide weir in the side of a stilling box simulating the reservoir. Two flow rates were used, 15 l/s/m and 30 l/s/m, with a flow velocity through the poles of 0.25m/s and 0.3m/s respectively. This data would be scaled by Froude similarity and so for the 10m log described above this would be simulating flow rates of 15 m^3/s/m and 30m^3/s/m at velocities of 2.5 and 3m/s.

A single row of poles were set up in front of the weir and the model was operated at a constant flow rate with model trees being placed in the reservoir and allowed to flow freely towards the weir. The percentage of trees that were trapped was noted and is recorded in Table 1. Trapped trees were removed from the model so that a 'log jam' could not form.

Table 1 – Percentage of trees trapped by a single row of poles

Spacing (mm)	Flow (l/s/m)	Tree shape				
		Log	1B	2B-3D	2B	3B-3D
100	15	10.0	10.0	30.0	83.3	80.0
	30	0	15.3	28.6	53.3	50.0
75	15	9.1	53.8	92.9	100	100
	30	8.3	60.0	86.7	100	100
50	15	50.0	100	100	100	100
	30	29.4	100	100	100	100

As the tree approached the row of poles the main 'trunk' section was observed to align with the flow and tended to pass through the middle between a pair of poles, it was therefore quite unlikely that the simple log shaped tree would become trapped. However, it is probable that this same flow orientation would happen through a bridge and thus the simple log would be less likely to become trapped and present a hazard.

For the 100mm spacing the 1B and 2B-3D shaped logs had a critical width less than the gap and thus were more likely to pass through than become trapped. Those that were trapped normally became lodged with the pole against the joint between the log and the branch (Photograph 1) although other wedge positions were possible (Photographs 2 and 3). The 2B shaped log could also become wedged between a pair of poles held by both branches (Photograph 4). The 3B-3D tree was a more complex shape and it was observed that this could become trapped by branches trailing under water and becoming wedged on the bed at the lower flow. In all cases increasing flow rate meant that more trees passed through the row of poles.

Photograph 1 – A 2B type log trapped on a single pole.

Photograph 2 – A 2B-3D type log trapped between two poles.

Photograph 3 – A type 1B log trapped on a pair of poles.

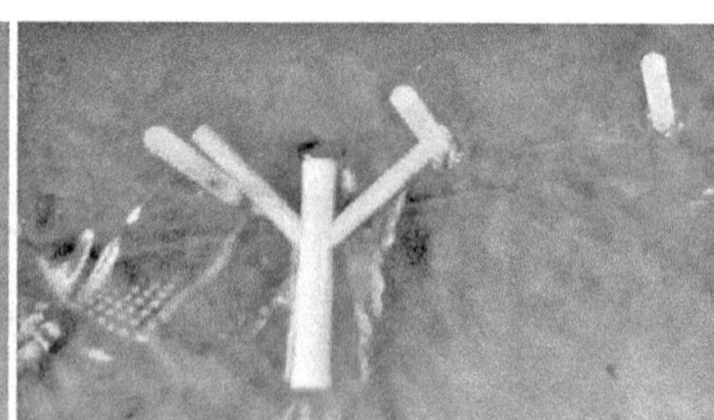

Photograph 4 – A type 2B log trapped between two poles.

As could reasonably be expected, the efficiency of the operation was improved by reducing the pole spacing relative to the size of the tree that was to be trapped. For example, a 2B type tree could not physically pass through a spacing less than 75mm. At 50mm spacing, which was half the length of the object being trapped and less than the critical width of all but the straight log, 100% of all trees and some of the straight logs were trapped.

If a log jam was allowed to form (Photograph 5) then a greater trapping efficiency than that noted in Table 1 was achieved and there was some build up of water level behind the log jam.

Photograph 5 – Log jam

The experiments continued with the poles in two rows spaced so that a pole on the second row lay exactly in the middle of two poles on the first row. The two rows were spaced apart so that the triangular distance from the upstream pole to the downstream pole was equal to the pole spacing, thus any three adjacent poles formed the corners of an equilateral triangle.

This arrangement was found to be much more efficient, as shown in Table 2. The logs continued to align with the flow to pass between the middle of two poles on the first row, which caused then to impact onto the pole in the second row. Sometimes the log bounced off or rotated and passed through but normally it rotated and trapped across the two rows of poles (Photograph 6 and 7).

Table 2 – percentage of trees trapped by a double row of poles

Spacing (mm)	Flow (l/s)	Tree shape				
		Log	1B	2B-3D	2B	3B-3D
100	15	33.3	38.9	81.3	100	100
	30	20.0	16.6	71.4	100	100
75	15	50.0	100	100	100	100
	30	0	100	100	100	100
50	15	80.0	100	100	100	100
	30	78.6	100	100	100	100

Photograph 6 - Log trapped across two poles on the second row

Photograph 7 - Log trapped between the two rows of poles

When installing poles of this nature the designer should carry out structural checks to ensure that the poles have adequate foundation to resist the turning moment generated by the head of water that could build up behind a log jam.

Many old earth embankment dams in the UK have thin upstream clay blankets that were either part of the original construction or else were added later in an attempt to make the dam watertight. Therefore, if barrier poles are to be installed at the weir approach, then care should be taken to ensure that the integrity of any waterproofing element directly beneath the surface is not compromised. Failure to do so could lead to leakage, which may be difficult to detect.

Poles will also collect quantities of smaller items of debris during normal operation and a suitable regime should be devised for clearing this debris, possibly during summer months when reservoir levels are low.

CONCLUSIONS

Model tests have shown that a tree trunk will tend to align itself with the flow and pass between a pair of poles designed to trap it: the higher the flow rate and hence velocity, then the greater the number of trees that will pass through a row of poles.

A log can hit a pole and rotate to pass through the gap. As could reasonably be expected the smaller the gap size relative to the tree the more chance there is of trapping it. If the gap is half the length of the tree and less than the critical width of the branches, then there is complete certainty that the tree will be trapped.

If a second row of poles is placed downstream, then a log passing through the gap in the first row of poles is likely to hit a pole in the second row, rotate and become trapped. With two rows of poles, a 100 percent trap can be achieved with a spacing three quarters of the length of the tree even if the critical width is slightly less than the pole spacing.

REFERENCES.

Godtland K, Johnsson N and Cederström M (1996) *Floating debris and dam safety under extreme floods.* Proceedings of the European Symposium on Repair and Upgrading of Dams, Stockholm, pp 374-382.

Wallerstein C.R and Abt S.R (1997) *Debris Control at Hydraulic Structures in Selected Areas of the United States and Europe.* US Army Corps of Engineers Waterways Experimental Station, Contract Report CHL-97-4.

Recent advances in the numerical modelling of embankment dams

N. KOVACEVIC, Geotechnical Consulting Group, London, UK
D. M. POTTS, Imperial College, London, UK
P. R. VAUGHAN, Imperial College, London, UK

SYNOPSIS. The main concern when designing or analysing embankment dams is usually their stability. However, assessing their deformation is also important. This paper describes the recent use of advanced numerical analysis to predict the latter. Examples of its use in assessing likely movements of six embankment dams constructed in the UK are provided. Various phases in dam life are considered: embankment construction, first reservoir impounding, subsequent reservoir operation and raising the crest level. The use of adequate constitutive laws and, in particular, the importance of modelling the plastic behaviour of embankment fills during loading, unloading and re-loading is clearly demonstrated.

INTRODUCTION

Embankment dams are complex geotechnical structures. The materials involved in their construction vary widely, involving foundations ranging from soft clays to hard rocks and fills ranging from compacted clays to rockfills. In the beginning their design and construction were largely empirical and based on past experience rather than on theory. With no intention of diminishing the important role of experience, the advances made in the theory of soil mechanics, together with the development of advanced numerical analyses in the last 40 years, have made it feasible to predict stresses, strains and displacements in dams during construction, first impounding and subsequent operation.

Many attempts have been made to predict the behaviour of embankment dams. The finite element (FE) method is a powerful tool for such predictions. A large number of FE analyses of actual dams have been reported in the literature and these have been extensively summarised by Duncan (1994).

The FE analysis of stresses and movements in embankment dams is an exceedingly complex problem. The techniques for modelling construction, first impounding and subsequent operation are covered in various publications (see e.g. Naylor, 1991). Many factors should be considered, such as construction in layers, the stiffness of the simulated layer, compaction stresses, etc. However, the most influential factor is the modelling of the stress-strain behaviour of the fill by an appropriate constitutive law (Duncan, 1992). A review of the different constitutive laws used in the numerical analysis of embankment dams can be found in various publications (see e.g. Potts & Zdravkovic, 2001). Both elastic and elasto-plastic formulations are available. The former are simpler to use, but the latter are far superior.

All FE analyses reported in this paper were carried out using the computer code ICFEP (Potts & Zdravkovic, 1999). The available space does not allow for a detailed description of the constitutive models used, laboratory testing performed and model parameters derived. However, references are provided wherever appropriate.

ROADFORD DAM

Roadford Dam in Devon is a homogeneous 41m high rockfill dam completed in 1990. Unusually conservative slopes were adopted for this dam (Figure 1) because the Carboniferous rockfill used in dam construction was prone to weathering. An important design feature of the dam was the presence of a less compressible sand-waste fill to reduce differential movements between the asphaltic membrane and the stiff concrete block at the upstream toe.

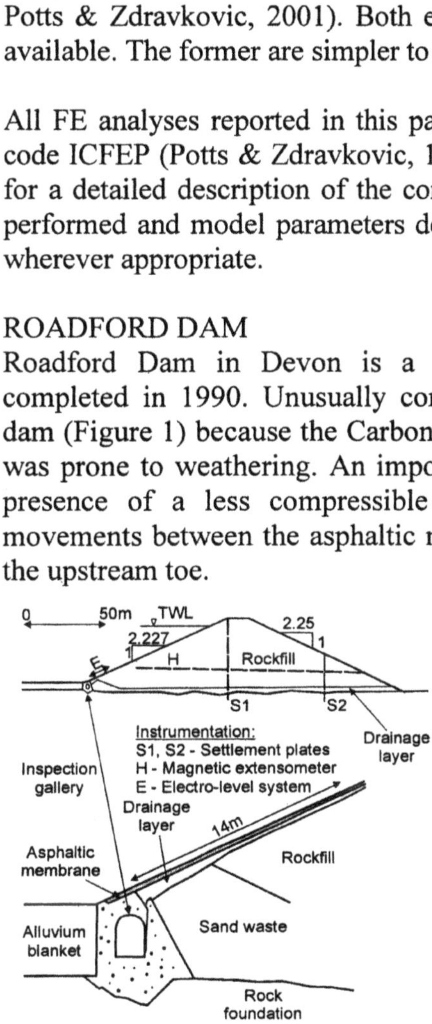

Figure 1. Cross-section of Roadford Dam.

The properties of the rockfill were measured in 250mm diameter triaxial and 1m diameter oedometer equipment by the British Research Establishment (BRE). Results of a suite of drained triaxial tests in compression at different confining stresses are shown in Figure 2. The fill was initially characterised by a relatively simple non-linear elastic perfectly plastic model of the Mohr-Coulomb (MC) type (Figure 2a). As a research exercise (Kovacevic *et al*, 1994), the rockfill behaviour was also predicted (Figures 2b and 2c) by using two more complex elasto-plastic models of the Lade's type (Lade, 1977; Lade & Kim, 1988), both of which account for the plastic behaviour before and after the peak strength of the rockfill is mobilised.

The measured settlements near the dam centre line and mid downstream slope during construction are presented in Figure 3. The magnitude of the

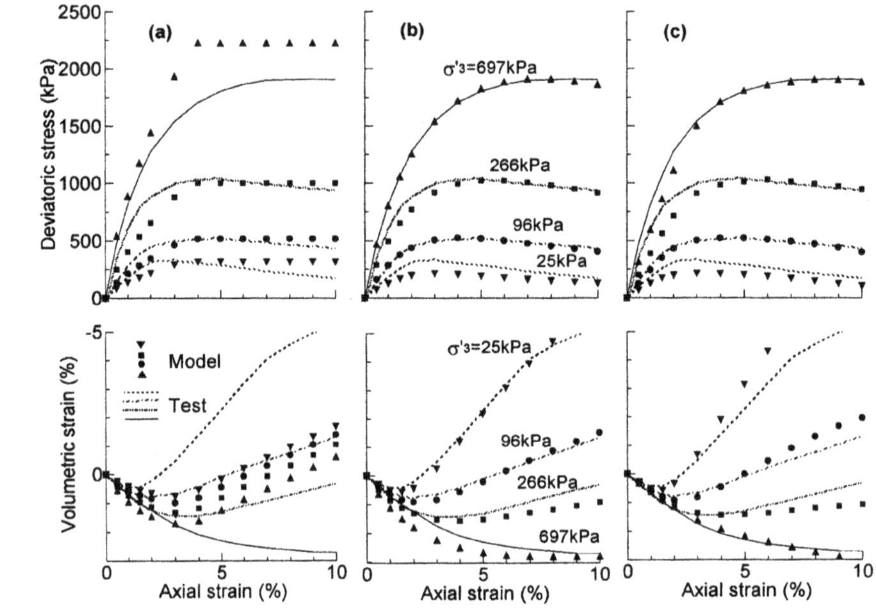

Figure 2. Predicted and measured behaviour in triaxial tests by (a) nonlinear elastic MC and (b) & (c) Lade's elasto-plastic models – Roadford Dam.

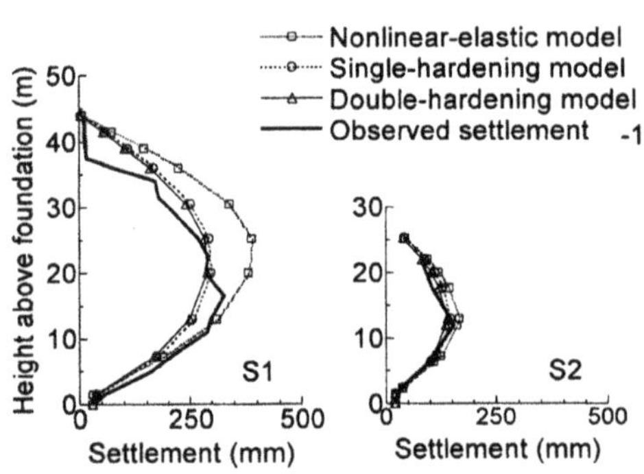

Figure 3. Predicted and observed settlements during construction - Roadford Dam.

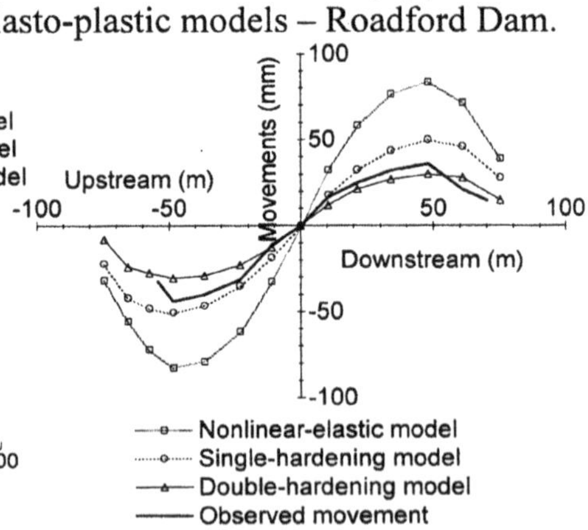

Figure 4. Predicted and observed lateral movements during construction - Roadford Dam.

predicted settlements is in reasonable agreement with the measurements. Whereas the non-linear elastic MC model overpredicted the settlements slightly, the Lade's elasto-plastic models underpredicted them.

The measured and predicted lateral movements along the extensometer H (see Figure 1) are shown in Figure 4. It can be seen that the non-linear elastic MC model overpredicted lateral movements by a factor of two. The prediction could have been improved but only by using soil parameters which no longer fitted the triaxial test results. However, both Lade's elasto-plastic models predicted the horizontal movements reasonably well, so the behaviour of the embankment was recovered from the available laboratory data. The likely cause for the difference between the above predictions is the ability of the Lade's models to account for the plastic strains pre-peak, which are qualitatively quite different from the elastic strains predicted by the non-linear elastic model.

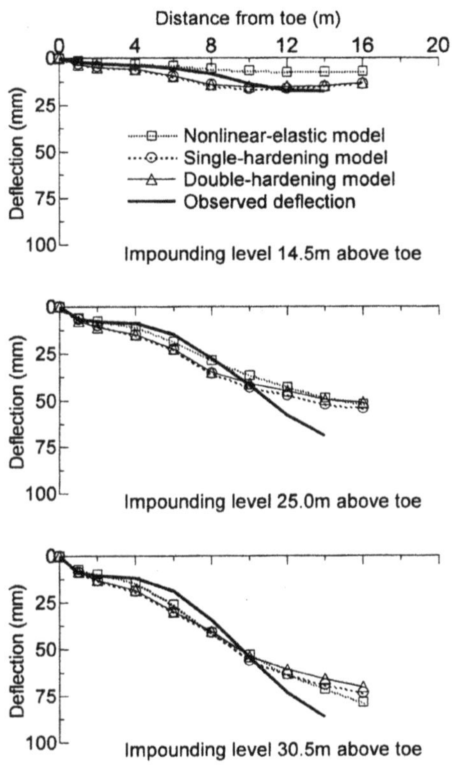

Figure 5. Predicted and observed deflections during first impounding - Roadford Dam.

Measured membrane deflections at three reservoir impounding levels are shown in Figure 5 where they are compared with the predictions given by the various models considered. There is no doubt that the FE analyses were useful in confirming the effectiveness of the sand-waste fill in reducing differential movements at the upstream toe of the dam.

WINSCAR DAM

A similar exercise has been carried out on the 53m high Winscar Dam (Kovacevic *et al*, 2002), which was the first dam in the UK to use an upstream asphaltic concrete membrane as the watertight element. Three horizontal plate gauges were installed in the 50m high section of the dam during construction at positions A, B and C as indicated in Figure 6.

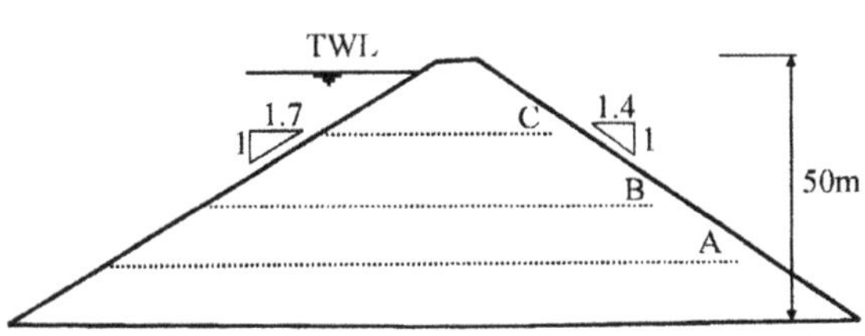

Figure 6. Cross-section of Winsacr Dam.

A comparison of the predicted and observed movements at the end of construction is shown in Figure 7. The first prediction was made by Penman & Charles (1985) using a 'simple' linear elastic FE analysis. The second prediction was obtained using the Lade's (1977) elasto-plastic model. The FE analyses using the two different constitutive models underestimated the dam movements during construction. This was particularly so at the higher dam elevations where the large observed settlements may well be due to creep in the sandstone rockfill (Penman & Charles, 1985)

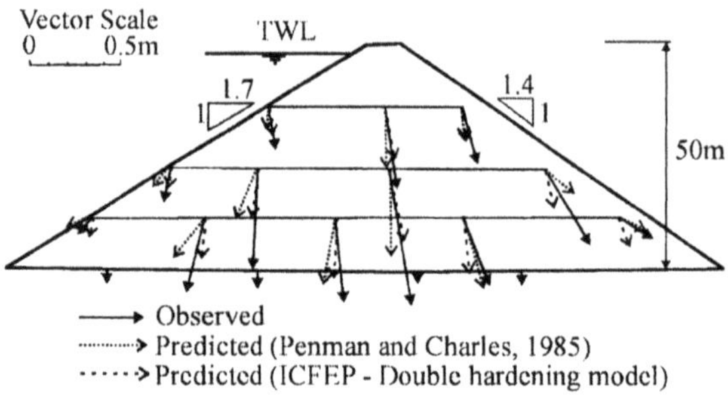

Figure 7. Comparison of predicted and observed movements during construction of Winscar Dam.

A comparison of the predicted and observed deflections of the upstream membrane during reservoir impounding is shown in Figure 8. Predictions presented in Figure 8a were made by Penman & Charles (1985). Initially, they significantly over predicted the membrane movements. Subsequently, they modified the model parameters in their analysis to simulate the stress changes during reservoir impounding more closely and, as a result, smaller and more realistic membrane movements were obtained. However, the 'new' parameters could

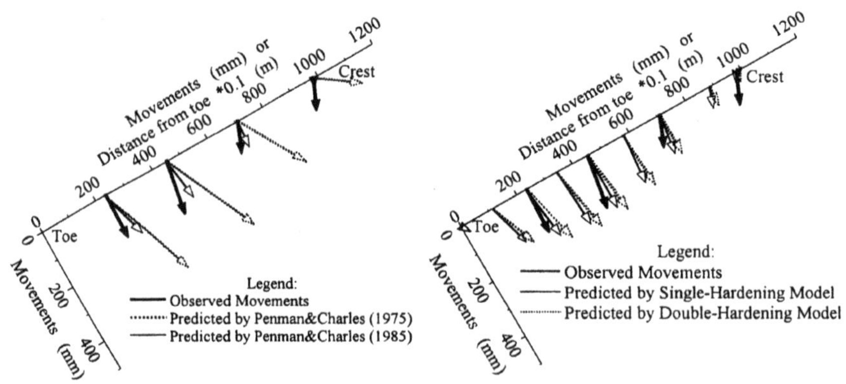

a) predictions by Penman & Charles (1975, 1985) b) predictions by ICFEP using Lade's elasto-plastic models

Figure 8. Predicted and observed movements of upstream membrane during reservoir impounding of Winscar Dam.

not recover the observed movements during dam construction any longer. No such 'problems' were encountered in the ICFEP analysis using the Lade's elasto plastic constitutive models (Figure 8b) which were able to predict reasonably well dam movements during both construction and reservoir impounding without adjusting the input parameters derived from available laboratory tests.

OLD EMBANKMENT DAMS

There are a large number of embankment dams with narrow central cores of puddle clay which have been built in the U.K. since the early 19th century. Detailed inspections of these dams have shown that they developed significant crest settlements during operation. These settlements could not be accounted for by 'standard' creep theories. The BRE undertook a detailed programme of field observation of their behaviour under operational cycles of reservoir drawdown and impounding (Tedd *et al*, 1997). It was concluded that the dams upstream fill was subject to substantial increases and decreases of effective stress and that the strains due to drawdown were not fully recovered during re-impounding resulting in permanent deformation, predominantly settlements.

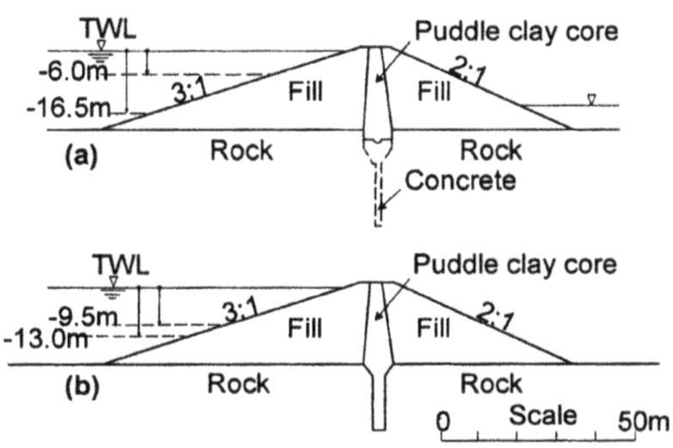

Figure 9. Cross-sections of (a) Ramsden and (b) Walshaw Dean Lower Dam.

To gain further insight into the mechanism of their behaviour during operational cycles of drawdown and re-impounding, several old dams were analysed by the FE method of analyses at Imperial College (Kovacevic *et al*, 1997). Results are now presented for two puddle clay core dams, whose sections are presented in Figure 9.

Ramsden Dam

The 25 high Ramsden Dam has a central puddle clay core and a concrete filled cut-off trench. It was analysed using one of the Lade's elasto plastic models (Lade & Kim, 1988), given its success in reproducing the rockfill behaviour both during dam construction and first reservoir impounding. The laboratory test data on the various materials involved were provided by Horton (1992). The results of cyclic loading in an oedometer test on the dam shoulder fill are presented in Figure 10a. In order to account for the observed permanent deformation after each operational cycle of draw-down and re-impounding, the non-linear elastic moduli in the Lade's model during unloading were made to be stiffer than during re-loading (Figure 10b).

The predictions were in reasonable agreement with observations during two reservoir draw-downs, particularly the smaller second one (Figure 11). However, the analysis showed that there was a tendency for the predicted permanent displacements to be towards upstream, rather than downstream as the observations suggest. It was suspected that a part of the problem was in the inability of the Lades's elasto plastic models to account for the hysteretic behaviour during unloading/ re-loading cycles. Models accounting for the observed hysteretic behaviour exist, and the 'bubble' model due to Al-Tabbaa & Wood (1989) successfully predict the observed fill behaviour in the oedometer test as shown in Figure 10c.

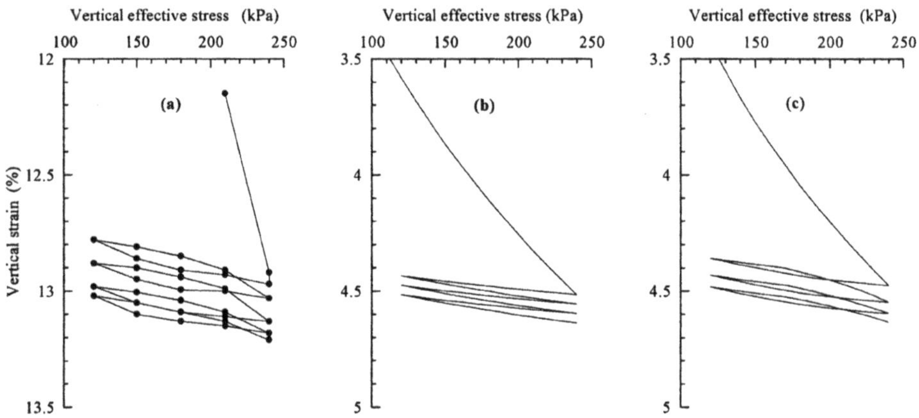

Figure 10. Behaviour of shoulder fill during cyclic loading in oedometer test: (a) observed and predicted by (b) Lade's and (c) 'bubble' model.

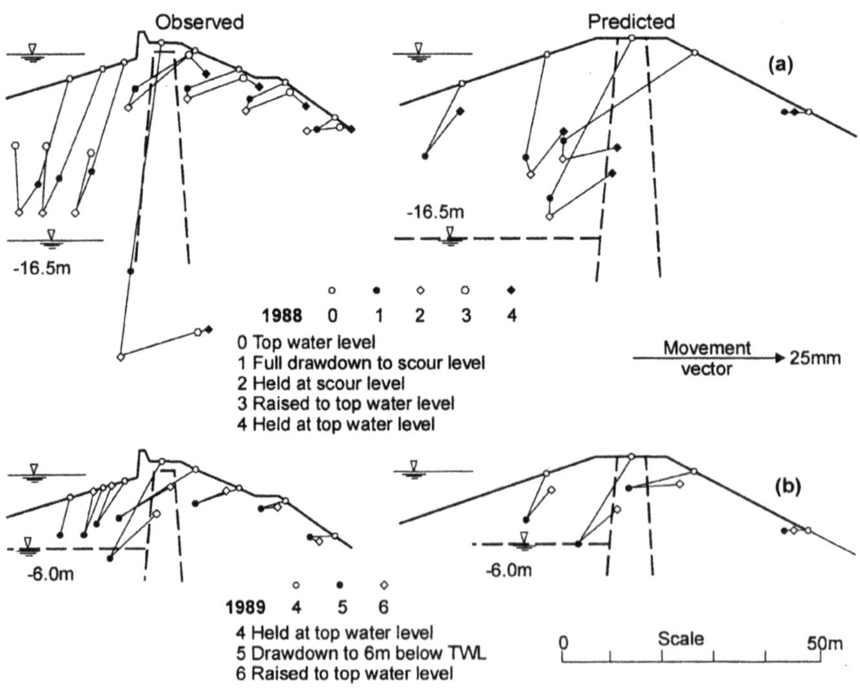

Figure 11. Observed and predicted behaviour of Ramsden Dam during (a) 1988 and (b) 1989 drawdowns – Lade's (1988) model.

Walshaw Dean Lower Dam

Walshaw Dean Lower Dam is the lowest in a chain of three dams completed in West Yorkshire at the turn of the last century. It has a maximum height of 22mm, free draining rockfill shells, a narrow central puddle clay core and a 20m deep puddle clay filled cut-off trench (see Figure 9b).

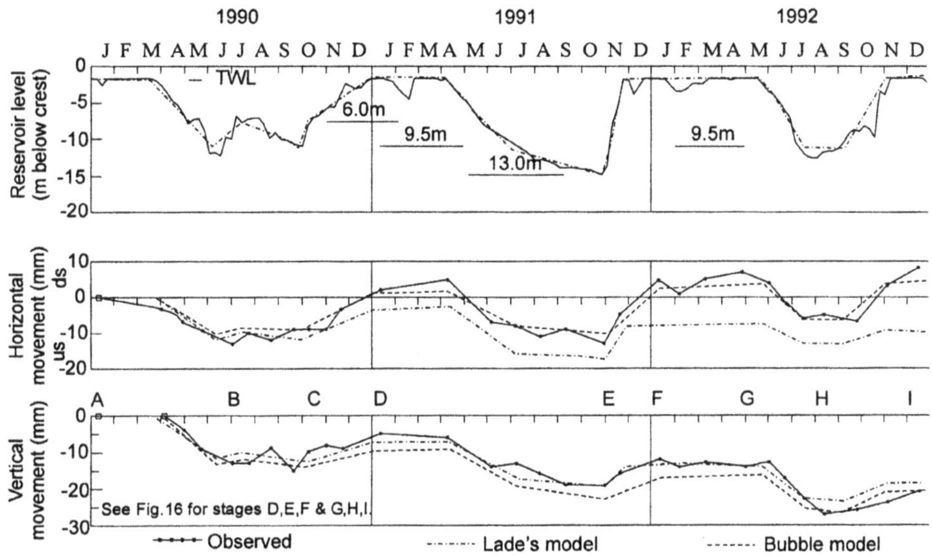

Figure 12. Observed and predicted displacements of the crest during three drawdowns – Walshaw Dean Lower Dam.

The dam was analysed using both Lade's (1988) and 'bubble' models. The measured horizontal and vertical displacements at the dam crest are compared with those predicted by the two models in Figure 12. It can be seen that while the settlements were reasonable predicted by both models, only the 'bubble' model successfully predicted the observed movements downstream. Thus, it appears that an overall downstream movement of the dam crest is a result of the plastic behaviour during unloading/ re-loading loops which the Lade's model cannot account for.

LADYBOWER DAM

FE analyses have been utilised more recently in the design and reconstruction of Ladybower Dam (Vaughan *et al*, 2000). This is a 43m high dam built in Derbyshire during the Second World War. It is one of the last puddle clay core dams constructed in Britain and certainly it is the highest one. The dam settled significantly, at least 1.5m to date. The narrow crest was raised on three occasions to maintain freeboard, steepening further the already steep downstream upper slope. The apparent increase in settlement rate due to rigorous reservoir operation in the last 20 years meant that the freeboard would have become inadequate to meet modern flood safety requirements. Studies showed that the best solution was to raise the dam again by constructing new fill on the downstream shoulder (Figure 13).

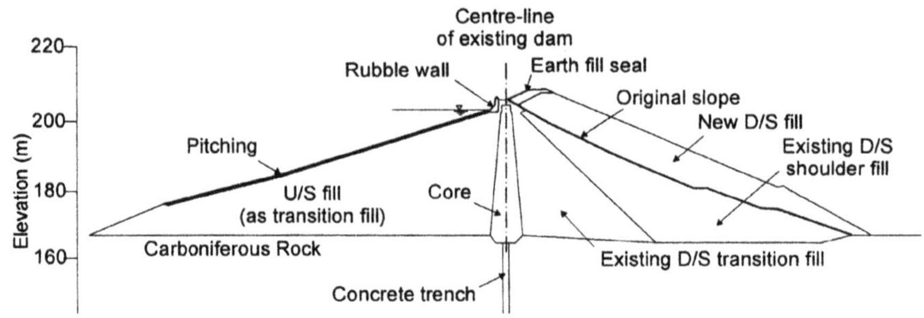

Figure 13. Typical cross-section of old and reconstructed Ladybower Dam.

Before doing so, it was decided to analyse the Ladybower dam using the same procedures as used before for the BRE study. Both the original dam and its raising were analysed. This provided an independent check on the cause of settlement for the original section, verified the stability of the reconstruction, and enabled a settlement prediction to be made for the modified section. However, it also provided predictions of the movements of the existing dam as the new downstream fill was placed, for comparison with observations.

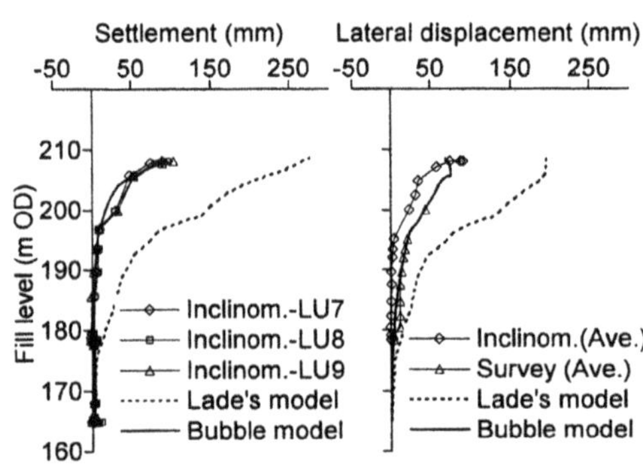

Figure 14. Observed and predicted (a) settlements and (b) lateral displacements at 'old' dam centerline - Reconstructed Ladybower Dam.

Observed settlements and lateral displacements along the axis of the original dam are plotted against the level (height) of the new fill in Figure 14. The observed movements only started to develop after 30m of fill had been placed. However, rather unexpectedly, the axis of the original dam moved downstream, being dragged down by the new fill. The 'before-the-event' prediction using one of the Lade's models over predicted the observed movements by a factor of 3. The 'after-the-event' prediction by the 'bubble' model was quite reasonable (Vaughan *et al*, 2004). Although this was a surprising result, bearing in mind that the parameters for both models were derived using the same available laboratory tests, it proved that the 'bubble' model can be successfully used in characterising the behaviour of dams during various phases of their life, such as construction, first reservoir filling, operational cycles of draw-dawn and re-impounding and also crest raising.

TRIAL BANK AT EMPINGHAM DAM

The examples of FE analyses of embankment dams presented so far involved the modelling of free draining rockfill materials (with the exception of the puddle clays in the case of the old dams). Thus, it was decided to test the capabilities of the 'bubble' model in characterizing the behaviour of modern, compacted clay fills (and even of in-situ clays in a dams' foundation) which, because of their low permeability, behave largely in an undrained manner, at least during dam construction.

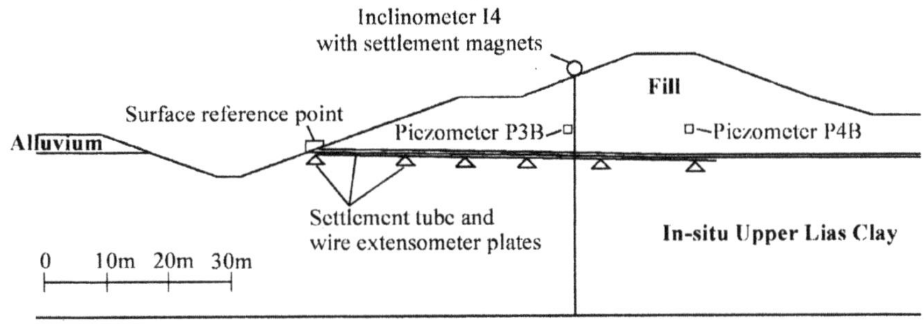

Figure 15. Section of trial bank analysed with installed instrumentation.

The 20m high instrumented trial bank (Figure 15) was built inside the footprint of Empingham Dam to examine and prove the strength of the stiff plastic Upper Lias clay (ULC) in the dam foundation (Bridle *et al*, 1990). The FE analyses using the 'bubble' model to characterise the behaviour of both the in-situ ULC and the fill derived from it were reported by Kovacevic *et al* (2007).

The model parameters for the ULC in the foundation were estimated from slow undrained triaxial tests on 260mm diameter samples. The results of two such tests involving post-peak unload/ re-load loops are shown in Figure 16 where they are compared with predictions from the 'bubble' model. The model cannot allow for strain-softening due to dislocation, shear surface formation and particle orientation at large strains; however, it accounts for pre- and post-peak plasticity, and the observed non-linearity of the stress-strain response during initial loading and subsequent unloading/ re-loading is predicted reasonably well.

The results obtained during undrained triaxial compression tests on the 100mm diameter field compacted samples of the ULC are shown in Figure 17, where they are compared with the predictions from the 'bubble' model. The matching of the observed behaviour is reasonable. In particular, the compressibility of the pore fluid was introduced in the model resulting in a realistic pore water pressure response and agreement with the volumetric strain changes observed in the laboratory tests.

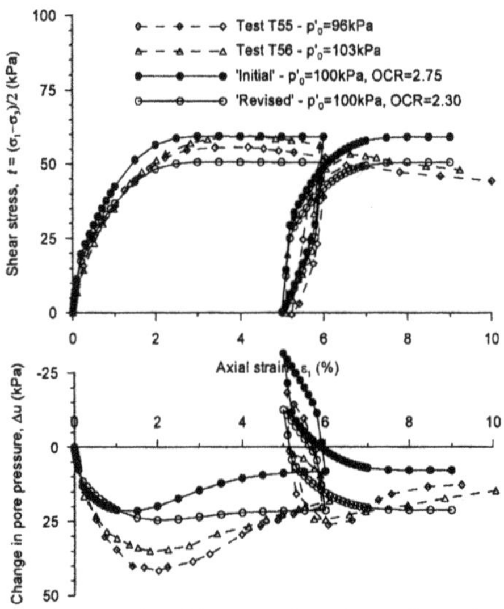

Figure 16. Observed and predicted behaviour of the in-situ ULC in undrained triaxial compression test.

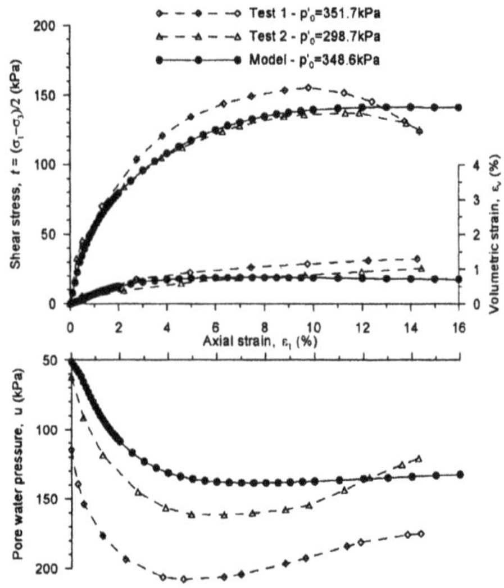

Figure 17. Observed and predicted behaviour of the ULC fill in undrained triaxial compression test.

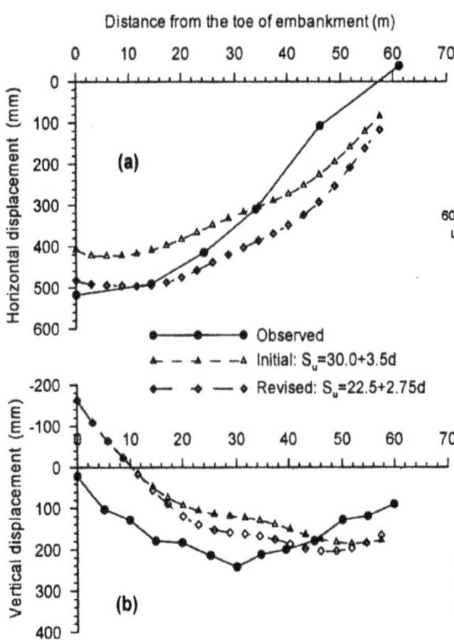

Figure 18. Observed and predicted (a) horizontal and (b) vertical displacements along the top of foundation.

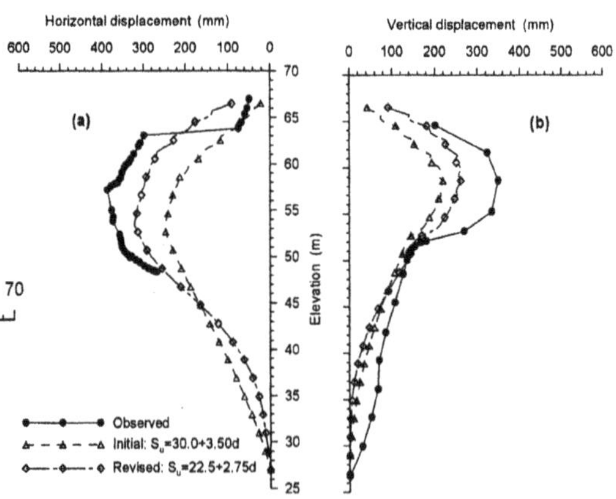

Figure 19. Measured and predicted (a) horizontal and (b) vertical deformations at the end of construction - Inclinometer I4.

The horizontal and vertical displacements at the top of the foundation (along the extensometer and settlement tube in Figure 15) upon completion of the trial bank are shown in Figures 18a and 18b respectively. The horizontal and vertical displacements measured by inclinometer I4 with the corresponding settlement magnets (see Figure 15) are presented in Figures 19a and 19b for the same stage of the analysis. Although the 'revised' analysis, which modelled a lower initial undrained shear strength (S_u) profile of the ULC in the foundation, gave better predictions of the observed trial bank behaviour, the 'bubble' model proved to be successful in characterising the behaviour of the ULC in both the foundation and the fill derived from it.

CONCLUSION

Embankment dams are important geotechnical structures which have often been analysed using advanced numerical analyses. However, to obtain predictions of deformation which agree well with those observed in the field, it is necessary to characterise the behaviour of the various fill (and foundation) materials using advanced constitutive models based on elasto-plastic theory. It has been shown that such models with their parameters derived from the available laboratory test data are capable of reproducing the full range of the observed behaviour of six embankment dams constructed in the UK.

REFERENCES

Al-Tabbaa, A. and Wood, D.M. (1989). An experimentally based 'bubble' model for clay. *Proc. 3rd Int. Conf. Numerical Models Geomechanics*, Niagara Falls, Canada, 91-99.

Bridle, R.C, Vaughan, P.R.and Wernek, M.L.G. (1990). The instrumented trial slope at Empingham Dam. *Proc. Conf. Geotech. Instrumentation in Civil Engineering Projects*, Inst. Civ. Engrs., London, 415-427.

Duncan, J.M. (1992). Static stability and deformation analysis. *Proc. Spec. Conf. Stability and Performance of Slopes and Embankments – II*, Geot. Rng. Div. ASCE, Vol.1, 222-266.

Duncan, J.M. (1994). The role of advanced constitutive relations in practical applications. *Proc. 13th Int. Conf. Soil Mechs & Found. Eng.*, New Delhi, India, Vol.5, 31-48.

Holton, J. R. (1992). In-service deformation of a puddle clay core dam', *Dams & Reservoirs*, Vol.2, No.1, 12-18.

Kovacevic, N., Potts, D.M. and Vaughan, P.R. (1994). Finite element analysis of a rockfill dam. *Proc. 8th Int. Conf. Computer Methods & Advances in Geomechs,* Morgantown, West Virginia, Vol.3, 2459-64.

Kovacevic, N., Potts, D.M., Vaughan, P.R., Charles, J.A. and Tedd, P. (1997). Assessing the safety of old embankment dams by observing and analysing movement during reservoir operation. *Trans. 19th Int. Cong. Large Dams,* Florence, Q.73, R.34. 551-566.

Kovacevic, N., Vaughan, P.R. and Potts, D.M. (2002). A comparison between observed and predicted deformations of Winscar dam. *Proc. 8th Int. Conf. Numerical Models in Geomechanics,* Rome, Italy, 565-571.

Kovacevic, N., Vaughan, P.R. and Potts, D.M. (2004). A comparison between observed and predicted deformations of an old dam during reservoir operation. *Proc. 9th Int. Conf. Numerical Models in Geomechanics,* Ottawa, Canada, 587-593.

Kovacevic, N., Higgins, K.G. and Potts, D.M. (2007). Finite element back-analysis of trial bank at Empingham Dam. *Proc. 10th Int. Conf. Numerical Models in Geomechanics*, Rhodes, Greece, 587-593.

Lade, P.V. (1977). Elasto-plastic stress-strain theory for cohesionless soil with curved yield surface. *Int.Jour. Solids & Structures*, Vol.13, 1019-35.

Lade, P.V. and Kim, M.K. (1988). Single hardening constitutive model for frictional materials, III. Comparisons with experimental data. *Computers & Geotechnics*. Vol.6, 31-47.

Naylor, D.J. (1991). Finite element methods for fills and embankment dams. *Advances in Rockfill Structures*, 291-340.

Penman, A.D.M. and Charles, J.A. (1985). Behaviour of rockfill dam with asphaltic memebrane. *Proc. 11th Int. Conf. Soil Mechs. Found. Eng.,* San Francisco, Vol.4, 2011-2014.

Potts, D.M. and Zdravkovic, L. (1999). Finite element analysis in geotechnical engineering: theory. *Thomas Telford,* London.

Potts, D.M. and Zdravkovic, L. (2001). Finite element analysis in geotechnical engineering: application. *Thomas Telford,* London.

Tedd, P., Charles, J.A., Holton, I.R. and Robertshaw, A.C. (1997). The effect of reservoir drawdown and long-term consolidation on the deformation of old embankment dams. *Geotechnique*, Vol.47(1), 33-48.

Vaughan, P.R., Chalmers, R.W. and Mackay, M. (2000). Ladybower Dam: Analysis and prediction of settlement due to long term operation. *Dams 2000, 11th Biennial Conference*, British Dams Society, Bath, 360-376.

Vaughan, P.R., Kovacevic, N. and Potts, D.M. (2004). The and now: some comments on the design and analysis of slopes and embankments. *Proc. Conf. Advances in Geotech. Engng*, Vol.1, Thomas Telford, 241-290.

Development of an integral system for dam and landslide monitoring based on distributed fibre optic technology

R. HOEPFFNER, Technische Universität München, Hydraulic Engineering – Germany
J. SINGER, Technische Universität München, Engineering Geology – Germany
K. THURO, Technische Universität München, Engineering Geology – Germany
M. AUFLEGER, Universität Innsbruck, Hydraulic Engineering – Austria

---

SYNOPSIS. A large benefit of distributed fibre optic sensing technology is the opportunity to receive information on temperature (Distributed Fibre Optic Temperature sensing – DFOT) and strain (Distributed Fibre Optic Temperature and Strain sensing – DFOST) along tens of kilometres of a rather simple fibre optic cable in a spatial resolution of roughly one metre.

As part of the development of an integral monitoring setup for dams and landslides, tests carried out on the slope of "Aggenalm Landslide" in the German Alps are described in the paper. Fibre optic strain sensing and TDR cables (Time Domain Reflectometer) were installed vertically parallel to inclinometer casings into the landslide in order to receive information on the location, thickness and movement rate of the sliding plane and at the same time making a direct comparison of the different measurement techniques possible. Horizontally installed fibre optic strain sensing cables can provide information on the development of the primary and secondary cracks on the surface of the landslide.

In the next stage of the project, the installation of strain and temperature sensing cables into an embankment dam is planned. Settlements will be measured using strain sensing cables and seepage can be evaluated by heating up a temperature sensing cable (heat-up method), a method that has successfully been used in canal facings and dykes for leakage detection.

INTEGRAL CONCEPT

The aim of the present work is to develop an integral measurement system for hydraulic structures and their reservoirs with top of the line monitoring and data evaluation techniques.

Distributed temperature sensing in concrete

For more than a decade, fibre optic technology has been used for the monitoring of hydraulic structures. Massive concrete dams have been equipped with fibre optic cables for distributed temperature sensing. The spatial resolution in which the Raman-shift based distributed temperature sensor DTS (or ROTDR for Raman optical time domain Reflectometer) is able to evaluate the temperature along the fibre is roughly 1 m, thus 1,000 measuring points along 1 km of optical fibre can be obtained. The sensing length can today be up to 30 km and more. The large data volume allows the production of detailed temperature cross sections through concrete dams in order to evaluate them for temperature gradients. They mainly occur due to the development of hydration heat of the hardening concrete and such areas are at particular risk of cracks forming within the structure. The comparison of the performance of singe ended temperature and strain sensing devices is summarized in the following table.

Method	model	value	temp. accuracy	strain accuracy
Raman sensor	DTS, ROTDR	temperature only	up to 0.01°C	-
Brillouin sensor	DTSS, BOTDR	temperature, strain	up to 1°C	up to 20με

Distributed seepage and strain monitoring in embankment dams

In soil, a combination of optical fibres and copper wires arranged next to each other in the same cable allows the distributed evaluation of the saturation and under specific circumstances the velocity of water in soil by heating up the copper wire to a certain amount and measuring the heat transfer. This technique has been successfully applied for leakage monitoring in open water canals and research is going on to apply this technology for the monitoring of embankment dams [Perzlmaier et al. 2007]. Recent research is dealing with the application of fibre optic strain sensing in embankment dams [Hoepffner, 2008].

Landslide monitoring

The monitoring of landslides is an important factor in relation to the safety of dams and reservoirs [ICOLD, 2002]. Especially during snow melting, heavy rainfall, first impoundment and quick water level variations, the banks of a reservoir are strongly affected. Depending on the geology, ancient sliding planes can be reactivated and endanger the reservoir and the dam. A combination of different monitoring techniques could help to monitor those movements and evaluate the size, cubature and velocity of the potential landslides. A new method has been developed by the authors to evaluate the shear zone and subsoil movements using inclinometers, TDR cables and optical fibres at the same time to identify large and small shear zones. For the evaluation of surface movements, fibre optic strain sensing cables were installed on the surface of the potential landslide.

The basic idea of monitoring landslides by means of fibre optic cables is not new. Higuchi et al. 2007 proposed a setup with extrinsic sensors to monitor the crack formation in landslides and Facchini (2001) proposed to measure the strain by attaching optical fibres to telephone poles on a landslide. What is new is the installation of the strain sensing cables directly into the ground as well as the vertical shear zone measurements.

TEST SITE AT THE “AGGENALM LANDSLIDE”

The Aggenalm landslide is situated in the Bavarian Alps about 3 kilometres southeast of the town of Bayrischzell. In the north it borders to the “Sudelfeld”, Germany’s largest contiguous skiing resort. One of the main access routes to the skiing resort crosses and is therefore affected by the landslide.

In 1935, after being triggered by heavy rainfall, the Aggenalm landslide destroyed three bridges and the road to the Sudelfeld area. Again after extreme precipitation in 1997 a debris flow originated from the landslide area and blocked the road. Since 2001 the landslide has been surveyed periodically twice a year by the Bavarian state office for the environment (Bayerisches Landesamt für Umwelt), showing average movement rates of about two centimetres per year.

Due to the growing economic relevance of the Sudelfeld access road for the tourism of the region, a detailed engineering geological investigation of the Aggenalm landslide was carried out in order to assess the underlying processes. Additionally, in the course of the research funding programme “Geotechnologies” of the German federal ministry of education and research, the Aggenalm landslide was selected as the test site for the installation of an innovative early warning system for alpine instable slopes [Thuro et al. 2007].

Geology

Tectonically the area of the Aggenalm landslide is part of the Lechtal nappe of the Northern Calcareous Alps, which is built up by various sedimentary rocks of mainly Triassic to Cretaceous age. Due to the alpine orogeny, the rock mass is folded into several large synclines and heavily faulted. In the last ice age the area was covered by glaciers, which resulted in typical glacial morphology and the abundance of various glacier deposits.

The Aggenalm slope is mainly built up by the "Kössner Schichten", an alternating sequence of limestone and marl, and the overlying "Oberrhätkalk" – massive limestones and dolomites. The whole sequence dips parallel to the slope with an average angle of 22°.

The marls which underlie most of the slope are sensitive to weathering and with time are decomposed to a clay-rich residual rock. This process coincides with a distinctive reduction of the rock mass strength [Nickmann et al. 2006] and is mainly responsible for the instability of the slope. In the upper part, the Aggenalm landslide can therefore be classified as a rock spread according to Cruden & Varnes (1996). Further downhill, with increasing deformation, the rock mass continuously disintegrates and the mechanism of the landslide changes into a very slow debris flow. As the events of 1935 and 1997 have shown, the Aggenalm landslide is sensitive to heavy precipitation and the accompanying rise in ground water levels.

CONVENTIONAL INSTRUMENTATION

Essential for the hazard assessment of a landslide is detailed information about the distribution, orientation and amount of deformation in four dimensions (on the surface and in the depth alongside boreholes and with high temporal resolution). Additionally the effect of triggering mechanisms (as e.g. precipitation) to the movement and their temporal relation is of great importance.

When selecting measurement techniques for the development of an early warning system, the ability to perform continuous measurements is essential. Time Domain Reflectometry and Distributed Fibre Optic Strain Sensing were chosen for economic reasons to determine subsurface deformation. In order to be able to assess the quality of the measurements, conventional sporadic inclinometer measurements were performed.

Monitoring of the surface deformation can be achieved economically and continuously e.g. by GPS measurements. Additionally the Aggenalm Landslide will be monitored by means of reflectorless tacheometry which will produce deformation data with high spatial resolution [Thuro et. al 2007]. However the main setback of these techniques is that they often cannot be used efficiently in forested areas. Here the measurement of surface movements with Distributed Fibre Optic Strain Sensing is an interesting alternative.

Time Domain Reflectometry (TDR)

Time Domain Reflectometry can be described as "cable-based radar": The TDR cable tester emits electric pulses which are sent through a coaxial cable. When these pulses approach a deformed portion of the coaxial cable a signal is reflected to the cable tester. As with radar, due to the known propagation velocity of the electromagnetic wave, by measuring the time span between emission and reception of the electric pulse, the distance to the deformation can be determined. Furthermore the analysis of the reflected signal (amplitude, width, form etc.) can reveal information about the type and amount of deformation.

For landslide monitoring the coaxial cable is installed into a borehole and connected to the rock mass with grout. When the rock mass starts to move, the coaxial cable is deformed (e.g. altering the distance between inner and outer conductor of the cable). This results in a change of the electric properties (impedance) of the cable, which can be measured with TDR.

The monitoring of deformations in rock/soil using TDR is therefore an indirect measuring method: the deformation itself is not measured (as e.g. when using inclinometers) but a dependant value: the change in impedance of a coaxial cable due to deformation. A detailed description of the underlying physics and the setup for landslide monitoring is given in O'Connor & Dowding (1999).

Several different TDR devices with different inducted voltage pulses are available. However, so far there is no complete system for deformation monitoring including analysis software available. For the current project a Campbell Scientific TDR100 device and newly developed analysis software was used [Singer et al. 2006].

INSTRUMENTATION FOR DISTRIBUTED FIBER OPTIC STRAIN SENSING

Distributed strain sensing theory

By evaluating the power and frequency shift of the Raman and/or Brillouin peaks in the backscattered frequency spectrum, the temperature and strain along an optical fibre can be evaluated under certain circumstances.

Within the last decade, the backscattered light in the Brillouin frequency band gained interest for distributed sensing as it is sensitive to both temperature and strain along the fibre. Depending on the method for distributed strain sensing, the spatial resolution can be up to several centimetres for a loop configuration using the Brillouin optical time domain analysis (BOTDA) or Brillouin optical correlation domain analysis

(BOCDA) or roughly 1 m for the Brillouin optical time domain reflectometry (BOTDR) which only needs access to one end of the fibre. As temperature compensation is a major task, a Distributed Temperature and Strain Senor (DTSS) which compensates for temperature without additional temperature measurements and cables was introduced by Parker et al. (1997). The latter method, which is based on spontaneous Brillouin scattering, has the great advantage that measurements can also be taken even if the fibre is damaged within the structure as only access to one end is needed at a measuring range of more than 20 km. Field tests showed that the resolution of 1 m is suitable for crack detection in concrete even at crack widths of less than 1 mm [Hoepffner et al. 2007].

A measuring system for distributed fibre optic strain measurements involves basically three parts. A laser sends coherent, monochromatic impulses of light into an optical single mode fibre. As the light propagates down the fibre, it is reflected at any position within the fibre. The photons of the light interact among others with density variations in the fibre core caused by acoustic phonons that propagate along the fibre. The information on both temperature and strain is included in the frequency shifted Brillouin peaks. The third part is the detection and evaluation unit, where the backscattered frequency pattern are detected by a photodetector and processed in a computer. The stimulated Brillouin scattering process additionally involves a continuous wave laser signal which is coupled into the other side of the fibre which is used to stimulate the backscattered signal. Commercially available sensing devices include the laser(s) and the processing unit with software for user friendly data visualization.

Sensing devices

Two methods for distributed strain sensing are most commonly used: BOTDA and BOTDR. As BOTDA involves two counter-propagating light waves in the fibre, the fibre has to be continuous and accessible from two sides. The spatial resolution can be up to a few centimetres. This method is, as far as spatial resolution is concerned, particularly suitable for the subsoil sensing of a shear zone. In harsh environments or in applications where great mechanical impact on the optical fibre cables can not be ruled out, BOTDR is favoured as the measurements are taken single ended which means that the local strain in the fibre can still be evaluated, even if a fibre ruptures. However, due to physical restrictions, the spatial resolution can not be greater than around 1 m. For the monitoring of surface movements and any application inside of a dam structure, BOTDR is most favourable. As the goal should be the integral measurement of the dam and landslides with only one fibre optic strain sensing device, the BOTDR method is favoured.

The strain sensing device being used in the current tests is the Sensornet DTSS which has the unique ability of evaluating simultaneous distributed temperature compensated strain measurements from one single mode fibre.

Cable

The specification for a DFOST sensing cable can easily be summarized by "as tough as necessary, as strain sensitive as possible". Unlike standard telecom cables that are commonly used for distributed temperature sensing, strain sensing cable need a good bond between the very fragile optical fibre in the centre and the outer sheath, as well as to the matrix where the fibre is installed in. The goal is an excellent strain transfer but on the other hand a good protection of the cable against mechanical forces that mainly occur during installation.

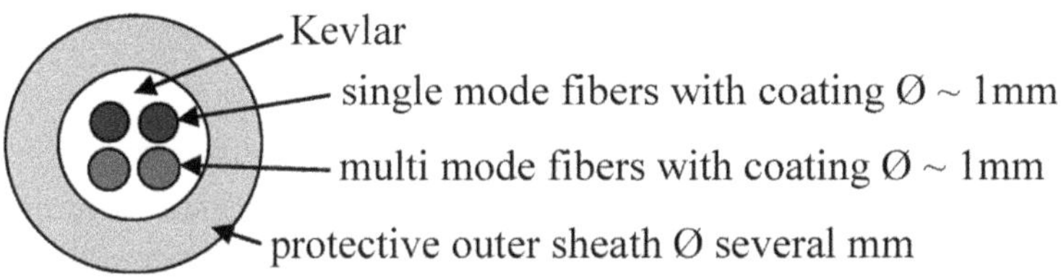

Figure 1: Cross section of a strain sensing cable.

The cables used in the present tests is a tight buffered especially designed strain sensing cable with two single mode fibres for redundant strain sensing and two multi mode fibres for independent temperature measurements, inside (Fig. 1). A standard cable with gel embedded fibres and thus no load transmission was used as extension for the surface strain sensing cable as the actual measurements were taken in an area difficult to access with the sensing device.

CABLE INSTALLATION AT THE LANDSLIDE

For the first test run, three cables were installed in fall 2008, 44 m optical strain sensing cable in the upper part of the landslide and 46 m optical stain sensing cable in a loop configuration in a 23 m deep borehole next to a 23 m long TDR cable, both attached on the outside of an inclinometer casing.

Measurement of surface movements

An appropriate section where considerable surface movements can be expected was visually identified, supported by geological movement indicators. The cable was installed into an excavated trench with a depth of at least 15 cm where possible, following the slope line. The cable in the trench was protected from people and animals that might pass the test area. The transmission of soil movements into the cable without applying further means could not be expected. Thus in order to guarantee a good transmission of movements to the cable, special anchors of up to 50 cm length were designed and hammered into the ground. The top of the anchors is designed in a way that the optical cable can easily be attached without applying excessive mechanical forces on them. As depicted in Fig. 3, the cable was attached in total to 30 anchors of different length and varying spacing.

Fig. 2 shows the installation work which was carried out during strong snowfall. The cable was a little pre-strained in order to be able to locate and measure even very small movements. The readily installed cable passes a secondary rotational sliding plane and an area of probable rapid surface movements. It is planned to measure strain in the rift at the beginning of the secondary slide plane, and contraction in the lower part of the stretch. Additional strain zones can be expected in the steep part of the cable layout, where surface movements up to a depth of some hundreds of millimetres are expected.

Measurement of subsoil movements and shear zones

Shear zones are vital parameters in order to obtain information on the thickness of a landslide and thus its cubature. It is generally not possible to evaluate the depth of a shear zone from boring profiles only. For more than a decade, the TDR measuring technique proved to be a very appropriate method for the detection of localized shear planes. However the deformation detection reaches its limit, resulting in a gradual bending of the coaxial cable with no significant change of the cable cross section geometry (see Fig. 3).

On the other hand, optical fibres have a maximum elongation of around 3 % and are sensitive to buckling. Very distinct shear zones would lead to a break of the fibre after only little movement, where movements within a thick shear zone would be measured as strain within the moving zone. Therefore the combination of the two systems could provide very valuable information on the situation and the movements at the shear plane. In our tests, the cables were attached onto the outside of an inclinometer casing. The inclinometer data can later be used to validate the measurements.

TDR

A single semi-rigid aluminium coaxial cable with a diameter of about 12 mm was attached to the outside of the inclinometer casing with duct tape. In order to prevent ingress of moisture into the cable, its end was sealed with heat shrink tubing and a cap. When installed below the groundwater table the aluminium cable should be protected against corrosion either by using a jacketed cable or by applying a corrosion inhibiting paint.

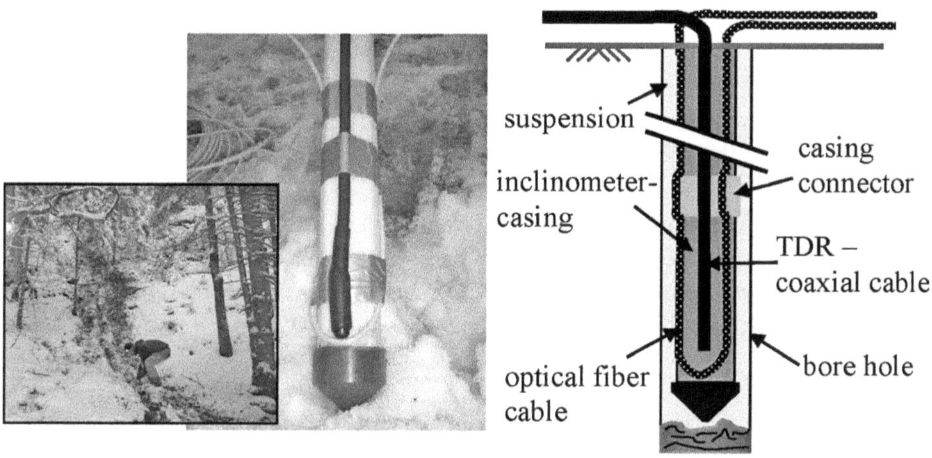

Figure 2: Equipment in the vertical borehole: TDR cable, optical cable and inclinometer and near-to-surface cable installation during snow fall (left)

Fibre optic cable

The fibre optic cable was also attached onto the outside of the inclinometer casing with duct tape. On the low point of the casing, the cable was looped back up, to be able to measure the cable from two sides which could also be essential in case of a damaged fibre. To have the theoretical opportunity to measure the direction of the subsoil movement, the two parts of the cable were attached on the casing with 90° spacing thus at 12 and 3 o'clock (see fig. 2).

FIRST DATA EVALUATION

TDR

Until the installation of the continuous measurement system is completed in the spring of 2008, only sporadic TDR measurements were carried out. In the three measurements since December 2007 so far no deformation could be detected. This is not surprising since the installation of a coaxial cable parallel to an inclinometer casing prohibits an effective transmission of the rock mass deformation to the cable and results in a rather gradual deformation of the coaxial cable, which can not be measured. Only after a considerable amount of deformation (several centimetres) the measurement will start to change the cable geometry and therefore produce results. At this time the inclinometer will almost certainly be near the end of its lifetime, so that the TDR measurements can considerably prolong the lifetime of a subsurface deformation measurement site. This could, for example, enable identification of additional shear planes below a main shear zone [Singer et al. 2006].

In principle, the installation of a coaxial cable parallel to an inclinometer casing into the same borehole is not advisable for the above reasons. It is better to install the coaxial cable into a borehole of its own and to connect the cable to the rock mass with a rather stiff grout, thereby making the transmission of the rock mass deformation to the coaxial cable possible.

In 2008 the installation of several boreholes solely instrumented with coaxial cables for TDR deformation measurements are planned. Some of these are placed near to inclinometer measurement sites, so that a direct comparison of the collected data will be possible.

Optical cable

Right after installation, OTDR measurements were taken from the installed cables in order to assure the fibres are continuous, not excessively bent, damaged or broken. As the DTSS device was not available since the time the installation was made, the measurements were postponed and are estimated to be carried out in spring or early summer 2008 after snow melt.

INTEGRAL DAM AND LANDSLIDE MONITORING SETUP

An integral monitoring system based on optical fibres for several safety concerns of a hydraulic structure and its reservoir bears many advantages. The sensors are cheap while collecting a huge amount of data due to a spatial resolution of up to 1 m. Optical fibres are not affected by electrical interference, weather conditions or vegetation. The data collection and processing can be automated and remote controlled via modem. A permanently installed temperature and strain sensing device could provide information on the state of the dam and at the same time monitor landslides that might affect the reservoir, even if located many kilometres away.

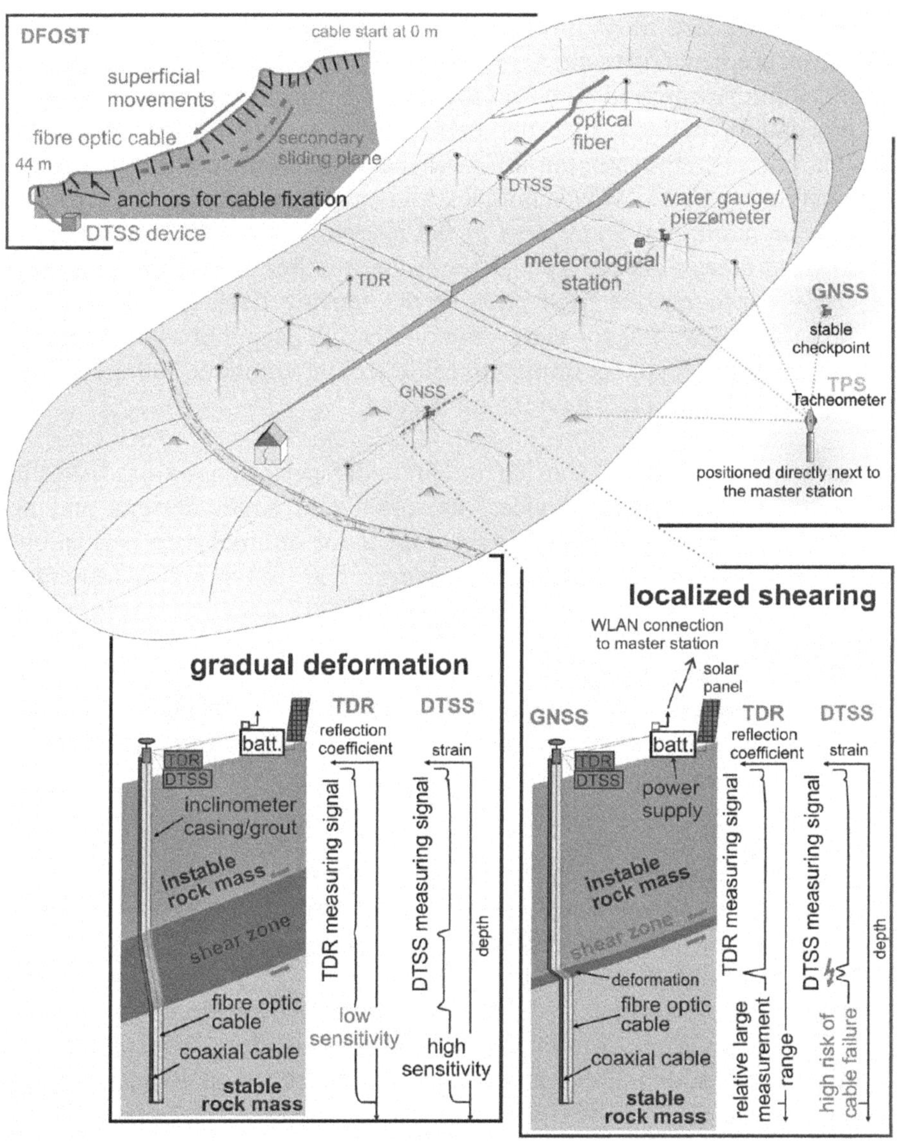

Figure 3: Measuring setup and expected measurement data at the Aggenalm Landslide.

Despite all its advantages, a combination with other measurement systems is absolutely essential. For the monitoring of landslides, TDR seems to be the ideal match for the evaluation of shear zones (see fig. 3).

OUTLOOK

The versatility of optical fibres and the sensing technology leads to the concept of an integral measuring system based on fibre optics. Dam monitoring with optical fibres in connection with monitoring of the slopes of the reservoir are ideal in connection with conventional methods. The TDR measurements presented in this paper have been used in multiple landslides but research is going on towards the optimization of the software and evaluation methods in order to receive not only information on the location but also the amount of movement in distinct shear zones. A combination of those two techniques allows the monitoring of the entire structure including the shore area of the reservoir remotely even in remote and wooded areas. As the optical sensors are rather cheap, the advantages can easily overcome the high price of the sensing device. For TDR, the coaxial cables as well as the sensor are relatively cheap, but as the range of a single measuring device is limited to cable lengths of about 300 m, for large project areas several TDR devices are needed.

Further research is being carried out into the performance of distributed fibre optic strain sensing devices, use-orientated cable designs and new installation, layout and monitoring techniques for optimal data evaluation in landslides, concrete and embankment dams. The first strain measurements at the Aggenalm Landslide are scheduled for 2008.

ACKNOWLEDGEMENTS

The authors would like to express their gratitude to the German federal ministry of education and research for the financial support through the research fund "Geotechnologies".

REFERENCES

Aufleger, M.; Conrad, M.; Perzlmaier, S.; Porras, P.; Strobl, T. *Distributed Fibre Optics Temperature Measurement on its way to become an ordinary Tool in Dam Monitoring*, Hydro Review Wordwide, Vol. 13,N.4, Hci Publ., Kansas City, USA, 2005

Cruden, D. M., & Varnes, D. J. (1996). *Landslide types and processes*. - in A. K. Turner & R. L. Schuster (eds.): Landslides: Investigation and Mitigation, National Research Council, Transportation Research Board, Special Report 247, p. 36-75.

Facchini, M. (2001). *Distributed optical fiber sensors based on Brillouin scattering*. Lausanne.

Higuchi, K., Fujisawa, K., Asai, K., Pasuto, A., Marcato, G. (2007). *Application of new landslide monitoring technique using optical fiber sensor at Takisaka Landslide*, Japan. 1st North American Landslide Conference, Vail, Colorado

Hoepffner, R. (2008). *Distributed fiber optic strain sensing in hydraulic concrete and earth structures - measuring theory and field investigations on dams and landslides* (woking title). Doctoral thesis. Issue no.121, Lehrstuhl für Wasserbau und Wasserwirtschaft. München, Technische Universität München.

Hoepffner, R., Moser, D. E., Aufleger, M., Neisch, V. (2007). *Performance of distributed and single point systems for crack detection in concrete dams.* Dam Safety Management. Role of State, Private Companies and Public in Designing, Constructing and Operating of Large Dams, St. Petersburg, Russia. Vol. p. June 27th, 2007:

Horiguchi, T., Kurashima, T., Tateda, M. (1989). *Tensile Strain Dependence of Brillouin Frequency Shift in Silica Optical Fibers.* IEEE Photonics Technology letters VOL. 1 (No.5).

ICOLD (2002). *Reservoir Landslides: Investigation and Management – Guidelines and case histories.* Bulletin 124, ISSN 0534-8293

Nickmann, M., Spaun, G. & Thuro, K. (2006). *Engineering geological classification of weak rocks.* - In: Culshaw, M., Reeves, H., Spink, T. & Jefferson, I. (ed.): IAEG Engineering geology for tomorrow's cities. - Proceedings of the 10th IAEG International Congress, Nottingham, United Kingdom, 6-10 Sept. 2006, 8 p. (Paper number 492, CD-ROM).

O'Connor, K. M. & Dowding, Ch. H. (1999). *GeoMeasurements by Pulsing TDR Cables and Probes.* - CRC Press, Boca Raton.

Parker, T. R., Farhadiroushan, M., Handerek, V. A., Rogers, A. J. (1997). *A fully distributed simultaneous strain and temperature Sensor using Spontaneous Brillouin Backscatter.* IEEE Photonics Technology letters, Vol 9, No. 7, July 1997

Perzlmaier, S., Aufleger, M., Dornstädter, J. (2007). *Detection of Internal Erosion by Means of the Active Temperature Method.* Assessment of the Risk of Internal Erosion of Water Retaining Structures: Dams, Dykes and Levees. [14th German Dam Symp. and 7th ICOLD Europ. Club Dam Symp.], Freising, Germany

Singer, J., Thuro, K. & Sambeth, U. (2006). *Development of a continuous 3d-monitoring system for unstable slopes using time domain reflectometry.* - Felsbau 24 (3), 16-23.

Thuro, K., Wunderlich, Th. & Heunecke, O. (2007). *Development and testing of an integrative 3D early warning system for alpine instable slopes (alpEWAS).* – Geotechnologien Science Report, 12 p. (in press).

Nonlinear seismic assessment of lightly reinforced concrete intake towers

R. SABATINO, Lloyd's Register, London, UK (formerly KBR Ltd)
A.J. CREWE, University of Bristol, UK
W.E. DANIELL, University of Bristol, UK
C.A. TAYLOR, University of Bristol, UK

SYNOPSIS. Published guidance on the seismic analysis of reinforced concrete intake/outlet towers is limited, especially for their nonlinear response, due to limited knowledge on the nonlinear characteristics of existing and new towers. Proving the integrity of existing towers is an international problem for dam owners, and an industrial need exists for a rational, cost-effective and validated method for their assessment.
This paper describes a series of tests aimed at investigating the seismic performance of typically reinforced, non-seismically designed towers. Monotonic and cyclic push-over tests were performed on 1/6th scaled models. The results from the physical tests were used to validate a 3D nonlinear finite element model of the towers, using embedded steel reinforcement and a smeared crack model to simulate crack properties of the concrete material. The dynamic performance of the structures was investigated by developing a simplified single degree of freedom model and performing a number of simulations to obtain fragility curves of the system. This simplified model was capable of simulating the degrading, hysteretic properties of the towers and was used to perform nonlinear time history analyses using a range of parameters. A probabilistic approach was selected as the basis of the performance evaluation process using fragility analyses as a tool for modelling the uncertainty associated with the parameter selection. Based on the experimental and analytical results, a three-staged assessment procedure for the seismic performance assessment of the towers was proposed.

INTRODUCTION

Intake/outlet facilities (Figure 1) form part of the vital infrastructure of a dam as they regulate the outflow of water from the impounded reservoir. In the event of an earthquake occurring, it is therefore essential that any damage to the intake tower does not induce the catastrophic failure of the dam, and consequent release of water. Continued operation of the facility may also be required to allow controlled release of water to permit essential repair work to be carried out if damage occurred to the main barrage itself. The seismic risk to dams in the UK has been studied extensively, resulting in the publication of design guides (Charles et al., 1991; Institution of Civil Engineers, 1998). However, limited guidance is available for specifically assessing the seismic vulnerability of intake towers (ICOLD, 2002; USACE, 2003). Existing seismic design codes, such as Eurocode 8 (BSI, 2004) provide limited guidance for their application to intake towers.

The purpose of this research was to determine the characteristic nonlinear behaviour of typical UK lightly reinforced concrete intake towers under seismic loading. This was done through a series of experimental and theoretical investigations into the nonlinear behaviour of scaled intake tower models subject to monotonic and cyclic pushover loads, leading to the development of a simplified probabilistic tool as part of a rational method for the evaluation of their seismic performance. By establishing appropriate performance requirements for given limit states, the seismic response of the towers was evaluated in a probabilistic context.

Figure 1: View of Errochty tower and access bridge

EXPERIMENTAL MODELLING OF INTAKE TOWERS

A desk study of the typical characteristics of existing reinforced concrete intake towers in reservoirs in the United Kingdom was undertaken. By averaging the typical values for geometry, reinforcement steel and material properties, a prototype tower configuration was obtained. For the purpose of this project, a rectangular hollow free-standing tower was selected. The area of reinforcement steel to area of concrete ratio (ρ) was chosen as 0.25% for both vertical (longitudinal) and horizontal (secondary) reinforcement, representing typical UK values. The control house, access bridge and other appendages, as well as the water-structure and soil-structure effects, were not considered as the primary purpose of the experimental programme was the understanding of the structural response to earthquake type loading only. Table 1 summarises the prototype and scaled model geometrical properties. The choice of the 1:6 scale was based on a compromise between practical size for testing, cost and the ability of adequately replicating the failure behaviour of the intake tower.

Table 1: Prototype and scaled model geometry

Tower geometry	Height (m)	Width (m)	Wall thickness (mm)
Prototype	18	6	600
Scaled model	3	1	100

Two intake tower specimens (NSD-R-1 and NSD-R-2) were constructed in the Earthquake and Large Structures Laboratory (EQUALS), part of the Bristol Laboratories for Advanced Dynamics Engineering (BLADE) testing facilities at the University of Bristol. The specimens were constructed as ultimate strength, or replica, models (Harris and Sabnis, 1999) using model concrete and model reinforcement materials which satisfied the similitude conditions for the prototype materials. Full details are given in Sabatino (2007).

Model material characteristics

The correct modelling of the materials ensured the performance of the model under quasi-static loading to adequately replicate the behaviour of the prototype. For successfully modelling the correct failure mode of the structure, and in particular distinguishing between brittle and ductile failure, it was necessary to develop model materials which would satisfy the similitude requirements of cracking, bond and strength – the parameters which govern the nonlinear response at a local level. Therefore, stress and strain characteristics of the materials were not scaled down.

The model concrete was developed using typical constituent materials for ordinary concrete: cement, sand, grit, chippings and water. However, the a reduced aggregate size was used. Steel reinforcement was modelled using 4mm cold-rolled threaded bars, or studding, which were heat treated to obtain suitable constitutive stress-strain characteristics.

Monotonic and cyclic push-over tests of scaled models

Two quasi-static push-over tests on 1/6th scale intake tower models were carried out in order to determine their load-displacement properties, in particular their capacity and cyclic degradation characteristics (Figure 2). The test specimens were mounted onto a purpose built reaction frame and subjected to either monotonic (specimen NSD-R-1) or cyclic (specimen NSD-R-2) push-over loads. Four servo-controlled hydraulic actuators were used to impose horizontal and vertical loads. The lateral loading was representative of earthquake loading, whereas the vertical loading was used to simulate approximately the added mass for gravity similitude. The towers were instrumented with a range of strain gauges and displacement transducers designed to record data describing the response of the structure to the applied loads. Approximately 80 channels of data were recorded.

Figure 2: Tower specimen setup

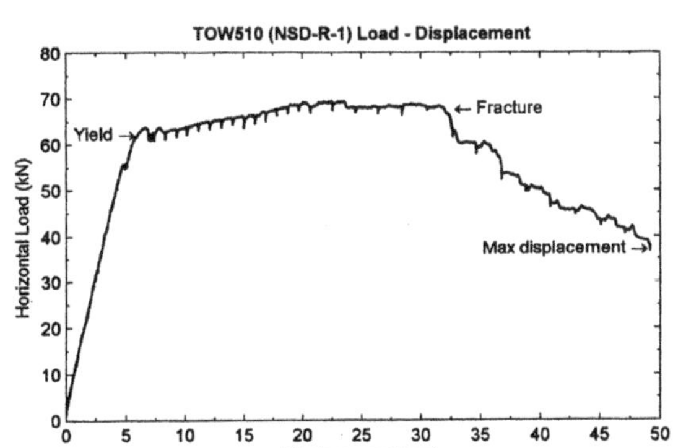

Figure 3: Load-displacement plot for monotonic test (NSD-R-1)

Test results

Monotonic test results indicated that the response of specimen NSD-R-1 was characterised by a tension failure of the under-reinforced section, with a single large crack forming at the base of the tower. With reference to the load-displacement plot (Figure 3), three distinct phases were observed:

1. The response of the specimen was stiff until first cracking was observed; first yielding of the reinforcement followed.
2. As further displacement was imposed, some hardening was observed until the peak load was reached.
3. Once fracture was initiated, the reaction force to the applied displacement dropped rapidly until the test was terminated.

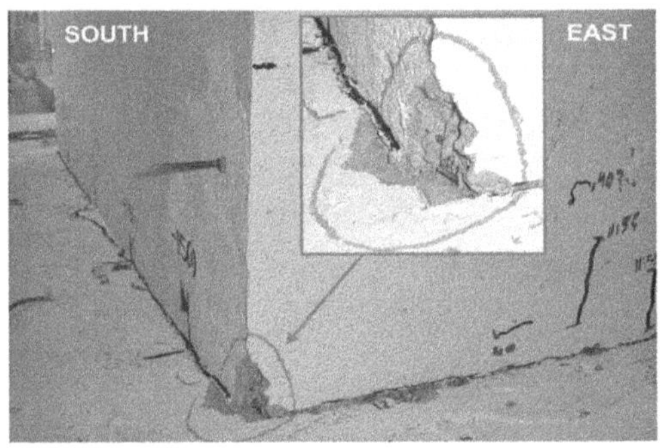

Figure 4: View of South-East corner crushing (NSD-R-2)

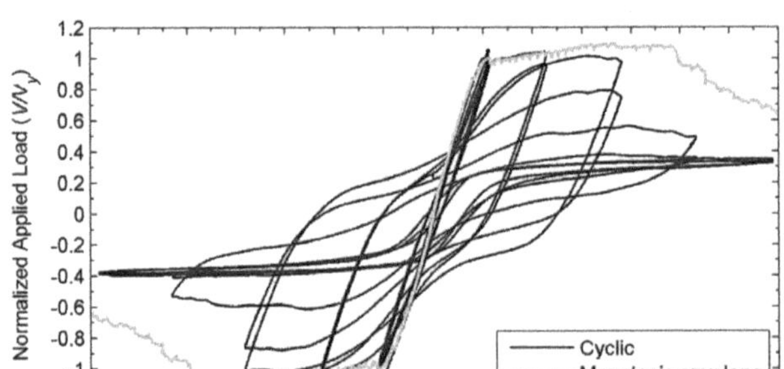

Figure 5: Normalized load-displacement plot.

The cyclic response of specimen NSD-R-2 confirmed the rigid body behaviour, with localised damage at the base including spalling of the concrete cover due to bar buckling (Figure 4). The following additional conclusions can be obtained from the cyclic test results:

1. Considerable strength and stiffness degradation was observed with increasing displacement amplitudes, making the monotonic envelope (Figure 5) a non-conservative estimate of the tower capacity.
2. Extreme pinching was likely to be caused by bond deterioration between the steel reinforcement and the concrete.

3. Yielding of the reinforcement was localised along the critical section, with small (elastic) strains being measured at other locations along the height of the tower.

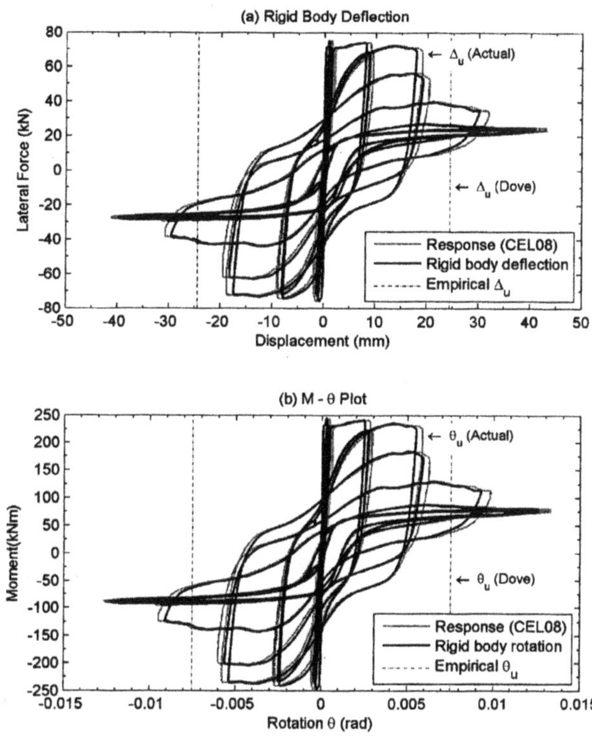

Figure 6: Experimental and empirical (a) ultimate deflection and (b) ultimate rotation values

Table 2: Comparison between experimental and predicted ultimate deflection and rotation values

	Actual (experimental) capacity	Theoretical capacity	Error
Ultimate deflection Δ_u	18.9 mm	24.4 mm	29.1%
Ultimate rotation θ_u	0.005832 rad	0.007514 rad	29%
Ductility ratio μ	21	28	33%

Figure 6 shows the ultimate deflection (Δ_u) calculated using the empirical expression proposed by Dove and Matheu (2005). The ultimate rotation (θ_u) was calculated, based on the rigid body assumption, by dividing Δ_u by the height. Both plots (a) and (b) show that the actual ultimate deflection and rotation of the specimen, obtained from the pushover tests, are less than the calculated empirical value. For this configuration, the predicted ultimate deflection (or rotation) is over-estimated using the empirical relation. This has considerable consequences on the ductility ratio of the tower (Table 2).

NUMERICAL MODELLING OF INTAKE TOWER SPECIMENS

Nonlinear finite element model

A Finite Element Analysis (FEA) of the intake tower specimens was undertaken to investigate the suitability of the method for modelling typical lightly reinforced concrete intake towers under seismic loading by comparing the numerical results with the benchmark experimental results. The modelling of the intake towers was performed using DIANA (release 9) finite element package, a general purpose commercial finite element code based on the displacement method (DIANA, 2006).

3D model definition

The finite element model geometry was based on the three-dimensional properties of the physical specimens, with the actual steel reinforcement layout modelled. The horizontal (monotonic and cyclic) load was applied in the form of explicitly specified load steps through the definition of a time curve. The vertical added mass was modelled as a point load applied at the top of the model. The concrete cracking was modelled numerically using the smeared crack approach to allow for a more versatile finite element model. The total strain fixed crack model (DIANA, 2006) was selected for its ability to formulate a single model with tensile and compressive constituent laws. The steel reinforcement was modelled as an elasto-plastic material with no ultimate strain defined.

Monotonic and cyclic analysis results

The duration of the monotonic analysis was 24 hours for 178 load steps, using a dedicated Windows server. The analysis results indicated a localised crack occurring at the base of the tower (Figure 7), with the rest of the model remaining within the elastic range. Figure 8 shows the load displacement results compared to the experimental results and the rigid body motion calculated empirically. The FEA results coincide closely with the rigid body motion, whereas the discrepancy with the experimental results was due to the error in the measurement of the lateral displacement of the tower, affected by the flexibility of the support frame.

The duration of the cyclic analysis was 24 days for 1279 load steps. The cyclic load history was defined to simulate the experimental cyclic loading. Figure 9 (a) shows the horizontal load against applied displacement plot for the top of the model. By comparison to the experimental results (grey curve), plotted for the same amplitude displacements, the numerical results slightly under-estimate the capacity of the tower, probably due to some difference between the numerical concrete strength and the actual tower concrete properties. However, the magnitude of the displacement and the unloading/reloading stiffness of the tower appear to coincide with the experimental results.

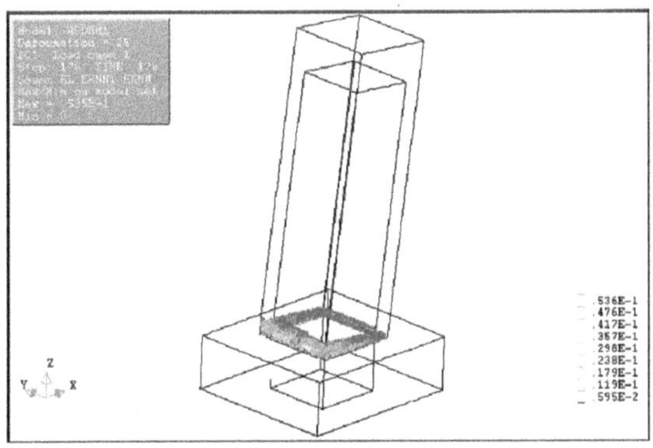

Figure 7: Monotonic crack distribution at time step 178 (f)

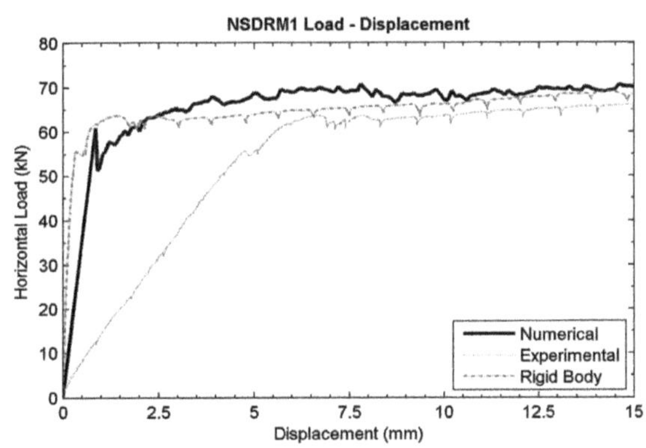

Figure 8: Monotonic load-displacement curve

Figure 9 (b) shows the plots, for both numerical and experimental results, of the normalized load, obtained by dividing the load by the yield value, and the normalized displacement, calculated by dividing the displacement by the yield value. The figure shows a good relationship between numerical and experimental results, suggesting that the FEA model is capable of adequately simulating the capacity of the structure, although strength degradation during the final cycle is more apparent for the experimental results.

Once fracture of the steel occurs in the physical model, this correlation is reduced as the numerical steel model does not allow for fracture of the steel to occur, emphasising the importance of defining the actual constitutive material characteristics in order to predict the full nonlinear response
Overall, the following conclusions may be drawn from the FEA results:

- The similitude between experimental and numerical results, in particular the crack distribution, indicate that the smeared crack approach is suitable for investigating the nonlinear response of lightly reinforced concrete towers.
- Shear was modelled explicitly using a shear retention factor. The analysis results were sensitive to the choice of the shear retention factor, even though the failure mode of the structure was tension. Careful consideration needs to be taken in selecting the crack model parameters to avoid a stiff response where it does not occur.

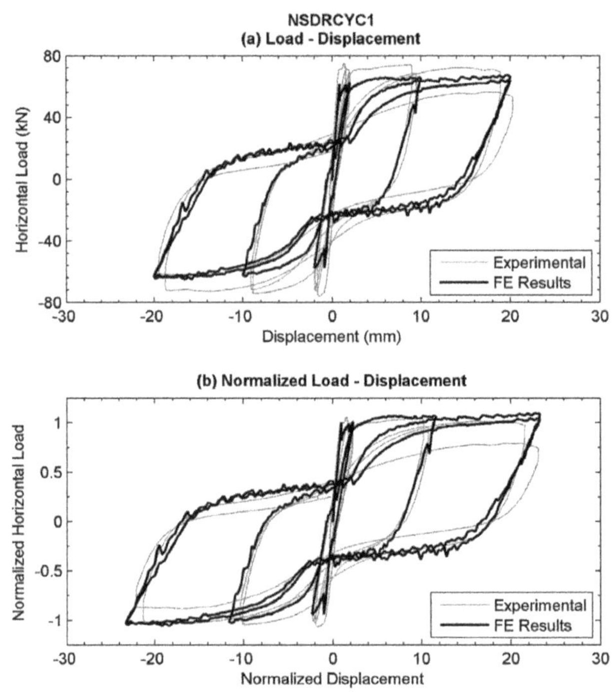

Figure 9: Cyclic load-displacement plots for numerical and experimental results: (a) actual; (b) normalized to yield values

Overall, the FE model showed a good simulation of the cyclic behaviour of the tower, but was impractical for use in probabilistic context due to the computational effort required. However, the results were used to validate the experimental results.

<u>Simplified model for dynamic time history analysis</u>

The response of the tower specimens described above suggests that the squat, lightly reinforced concrete towers may be approximated to a rigid block. The shaft section may therefore be modelled as a concentrated point mass, connected to a rotational spring modelling the crack opening at the critical section. The system lends itself to a single degree of freedom

(SDOF) idealisation, which is particularly advantageous for dynamic analysis of a nonlinear system, where the seismic action is represented as an acceleration time history. The hysteretic behaviour of a system can then be modelled by defining the characteristics of the SDOF spring to represent the nonlinear constitutive response of the structure. The SDOF idealisation (Figure 10) was used as the basis of a probabilistic approach to investigate the seismic vulnerability of various intake tower structures, thus allowing for the uncertainty associated with the selection of the parameters defining both structure and earthquake ground motion to be adequately represented.

SDOF model definition

An adapted Bouc-Wen model (Bouc, 1967; Wen, 1976) was chosen as a suitable mathematical representation of this relationship due to its versatility in defining the governing parameters. The model was adapted to simulate accurately the nonlinear response of the towers to earthquake-type loading, including structural degradation and pinching effects, and was calibrated against the experimental test results of the specimens. Details of the simplified model derivation are given in Sabatino (2007).

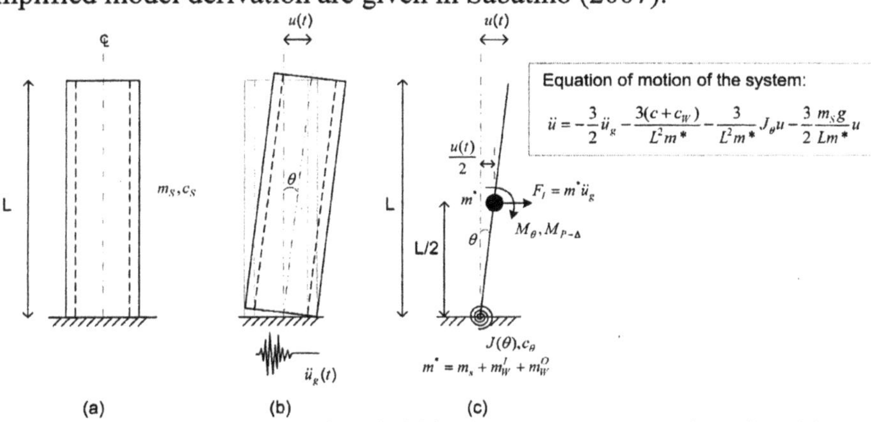

Figure 10: (a) Intake tower; (b) Rigid body response to earthquake; (c) SDOF idealisation

Earthquake ground motions

For the purpose of demonstrating the validity of the concept, synthetic earthquake ground motions were generated using the Kanai-Tajimi power spectrum (Tajimi, 1960) to represent typical UK, short duration earthquakes. The time histories generated for this study were deficient in energy content at very low frequencies. The results obtained should therefore not be viewed as meaningful in describing the real seismic performance of the intake towers.

Fragility analysis

The limit states considered in assessing the seismic performance of the towers were based on damage limitation (limit state 1 – LS1) and near collapse (limit state 2 – LS2) performance levels, as prescribed by Eurocode 8 Part 3 (BSI, 2005). By using the displacement ductility (μ) as a measure of the performance level, values of μ were determined for first yield ($\mu_{LS1} = 1$) and ultimate displacement at fracture initiation of the reinforcement steel ($\mu_{LS2} = 20$).

Results from the dynamic analyses showed that the response of the towers was dominated by their inertial characteristics. The stiff structures dissipated little energy in their elastic range and only after cracking was initiated was the dominant response governed by plastic yielding, with the displacement of the SDOF system proportional to the force applied.

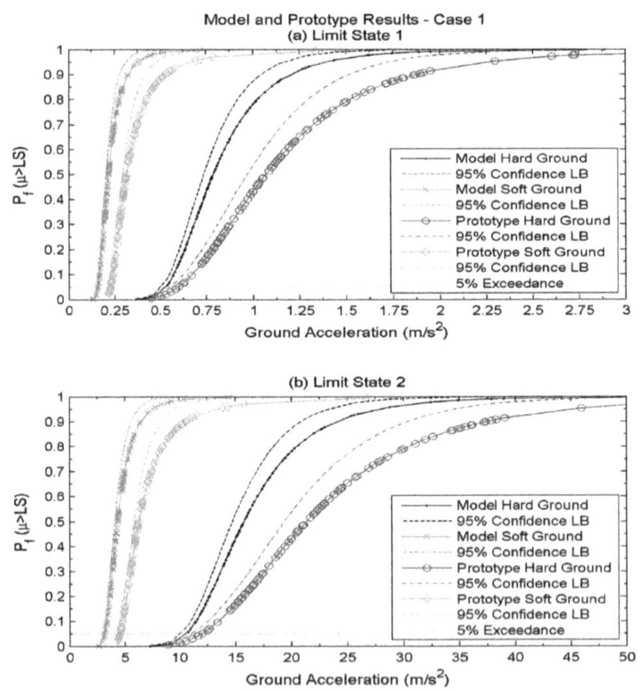

Figure 11: Typical fragility curves obtained for model and prototype towers (reservoir full conditions)

Indicative fragility curves describing the seismic performance of the structural systems considered were obtained (Figure 11). The main conclusions drawn from the fragility analyses are summarised below:

- The added mass contribution due to the reservoir water increases the seismic vulnerability of the system by approximately 15% for hard ground conditions.

- The response of the structures was sensitive to the shape and amplitude of the time histories used. The energy content of the earthquake, as well as the frequency content of the time history, greatly influenced the seismic performance of the structural models considered.
- The apparently large *PGA* response required to reach the collapse prevention limit state can be attributed to the geometrical characteristics of the towers, where large displacements are required before the centre of gravity of the structure is sufficiently displaced to induce overturning P-Δ effects. Although all the reinforcement steel may be fractured, the squat structure would respond in a rigid body rocking motion, requiring a large *PGA* to cause it to overturn.

Due to time constraints, the earthquakes generated for the purpose of this study were very crude. It was evident that a more precise evaluation of the seismic event, preferably through the generation of site specific time histories or at least by using more refined stochastic methods in simulating ground motions allowing for the appropriate ground characteristics to be modelled, would be required to obtain any significant performance assessment of a real structure.

These conclusions have been drawn from a simplified model which has been calibrated against quasi-static test results and a number of simplifying assumptions have been used. The validity of the dynamic response of the model requires further investigation, either through dynamic testing of tower models or, preferably, through comparison with time history analyses using more refined models over a range of parameters.

PROPOSED STAGED ASSESSMENT PROCEDURE FOR SEISMIC PERFORMANCE EVALUATION

The proposed staged assessment procedure, described below, allows for a gradual increase in the complexity of the analysis where necessary.

Stage 1: An initial performance assessment of the tower can be carried out using existing capacity spectrum methods. The capacity curve of the tower can be readily estimated using commercial packages, and the capacity spectrum method allows for an initial estimation of the whether a given seismic demand is likely to exceed the tower's capacity and therefore require a more rigorous nonlinear dynamic analysis.

Stage 2: Using a simplified model, a probabilistic, second stage analysis would follow if necessary. A number of simulations can be carried out to obtain fragility curves for the structure for various loading conditions and performance requirements. The Engineer can then determine the vulnerability of the tower for a given seismic hazard level, and establish whether there is the need for a more detailed, and costly, FE analysis.

Stage 3: A nonlinear FE analysis would only be required for those cases where, based on the fragility analysis results from Stage 2, the seismic demand exceeded the capacity of the tower. By selecting a few time histories, identified in the previous stage as having the most significant impact on the response of the tower, the dynamic time history analyses of the tower would provide a more accurate estimate of the structural performance.

CONCLUSIONS

Monotonic and cyclic pushover tests of scaled lightly reinforced intake tower models indicated that their response was characterised by a localized tensile failure at the base of the tower. The rest of the tower behaved as a rigid block, with negligible flexural response. Significant strength and stiffness degradation were observed for the cyclic test, with the monotonic envelope providing a non-conservative estimate of the capacity of the structure. The response of the towers indicated that the empirical method proposed by Dove and Matheu (2005) to calculate the ultimate deflection capacity of intake towers over-estimated the actual capacity of the specimens tested. Further experimental studies would be needed to clarify this issue.

The test results were compared to a FE model of the specimens, capable of simulating the crack propagation using a smeared crack material model for the concrete. The computational effort required to run the analyses made this approach not viable for use in a dynamic time history analysis. A simplified SDOF idealisation was developed and used to perform Monte Carlo type analyses of the towers for different loading conditions and performance levels.

Based on the results obtained, the proposed simplified model was shown to be a credible tool for the dynamic analysis of lightly reinforced concrete intake tower structures. However, gross simplifications have been assumed in defining both the structure and the synthetic earthquakes used in the analyses. Although it provides a useful tool for the seismic vulnerability assessment of the towers, it should be used as a first approximation. The good fit to the quasi-static cyclic response of the actual towers does not guarantee that the dynamic simulation accurately matches the behaviour of a real tower. The uncertainty associated with the variability of the parameters defining the structure and earthquake, as well as the sensitivity of the response to the energy content and shape of the time history, implies that a risk-based approach should be adopted for assessing the performance of the intake towers.

ACKNOWLEDGEMENTS

The authors would like to acknowledge the financial assistance provided by EPSRC under Grant Ref. GR/R93964/01, and KBR and Scottish & Southern Energy for their support as Project Partners.

REFERENCES

Bouc, R., 1967. *Forced vibration of mechanical systems with hysteresis.* Proceedings of the 4th Conference on Nonlinear Oscillations, Prague.

BSI, 2004. *BS EN1998-1: Eurocode 8: Design of structures for earthquake resistance. Part 1: General rules, seismic actions and rules for buildings.* British Standards Institution.

BSI, 2005. *BS EN 1998-3: Eurocode 8: Design of structures for earthquake resistance. Part 3: Assessment and retrofitting of buildings.* British Standards Institution.

Charles, J.A., Abbiss, C.P., Gosschalk, E.M. and Hinks, J.L., 1991. *An engineering guide to seismic risk to dams in the United Kingdom.* Building Research Establishment, Garston.

DIANA, 2006. *Finite Element Analysis – User's Manual. Release 9.* TNO, Delft.

Dove, R.C. and Matheu, E.E., 2005. *Ultimate Deflection Capacity of Lightly Reinforced Concrete Intake Towers.* ACI Structural Journal, Vol. 102, Issue 2, pp. 214-223.

Harris, H.G. and Sabnis, G.M., 1999. *Structural modelling and experimental techniques (2nd ed.)*, CRC Press, Boca Raton, pp.62-67.

ICOLD, 2002. *Seismic design and evaluation of structures appurtenant to dams, Guidelines.* Bulletin 123.

Institution of Civil Engineers, 1998. *An application note to An engineering guide to seismic risk to dams in the United Kingdom.* Thomas Telford, London.

Sabatino, R., 2007. *Seismic performance assessment of lightly reinforced intake towers in reservoirs.* Ph.D. thesis, University of Bristol.

Tajimi, H., 1960. *A statistical method of determining the maximum response of a building structure during an earthquake.* Proceedings of the 2nd World Conference on Earthquake Engineering, Tokyo, Japan, Vol. 2, pp. 781-798.

U.S. Army Corps of Engineers (USACE), 2003. *Engineering and Design - Structural Design and Evaluation of Outlet Works*, Publication No. EM 1110-2-2400.

Wen, Y.K., 1976. *Method of random vibration of hysteresis systems.* Journal of the Engineering Mechanics Division, ASCE, Vol. 116, No. 8, pp. 1798-1811.

4. 2007 storms and UK reservoir safety issues

The Defra Research Programme and its Management

AK HUGHES, Director of Dams & Water Resources, Atkins Ltd

SYNOPSIS. This paper describes the current Defra Research Programme. The programme currently includes a number of contracts including producing a new edition of the 'Guide to the Reservoirs Act, 1975', a reworking of the Guide to Emergency Planning and the production of a prioritised list of reservoir research topics covering a period of 5 to 15 years. The paper will also describe the formation, purpose and role of the Reservoir Safety Advisory Group (RSAG).

General Research Programme Call Off Contract to Defra and the 'Guide to the Reservoirs Act 1975'

In 2006 Atkins were awarded a three year contract to:

- give advice on reservoir safety issues to Defra and Government on an ad hoc basis
- produce a bibliography of published papers relevant to the UK
- seek the views of the profession to see whether it was timely to produce a new edition of the 'Guide to the Reservoirs Act 1975'.

The first two items are the 'normal' elements of the contract whilst the third element was the additional item introduced for 2006.

Approach to Revising the 'Guide to the Reservoirs Act 1975'

The views of the profession were sought principally via a questionnaire and informally via a number of British Dam Society meetings. The questionnaire to all members of the British Dam Society resulted in a small majority in favour of producing a new version of the document. Many positive suggestions were made which included additions to the Guide, areas where more explanation was required, clauses which needed clarity/explanation and requests for more examples. However, concerns were raised over the cost to the profession of producing a new Guide. Thus in producing the new Guide the greatest challenges were associated with the proposed format of the Guide – book, loose leaf file or CD, and the costs of the publication.

Having sought comments and questioned the profession a significant number of issues needed resolution. These were resolved by the team, by discussion with the Defra lawyer, or by discussion with the Environment Agency lawyer. Discussions were also held with some of those who responded to the questionnaire, the enforcement team within the Environment Agency as Enforcement Authority and a number of other interested parties.

At the time of writing this paper, the new Guide is being assembled with sections which will bring it up-to-date with respect to the Water Act 2003; provide new Appendices giving information on the Reservoirs Committee of the ICE how it operates; what it expects from candidates seeking appointment or reappointment to the Panel and the way that those candidates are assessed including the work and initiatives of the Enforcement Authority.

These initiatives include the incident reporting system and examples of its output, moves to improve the database of information on the dams subject to the Act.

Once the 'new' Guide has been assembled, decisions will be made by Defra on how the information in the 'new' Guide is to be disseminated.

The preferred method being suggested by the Contractor is by hard copy but an alternative is via a CD supplied free to all Panel Engineers and Engineers seeking appointment to the Panels constituted. It is hoped that this will be funded by the Environment Agency as the Enforcement Authority.

The Engineering guide to Emergency Planning

The draft Engineering Guide to Emergency Planning has received criticism from elements of the profession and in particular the Category 1 responders comprising what essentially are the emergency services (blue light organisations).

The criticisms centred on:

- unnecessary complexity
- undue costs likely to be incurred
- unsuitable for end user.

An Atkins led consortium formed with HR Wallingford approached the problem from a different direction whilst still following the principle of the 3 elements defined by the earlier Guide, namely:

- an on site plan
- dam break analysis and inundation mapping
- an off site plan

Atkins approach to the commission was to discuss the needs of the end user with the end users – the Category 1 responders and particularly the Local Resilience Forums (there are 47 of these covering England and Wales). One of the objectives of these discussions was to find out the format and information required, what detail was required and what accuracy was required of the mapping and hence the modelling.

Prioritised List of Research Topics

Atkins, in collaboration with HR Wallingford, Defra, Bristol University and Chris Binnie Consulting were commissioned by the Environment Agency to undertake research to establish the future research needs with respect to dams safety in the UK and identify sources of funding for future research.

Objectives

The main objectives of the project were therefore to:

- identify the issues that will impact dams safety in the future
- identify sources of funding including innovations for obtaining funding for future research projects
- provide a schedule of research projects over the planning period which sets out the order of priority with which projects should be done in order to ensure that the available funding in each year is spent to maximise the benefit to reservoir safety.

Guiding Principles

A number of guiding principles were identified as key to successful development of a research strategy. These were:

- identify the drivers that will shape the future use, and hence safety, of reservoirs.
- identify the likely impact of the drivers on dams safety.
- establish the research baseline to avoid duplication of effort.
- identify research that will address the issues arising from the drivers.
- identify projects which provide tools, guidance or new science deliverables of use to the profession.

Drivers

In order to develop an effective strategy for research supporting UK reservoir safety, it was first necessary to consider the drivers affecting reservoir design, maintenance and operation in the UK now, and in the future. Actual needs within the strategy may be grouped into theme areas such as supporting design, monitoring, emergency planning etc. Specific actions may then fall into categories such as science, guidance, tools/techniques etc. By analysing drivers and solutions in this way, a strategy was developed to meet industry needs and priorities both in the short and longer term (Figure 1). In addition to short and long term drivers (for example, legislation changes or climate change) specific industry needs and quick wins / opportunities will (and should) also directly have affected how the strategy was implemented.

Figure 1: Factors contributing to an overall strategy for reservoir safety in the UK

Short term drivers are more easily identified than long term drivers, since they typically already exist, or are likely to exist in the very near future. Identification of long term drivers requires careful consideration of trends and subsequently a prediction of factors that are most likely to affect UK reservoir operation in the future. The following analysis considers short term to be within the next 5 years and long term to be 15-20 years.

SHORT TERM DRIVERS

UK legislation (Water Act 2003).

The Water Act 2003 enhances reservoir safety by amending the Reservoir Act 1975 to:

- transfer enforcement powers from Local Authorities to the Environment Agency in England and Wales and
- empower the Secretary of State or the Assembly to direct owners of large raised reservoirs to prepare and maintain flood plans.

This requirement to prepare and maintain flood plans will be key to the management of the risks from dams in the future. An engineering guide which was commissioned by Defra has been developed by Jacob's to aide owners in developing these plans, but this is currently being revised. The final guide will be key to establishing the risks posed by large dams and the management of those risks. In addition, they will provide the EA and Emergency services with the information required to manage incidents at each dam thus meeting the requirements of the Civil Contingencies Act, as discussed below.

UK Civil Contingencies Act (2004)

The Civil Contingencies Act (2004) will deliver a single framework for civil protection in the UK. In the context of emergency preparedness, risks are those hazards (i.e. non-malicious events such as flooding) or threats (i.e. malicious events such as terrorist attacks) which could adversely affect an organisation and its ability to carry out its functions. The risk of major dam failure falls into both these categories of risk. There is therefore a requirement to identify and quantify the risks posed by individual dams and establish contingency measures for response to emergencies arising from a dam failure.

European legislation

The Water Framework Directive has established a new, integrated approach to the protection, improvement and sustainable use of surface waters. The Directive requires all rivers to achieve either good ecological status or, if classified as a Heavily Modified Water Body, good ecological potential. This is to be implemented in the UK. This will require variable flows to be released, set by the time of year and the required ecological conditions downstream. Thus specific variable flow releases will be required.

There are a number of reservoirs which have no facilities to release water when the reservoir is below top water level. In many others the current flow release is fixed throughout the year and is of limited size. It can be expected that in some cases new facilities for release will be needed, and these can be difficult to provide in dams which do not already have them. There may also be the need to provide the facility to vary the release to provide freshets. This may well require the provision of automatically controlled powered valves.

The aim of the EU Floods Directive is to reduce and manage the risks that floods pose to human health, the environment, infrastructure and property. According to the EU Floods Directive, Member States will have to reduce flood risk for those areas where the risk is deemed significant. This is done by first determining the extent of flood risk (through flood hazard mapping and flood risk mapping). The directive aims to map all sources of flooding including the risk from reservoirs, although it is not clear to what extent flood risk mapping from reservoirs will be required.

Succession
The maintenance of dam safety in the UK is heavily dependent on the availability of skilled and knowledgeable staff to carry out inspections and assessments of dams. There is a need to address the numbers as well as the quality of qualified professionals able to undertake these responsibilities.

LONG TERM DRIVERS

Climate change
Climate change in the UK is projected to result in wetter winters, more frequent and larger floods (magnitude and duration), and hotter and drier summers. Issues arising as a direct result of climate change include:

- more frequent extreme (in terms of magnitude and frequency) floods resulting in more severe and frequent overtopping of dams
- requirement for increased spillway capacity to accommodate larger floods
- requirement to reinforce spillways to prevent damage during long duration flood events
- requirement to raise dams to accommodate larger floods
- increased magnitude of winds on reservoirs, increasing the frequency and depth of overtopping due to waves
- increased vegetation on earth dams resulting in root damage and internal erosion. This will have implications for dam maintenance
- increased erosive power of rainfall
- dessication of earth dams under hotter drier summers leading to cracks and internal piping.

Indirect impacts of climate change include:

- shorter wetter winters and longer drier summers could lead to a change in the filling and drawdown sequence of reservoirs. This effect might be off-set to some extent by changes in the patterns of demand due to climate change, but there is the potential that the 'excess' water during the winter will need to be stored to meet deficits during summer. This could lead to a requirement to construct new or modify existing reservoirs for both flood retention during winter and water supply during summer
- operating regimes of existing reservoirs could also change due to changes in the timing of releases to meet changes in the timing of demand, and also to provide flood retention benefits and prevent overtopping. Hence reservoirs may have to be drawn down mid season to accommodate floods that may occur later on in the filling season. This could present operational difficulties, and could introduce further uncertainty to water supply in the summer.

Ageing Dams– deterioration

Many UK dams were constructed in the 1850s, 1880s and 1890s at the start of the Industrial Revolution. The average age of large dams in England is 110 years. As these dams continue to age, there will be increasing concerns over safety. Inspection intervals may need to be revised based on age of the dam, and monitoring measures put in place. A number of monitoring techniques are currently available and it would be important to examine these in order to determine which are the most suitable for UK dams.

Dams which are no longer needed may well be taken out of service by abandonment and discontinuance. The need for this may be accelerated by the likely requirements of the Water Framework Directive such as variable flow releases and also by the cost of operation including the preparation of emergency plans. There is also the question of the safety of decommissioned dams.

Dam removal/Discontinuance

One method of reducing the risks posed by dams is to abandon or discontinue them. In the US, dam removal exceeded dam construction for the last 10 years with 80 dams removed in the last 2 years. These were mainly for safety reasons but the regime does differ in as much as in the US the dams were licensed for a specific period with the intention of removing the dam at the end of the licence period.

So far there is little knowledge of the issues involved in dam removal and how best to cope with this requirement. There is therefore a need to provide guidance on the procedure for safe dam removal from both an engineering and environmental/ecological perspective.

Sustainability and Increased demand

There is the potential problem of more reservoirs being needed because of rising population and the reducing number of people in each home leading to higher per capita demand. While leakage levels are reported to be decreasing (see Ofwat report), new homes are to be constructed with a much lower water usage (see Defra documents and consultations), and Ofwat is setting targets for water companies to reduce the amounts of water delivered (see Ofwat letter to water companies). It is not clear whether these measures will offset the projected increase in demand due to increasing population.

In addition, the British Geological Survey study found that groundwater pollution is likely to reduce the amount of ground water that can be treated economically. Climate change is therefore likely to reduce the yield of direct water supply systems and is likely to reduce the yield of reservoirs. Thus, augmented water resources may well be needed, primarily in the south and east of the country. This could well require increased volumes of storage, possibly by raising existing dams, accepting treated sewage effluent into them, and possibly by a small number of new embankment dams.

Issues such as the planning and consultation horizons for the construction of new dams will need to be reviewed to ensure that the requirement for the construction of new reservoirs take full account of these issues.

Emergency Preparedness and Communication of Risks

With greater variation in flood events, greater understanding of how to handle these in real time is required. Growing concern about risk, safety and terrorism means greater need for emergency preparedness and operational response including: increased stakeholder involvement, management and monitoring of reservoirs, mechanisms of assessing potential dam breach and downstream inundation, disaster preparedness, emergency operational planning, emergency action plans, and remedial measures. An important aspect is the extent to which remotely possible breach impacts are placed in the public domain.

Renewable Energy

Demand for renewable energy will also lead to the construction of new dams, and the retro-fitting of existing dams, with hydroelectric power facilities. Hydroelectric dams may require new draw-off systems being installed, which may result in safety issues both during construction and operation. The Glendoe hydroelectric dam, currently under construction, is the largest to be built in the UK in 50 years. Further new hydroelectric schemes may well be built.

Advances/trends in Sciences

Modelling methods across a number of disciplines related to dams safety, as well as monitoring techniques and equipment, have advanced significantly and provide opportunities to assess and monitor dam safety with greater levels of accuracy. Numerical and physical modelling also enables the examination of unobservable scenarios. It will be important to examine input requirements, model assumptions and simplifications, and expected accuracy of outputs. In addition, guidance on appropriate methods, software and techniques is vitally important.

Risk and uncertainty

Dealing with risk and uncertainty is common in many, if not all, aspects of reservoir design, construction, operation and maintenance. The science of risk and uncertainty is rapidly evolving, and over the last decade a wide range of tools and techniques have been developed to support risk-based analyses.

In the process of making dam safety decisions, it is necessary to consider a plethora of factors including; public safety responsibilities, limitations on borrowings, market or other pricing constraints, business criticality, public opinion, loss financing, liability, and due diligence. Hence, engineers are being challenged to justify expenditure associated with one risk-reduction measure relative to another.

For a dam owner with a portfolio of dams, there is also a need to determine the priority with which to address safety issues related to his stock. Engineers are therefore being asked to provide risk-based assessments which serve as inputs to a business decision making process in which dams are a key asset. Dam safety risks are therefore being managed in a more open, informed, and holistic manner than ever before.

This trend towards a risk-based approach to dam safety management requires a shift in approach to dam assessment. For instance, currently used deterministic methods such as the PMF do not suit this risk-based approach. There are projects currently underway, dealing with this issue in the UK and elsewhere. It is important that the future research strategy ensures that appropriate methods and guidance are supported.

Specific Industry Needs and 'Quick Wins'

Specific industry needs may be identified through consultation with engineers and owners within the profession. 'Quick wins' are situations where projects may be initiated because circumstances offer a unique advantage for rapid implementation, whether this is scientific, financial or a combination of both.

DEFINING RESEARCH NEEDS

Establishing the baseline

For most issues that need to be addressed, there is already research that has been undertaken either as a precursor to more detailed work in the UK, or elsewhere in the world with direct relevance to the UK. In developing the strategy, limited research has been undertaken to establish the current state-of-play with respect to most aspects of dam safety. However, it is imperative that the baseline of each topic/issue is established before any detailed assessments are undertaken. For this reason, most projects proposed will start with a preliminary stage which will look at previous work.

The outcome of research undertaken in this phase is likely to be documents reviewing the existing state of the topic and also outline the scope for further work to address the problem during the main study. Some topics have already established baselines and can therefore move into detailed assessment. In addition, 'quick win' projects can also be started immediately.

Defining research Needs

Having identified the drivers contributing to the strategy, an initial assessment of the existing research was undertaken for all aspects of dam safety. To this end, the broad topic of dam safety was broken down into disciplines, and searches conducted and summarised on all research falling under each discipline. Given the time and cost constraints on this project this initial assessment was restricted to research undertaken in the last 2 to 5 years in most cases. In addition, early consultation was sought with dam owners and panel engineers, which resulted in a list of their requirements with respect to dam safety. The wider profession was also included in this initial consultation through a questionnaire posted on the BDS website requesting contributions.

A database of research projects and article references was developed to aide the searches, and could form part of the project deliverables. Once previous and current projects were identified, these were examined against the research needs arising from the drivers discussed above to determine whether they addressed any of the issues. From this assessment, research gaps were identified and specific research projects formulated including project scope, estimated cost and duration, expected outcomes and products (guidance, tools, methods etc.).

PRIORITISATION OF RESEARCH PROJECTS

The prioritised list of research projects was developed as follows:

- Projects were categorises based on the following criteria:
  - Long Term/Short Term/Industry/Quick Win
  - Likely Outcome – Science, Methods, Tools, Guidance
  - Level of Difficulty – Hard/Easy
  - Cost/Sources of Funding
  - Duration of Project
- 'Quick wins' were identified from the list of projects. As a general rule, projects which meet the immediate needs of owners and reservoir engineers generated from the list of research requirements dictated by Panel Engineers and Owners, which could be easily achieved, and which would provide tools or guidance (either interim or final) on a particular topic, were considered to be quick wins. Quick win projects were given high priority
- Second consultation – experts from all disciplines involved in dam safety were consulted during a 2-day workshop which involved discussions and voting on each project presented
- A final prioritised list of scheduled projects was then prepared and different spending scenarios considered
- Consultation was carried out via the BDS web site in the later stages to enable those who were not at the workshop to take part and secondly to seek contributions.

The prioritised list, is then used by the Reservoir Safety Advisory Group (RSAG). The RSAG is an Institution of Civil Engineers (ICE) managed group of Panel Engineers, owners, academics and representatives of the EA and Defra and is supported by a secretariat supplied by the ICE. The RSAG meets at 6 monthly intervals to ensure that existing projects are on time and in budget and also to consider new projects and the most appropriate source of funding to seek. The formation of this group will ensure the appropriate research is done on time and to budget.

The Implications of the 2007 Summer Storms for UK Reservoir Safety

A L WARREN, Halcrow Group Ltd
E J STEWART, CEH Wallingford

SYNOPSIS. This paper discusses the impact on reservoirs of the storms that affected north-east England and the Gloucestershire region in June and July 2007. The incident at Ulley Reservoir was well publicised and there were a large number of other dams that were overtopped and /or damaged by these storm events. The paper aims to:

- Evaluate the meteorological and statistical significance of the storm events experienced in the two regions;
- Summarise the impacts on reservoirs in both regions;
- Comment on the possible implications for reservoir safety operation, management and legislation; and
- Comment on how current research into extreme rainfall events should improve flood risk assessments for dam safety in the UK.

INTRODUCTION

Two separate storm events in England in the summer of 2007 led to widespread flooding which was widely covered by the national media. This paper aims to evaluate the meteorological and statistical significance of the storm events and to discuss the impacts on reservoirs. The incident at Ulley reservoir near Rotherham was particularly notable, where erosion adjacent to an old masonry spillway channel almost led to the failure of a large reservoir. The impact of a failure of this dam would have been significant both in terms of risk to life and damage and disruption to major infrastructure. However, all of the reservoirs in the two affected areas were affected by the storms to differing extents. It is appropriate, especially in the face of climate change, to review the impact of these floods with respect to current reservoir flood safety provisions in the UK.

Ensuring reservoir safety into the future 2008, Thomas Telford, London

METEOROLOGICAL SIGNIFICANCE OF THE 2007 STORMS

The late spring and early summer periods in 2007 were exceptionally wet throughout the UK, causing widespread and repeated flooding during June and July. According to Pitt (2007), this was the wettest ever May to July period since national records began in 1766. The most significant flood events followed two particularly extreme storm events, one on 24 to 25 June which affected a band across England from Worcestershire to the North York Moors, and the other on 19 to 20 July which gave rise to flooding in Gloucestershire, the Midlands and south-east England.

Following the first of these storms, the potential threat of the structural damage experienced at Ulley Dam hit the headlines when the M1 was closed for 40 hours and 1000 people were evacuated from their homes. The rainfall event which triggered the alert was caused by a deep area of low pressure moving gradually eastwards. Frontal systems around the northern side of the depression affected a vast swathe of north-east England, the Midlands, Wales and south-west England. Rainfall totals were exceptional, with more than 90 mm of rain in 18 hours being recorded in the Rotherham area (Environment Agency, 2007[1]). The Flood Estimation Handbook's rainfall depth-duration frequency model estimates the return period of such an event to be about 100 years (an annual probability of occurrence of 1%). Thus, while the rainfall event which triggered the damage to the spillway was reasonably rare, it was far from being a very extreme event in itself. However, it followed a remarkably wet spring period, and heavy rainfall had affected much of northern England on 14 to 15 June, leaving many catchments very vulnerable to further rainfall (Marsh and Hannaford, 2008). It has been estimated (Environment Agency, 2007[1]) that the return period of the flood flows at Sheffield and Rotherham was about 200 years.

The second major storm on 19 to 20 July was again caused by a slow-moving depression, this time centred over south-east England. Heavy rainfall moved northward and exceptional storm totals were recorded across the Cotswolds and the lower part of the Warwickshire Avon catchment. The rainfall totals recorded were somewhat higher than those of the June event in northern England, with 142.6 mm being recorded in 24 hours at Pershore College in Worcestershire (Marsh and Hannaford, 2008). The estimated return period of this event from the FEH model is over 800 years and the extreme rainfall was remarkably widespread, although the return period of the resulting fluvial flooding has been estimated to be about 200 years.

It is not unusual for extreme rainfall events to occur in the summer in the UK, but as they are generally caused by convective storms, they tend to be isolated, although their effects can be devastating as demonstrated by the Boscastle floods in August 2004. It is, however, very rare for summer flooding to be experienced on a regional rather than catchment scale. The distinguishing feature of the storms experienced in the summer of 2007 is that, although they were caused by embedded convection, they were associated with a succession of frontal systems which persisted for many weeks. Blackburn *et al.* (2007) attribute the unusual summer weather patterns to a southerly shift in the position of the jet stream steering Atlantic weather systems across the UK. It is interesting to note that current climate change projections point towards drier summers in the UK, associated with a poleward shift of the storm track; that is the opposite of the situation that pertained last summer.

VULNERABILITY OF DAMS TO EXTREME FLOODS

Flood-related damage to dams in the UK

The overtopping of dams in flood events is a major cause of dam damage and failure in many parts of the world. ICOLD (1995) report that, globally, floods account for 51% of dam failures. The Machhu II failure in India was attributed to the overtopping of a 26m high dam that failed resulting in the death of 2000 people (Charles, 2005). In the UK, many of the most notable dam failures have also been due wholly or partly to overtopping, whereby the hydraulic capacity of the spillway facilities and the available freeboard to the lowest point on the dam crest was insufficient to prevent the uncontrolled flow of water onto the downstream face of the dam. Significant flood-related failures include Darwen (12 lives lost in 1848), Cwm Carne (12 lives lost in 1875) and Skelmorlie (5 lives lost in 1925). There have been no lives lost due to dam failure in the UK since the introduction of reservoir safety legislation in 1930, but many dams have been overtopped. White (1994) lists four dam failures for the period 1960-71.

The failure of an embankment dam when overtopped is not assured and depends on various factors including the duration of the overtopping event, the nature of the ground cover on the downstream face and the velocity (erosive power) of the water flow. Hughes and Hoskins (1994) provide a useful summary on the wide range of factors that can contribute to the risk of damage to an embankment during an overtopping event and guidance on mitigating the risk. Erosion of the downstream face is not the only means by which dam integrity is threatened during floods. An embankment dam under flood conditions is at increased risk of damage due to:

- internal erosion during or at some time long after the flood event (refer to Defra, 2003);
- wave damage;
- unnatural flood inflows, for example due to the failure of an upstream dam.

Charles (2002) provides a summary of historical dam failure events in the UK. The Environment Agency intends to publish reports on reservoir incidents annually and an update on the progress of the voluntary reporting system has been provided (Hamilton-King *et al.*, 2008). Analysis of the national incident database administered by the Environment Agency was carried out for incidents at dams arising from flood inflow. Incidents were grouped as pre- and post-1985; 1985 being the year when the current UK reservoir legislation and safety regime came into force.

Table 1. Incidents where flood inflow was the primary threat to the dam

Incident Severity*	1799-1985	Post-1985
Level 1	10	1
Level 2	14	5
Level 3	1	7

*Level 1: Dam failure; Level 2: Emergency actions taken; Level 3: Precautionary actions taken.

Table 2. Incidents where the dam was overtopped

Incident Severity	1799-1985	Post-1985
Level 1	10	1
Level 2	11	4
Level 3	1	5

It should be noted that the incident statistics in Tables 1 and 2 do not accurately reflect the frequency of occurrence (particularly for Level 3 incidents) as the database has historically only captured well-publicized incidents. Recent developments in incident reporting have improved the level of reporting but reporting currently remains on a voluntary basis and hence the level of reporting is less than complete.

Guidance on reservoir flood safety provisions

When the Reservoirs (Safety Provisions) Act 1930 came into force, there was a diversity of opinion on appropriate levels of reservoir flood safety. In 1933, the Council of the ICE published guidance entitled *Interim report of the Committee on Floods in Relation to Reservoir Practice.* This guidance was updated in 1960 following the Lynmouth disaster of 1952. The publication of the *Flood Studies Report* (FSR) (NERC, 1975) and *Floods and Reservoir Safety* (ICE, 1978) provided new guidance on appropriate levels of flood safety and contributed to the large increase in spillway-related improvement works in Great Britain in the 1980's and 1990's. The ICE guidance on flood safety provisions was primarily aimed at panel engineers in relation to statutory reservoirs. Specific guidance for the design and maintenance of small reservoirs has since been provided (CIRIA, 1996) which draws from the ICE guidance. The Flood Estimation Handbook

(FEH) (Institute of Hydrology, 1999) did not aim to specifically address the estimation of very extreme flood events and as such was of limited relevance to the design of spillway facilities for reservoirs that pose a high degree of risk to downstream persons and property. Over recent years Defra has issued guidance notes for panel engineers in the use of the FSR and FEH in relation to spillway design and the Cox report (2003) reviewed the use of the FEH in the estimation of extreme rainfall. A revised model of rainfall frequency estimation and updated guidance are currently in preparation by the CEH and this is covered below.

Best practice in relation to spillway design or improvement is generally applied to large reservoirs through the provisions of the Reservoirs Act 1975. Qualified engineers who are familiar with the various flood safety guidelines are required to certify the adequacy of spillway provisions for statutory reservoirs. There are no such provisions for smaller, non-statutory reservoirs and the risk of small dam embankments being overtopped during floods is believed to be far greater than that for statutory reservoir embankments (Goff and Warren, 2008).

THE IMPACT OF THE 2007 FLOODS ON RESERVOIRS

Of the twelve reportable incidents held on the national database for incidents in 2007, eight were attributable to flood inflow.

Ulley Reservoir

A stepped masonry spillway failed at Ulley Dam near Rotherham on 25 June 2007 and was very similar to the failure of the stepped masonry at Boltby dam near Thirsk on 19 June 2005. The masonry spillway had been supplemented by a larger concrete spillway in 1943. However, the arrangement of the design meant that the older, stepped masonry spillway always operated preferentially. It appears that the masonry had lost much of its pointing and had little bedding mortar between the blocks. The unusually high flow conditions in the stepped channel might have led to the development of high hydrodynamic pressures within the walling, leading to the collapse of the wall and the subsequent erosion of the embankment fill material. The resulting loss of support to the dam core, and fear for the stability of the dam led to the closure of the M1 motorway and a widespread evacuation downstream of the dam. This particular incident has led many to question the performance of masonry stepped spillways and it seems clear that research is warranted.

Other reservoirs in the north-east

Reporting of the impact of the storms in the north-east is mostly anecdotal as not all the incidents would have been judged as reportable, and others were not formally reported to the Environment Agency for various other reasons. It is believed that more than 20 reservoirs were in some way damaged or endangered by the storm event in this region. The industry is also aware that there were several other examples of where masonry spillway channels were damaged by spillwater, but no details are currently available. This further underlines the need for research.

At two separate incidents, embankments were overtopped that featured sheet piles that had been installed to raise the crest level. Both suffered damage by erosion to the crest and downstream face.

A further incident arose at a reservoir which is located in a catchment area where the geology is predominantly chalk. The overflow spillway system was overwhelmed by the amount of groundwater issuing from a chalk escarpment and the dam embankment was overtopped.

Reservoirs in the Gloucestershire region

The number of reservoirs affected by the storms in the Gloucestershire area was far fewer than the number in the north-east, largely due to the fact that the areal density of reservoirs in this area is less than that in the north-east. It appears that the statutory reservoirs in the region were not seriously affected by the storms. This may be in part due to the fact that most of the larger reservoirs would have had available storage at this time of the year such that the flood runoff volume was partly or wholly absorbed. In addition, statutory reservoirs generally have well sized and designed spillway facilities capable of dealing with large floods. However, there were several non-statutory reservoirs that did not fare so well. Non-statutory reservoirs are often kept full or near full to serve some amenity function and therefore only the effectiveness of the spillway facilities can prevent overtopping in such circumstances when floods arise. To compound the threat to non-statutory reservoirs, these generally have spillway design capacities well below that of statutory reservoirs and are generally not so well maintained.

We are aware of two reservoir incidents in the Gloucestershire area that caused alarm, both at non-statutory reservoirs. In the first case (Incident 313) the dam crest had not been maintained at a uniform elevation and this led to overtopping of the embankment at the low spot and erosion steps to form on the downstream face. Monitoring of a seepage point at the toe over the following three months led to concern that internal erosion was occurring and the embankment was grouted in addition to repairs to the downstream face.

In the second case (Incident 321), a statutory reservoir had been discontinued but left with inadequate spillway provision. The dam was very nearly overtopped during the flood event. A panel engineer was called to the scene but was initially unable to reach the reservoir due to the transport disruption.

IMPLICATIONS FOR RESERVOIR OPERATION, SURVEILLANCE AND LEGISLATION

Flood events represent one of the principal threats to reservoir safety and this is widely acknowledged in the guidance and training given to reservoir operators and engineers. The great majority of statutory reservoirs are provided with spillway facilities able to withstand floods of return period far greater than that experienced in either the north-east or the Gloucestershire areas in 2007. It is therefore surprising perhaps that we continue to be able to learn so much about the safety of reservoirs when moderately severe flood events occur.

Operation

Ideally, where reservoirs have poor provisions for dealing with floods, operators should be encouraged to either address the problem or to maintain some flood storage below the spill level. Currently, there are many problems in promoting or enforcing such behaviour:

- the problem generally only arises at a proportion of non-statutory reservoirs;
- there is currently no comprehensive database of non-statutory reservoirs that pose a significant risk were they to fail and hence no database of reservoir owners from which to impart good advice;
- non-statutory reservoirs are often kept full to meet some amenity purpose such as fishing;
- non-statutory reservoirs often do not have operable draw-offs by which the reservoir level could be effectively maintained below the spill level.

Surveillance

Flood events can highlight deficiencies in surveillance, especially when ‘near miss’ incidents arise. Experience shows that surveillance at non-statutory reservoirs varies from excellent to none and usually the personnel involved in the surveillance have no formal training in dam safety. In contrast, statutory reservoirs are provided with a suitably trained Supervising Engineer that is able to check on the condition of the dam following a flood event and to take appropriate steps to rectify defects and monitor performance.

The Ulley incident has caused some concern for the owners of masonry-lined spillway channels who will need to take account of the findings of any future research to ensure that such channels are maintained appropriately and kept under surveillance to ensure adequate performance under flood conditions.

Legislation

It seems very likely that many reportable reservoir incidents arose during the 2007 floods that were not reported by major reservoir owners. We will never know whether the implementation of a mandatory system of incident reporting some years ago might possibly have prevented the Boltby and Ulley incidents. However it now seems very likely that mandatory incident reporting will be introduced in Great Britain (Environment Agency, 2007[2]). The Ulley incident also underlined the need for effective emergency planning and measures were already in preparation for the production of inundation maps and response procedures. These provisions are due to be implemented for statutory reservoirs from 2009 (Environment Agency, 2007[2]).

The flood events have also served to remind us that non-statutory reservoirs are far more likely to be adversely affected by flood events than statutory reservoirs (Goff and Warren, 2008). While the hazard that such reservoirs pose is generally small, many pose a significant risk to life and there is currently no effective legislation to assess the risk and make adequate provision for design, monitoring and surveillance as might be appropriate.

CURRENT RESEARCH ON EXTREME RAINFALL FOR RESERVOIR FLOOD SAFETY

An ongoing research project funded by Defra is reconsidering the basis of the rainfall frequency model developed as part of the FEH (Institute of Hydrology, 1999) with particular reference to the estimation of very long return period rainfall. The project aims to develop an alternative depth-duration-frequency model that is specifically designed to deliver frequency estimates for return periods of between 100 and 10,000 years with applications in reservoir flood safety assessment. A number of aspects of the current model are being examined, notably the way in which information from different sites is pooled together and the importance of spatial dependence in the estimation of rainfall extremes across the range of return periods. In addition, recommendations will be made about the applicability of the estimates of probable maximum precipitation (PMP) for the UK which, although not revisited since 1975, still forms part of the ICE guidance for reservoir flood assessment (ICE, 1996).

Although the study will deliver a revised procedure for the estimation of design rainfall depths for return periods of 100 years and above, the research does not include an evaluation of rainfall-runoff modelling for reservoir safety.

CONCLUSIONS

The summer 2007 floods served to highlight a diverse range of reservoir safety issues:

Response

- the benefits of formal emergency planning and inundation mapping;
- the difficulties of reaching reservoirs to provide advice or to undertake emergency measures during severe floods.

Robustness of intrinsic dam design

- there appears to be a need for further research into the performance of masonry-lined spillway channels;
- crest-raising works using sheet piles must take into account the impact of overtopping events;
- spillway systems for reservoirs in catchment areas with well developed groundwater systems should be checked to ensure an appropriate level of protection against overtopping;
- effective draw-off or bottom outlet works enable operators to effectively provide flood storage but many small reservoirs have no such facility and inadequate spillway/freeboard provisions.

Monitoring and surveillance

- the floods served to highlight the fact that the provision of monitoring and surveillance at non-statutory reservoirs varies greatly, and is not carried out at some reservoirs. The dams forming many non-statutory reservoirs would pose a risk to life were they to fail.
- Where unusually high reservoir levels have been experienced (with or without overtopping of the crest) the monitoring and surveillance frequency should be increased for embankment dams over the months following any severe flood events to check for signs of internal erosion.

Industry issues and legislation

- the flood events have served to support the need for legislation in relation to both emergency planning and mandatory incident reporting;
- there appears to be a need to appraise the risk posed by non-statutory reservoirs and to provide appropriate legislation and guidelines for panel engineers. It would appear reasonable that a person living downstream of a small reservoir is equally deserving of an

appropriate level of statutory protection as that for a person downstream of a large reservoir.

- there is current research being carried out to re-evaluate the risk of extreme rainfall events for application to reservoir safety;
- flood events do not always occur in a traditional manner according to the season and may not conform to postulated climate change trends.

REFERENCES

Blackburn, M., Hoskins, B. and Slingo, J. (2007) *Notes on the meteorological context of the UK flooding in June and July 2007.* The Walker Institute, University of Reading, July 2007, 3 pp. http://www.walker-institute.ac.uk/news/summer_2007.pdf

Charles, J.A. (2002) *"A historical perspective on reservoir safety legislation in the United Kingdom"* in *Reservoirs in a Changing World*, the 12th Conference of the British Dam Society, Trinity College, Dublin. Thomas Telford.

Charles, J.A. (2005) Use of Incident Reporting and Data Collection in Enhancing Reservoir Safety. *Dams and Reservoirs*, vol.15, no.3, November, pp29-35.

CIRIA (1996). *Small Embankment Reservoirs,* Report 161.

Cox, D.R. (2003). *Some comments on 10,000 year return period rainfall.* www.defra.gov.uk/environment/water/rs/pdf/coxreport.pdf

Defra (2003). *Early detection of internal erosion*, KBR Consultants. http://www.defra.gov.uk/environment/water/rs/pdf/internal-erosion-report.pdf

Environment Agency (2007[1]). *Review of 2007 summer floods,* Environment Agency Report. 60 pp.

Environment Agency (2007[2]) *Improving Safety, protecting lives* Biennial report. 25pp.

Goff, C.A. and Warren, A.L. (2008). The safety of small British reservoirs in *Ensuring reservoir safety*, proceedings of the conference of the British Dam Society, Warwick. Thomas Telford.

Hamilton-King, L, Hope, I.M. and Warren, A.L. (2008). Post-incident reporting: learning from experience to promote reservoir safety in *Ensuring reservoir safety*, proceedings of the conference of the British Dam Society, Warwick. Thomas Telford.

Hughes, A.K. and Hoskins, C.G. (1994). A practical appraisal of the overtopping of embankment dams, in *Reservoir Safety and the Environment*. Proceedings of the 8th conference of the British Dam Society, Exeter. Thomas Telford.

ICE (1978), *Floods and Reservoir Safety.* Thomas Telford (now in 3rd edition, 1996).

ICOLD (1995) *Dam failures: statistical analysis.* ICOLD Bulletin 99.

Institute of Hydrology (1999) *Flood Estimation Handbook (five volumes).* Centre for Ecology and Hydrology, Wallingford, UK.

Marsh, T. J. and Hannaford, J. (2008). *The summer 2007 floods in England and Wales. National Hydrological Monitoring Programme,* CEH Wallingford/BGS, February 2008. 36 pp.

Natural Environment Research Council (1975). *Flood Studies Report (five volumes)*, NERC, London.

Pitt, M. (2007). *Learning lessons from the 2007 floods.* Interim report of the Pitt Review, Cabinet Office, London. 165 pp.

Warren, A.L. and Hope, I.M. (2006) A new incident reporting system for UK Dams in *Improvements in reservoir construction, operation and maintenance*, the 14th Conference of the British Dam Society, University of Durham. Thomas Telford.

Wright, C. (1994). UK reservoir failures and safety legislation in *Dams and Reservoirs*, vol.4, no.3, October, pp20-21.

Ulley Reservoir and high velocity spillway flows

J L HINKS', Halcrow Group Limited
PJ MASON, Montgomery Watson Harza
JR CLAYDON, Independent Consultant

SYNOPSIS. In the small hours of Tuesday 26 June, 2007 Rotherham Metropolitan Borough Council requested the evacuation of areas downstream of Ulley Dam and the closure of the M1 motorway because of fears about the safety of the dam. The danger arose because of the failure of the masonry walls of one of the original spillways at the dam during a flood with a return period of about 200 years.

This paper discusses the background to the incident, the problems associated with high velocity flows in masonry spillways, the handling of the emergency and plans for the restoration of the dam.

INTRODUCTION

Ulley reservoir is located about 5 km to the south-east of the town of Rotherham in Yorkshire. The dam, which is earthfill with a puddle clay core, was completed in 1873 and was originally used for water supply. In 1986 it was sold by the Yorkshire Water Authority to Rotherham Metropolitan Borough Council for £1 plus outstanding debts. Since then it has formed the centrepiece of the Council's Ulley Country Park.

92.4 mm of rain (measured at the Maltby raingauge) fell on 24 and 25 June, 2007. This caused flooding in the Borough, most notably at Catcliffe, and also led to significant spill from Ulley reservoir. At 19.00 on Monday, 25 June the Manager of the Country Park visited the dam and decided that there was a risk of erosion adjacent to the south bywash channel that might endanger the dam. The site was then visited the same evening by officials of the council and by the Supervising Engineer.

At 01.20 on Tuesday 26 June the Chief Executive of the council requested that areas downstream be evacuated and that the M1 motorway be closed in both directions downstream of the dam. The total number of people evacuated was about 1,000 although it is not possible to say exactly how many of these were evacuated as a result of the flooding already being experienced and how many were evacuated as a precaution against the possible failure of Ulley dam.

Responsibility for the enforcement of the Reservoirs Act, 1975 in England and Wales passed to the Environment Agency on 1 October 2004. Since then, their Reservoir Safety Section based in Exeter, have been active in promoting various initiatives aimed at improving reservoir safety. One of these is their Post-Incident reporting system for UK dams which has been developed on their behalf by Halcrow Group Limited. The final report setting up the system was submitted in February, 2007.

Under the relatively new Post-Incident reporting system the Environment Agency requested Jonathan Hinks and Peter Mason to prepare a review report on the Ulley incident with the objective of ensuring that any appropriate lessons were learned by those responsible for reservoir safety in the UK. This report, which was submitted in November 2007, did not cover future actions to be taken to ensure the safety of the reservoir. These matters were addressed in a report under Section 10 of the Reservoirs Act prepared by Mr.J.Claydon (Independent All Reservoirs Panel Engineer engaged by Ove Arup & Partners Ltd. as a sub-consultant).

Figure 1 - Works in progress filling scour hole

BACKGROUND

Ulley dam has a puddle clay core, a crest length of 205 m and a maximum height of 16 m. It impounds a reservoir with a capacity of 580,000 m^3.

The catchment area to the north-east, east and southeast is 11.86 km^2. The reservoir has a surface area of 0.12 km^2.

The dam is located on sandstones of the Middle Coal Measures of the Upper Carboniferous period. Mine workings are known to have taken place in the area in at least eight seams. During the twentieth century mining within the area of influence of the reservoir is known to have taken place as follows:

Barnsley Seam in 1927 at a depth of 434 m
High Hazel Seam in 1953 at a depth of 357m
Thorncliffe Seam in 1964 at a depth of 677 m
Wathwood seam in 1967 at a depth of 300 m

In each of the above seams an area of coal is said to have been left unworked to protect the dam from the worst effects of subsidence, although the design of such pillars may not in all cases have been based on present day understanding of the real effects of mining subsidence.

In addition to the above an unspecified seam was mined in 1935. A proposal to purchase a pillar in this seam was turned down by Rotherham Council who preferred to let the dam settle before building the new spillway in 1943.

There was damage to the scour tunnel and pipe in 1968 which may possibly have been associated with mining in the Wathwood Seam the previous year. However this is not certain since a pillar is believed to have been purchased in the Wathwood Seam.

As originally constructed the dam had narrow spillways down both mitres with masonry retaining walls on either side. These narrow spillways take the form of cascades with a series of plunge pool type stilling basins to dissipate energy. In 1943 a new concrete spillway was constructed at the south end of the dam to discharge, via a chute, to a short stilling basin adjacent to the stream. It is remarkable that resources were allocated to the construction of a new spillway in the middle of the Second World War.

Figure 2 - Flow down 1943 spillway and original south spillway during the incident. Original south spillway is on extreme right of photograph.

The hydraulics at the south end of the dam are complicated with water entering two spillways at different levels, in one case round a sharp right-angled bend.

	2006 survey (m AOD)
Previous TWL (southern bywash)	51.75
Morthern Weir	53.56
1943 weir	52.98
Peak WL in 2007 flood	53.55

It will be seen from the above that the lowest of the three weirs was that at the top of the southern bywash channel and that the 1943 weir was 1.23 m higher. The Morthern (sic) weir was still higher and did not actually discharge any significant amount of water in the June 2007 flood.

Total outflow in the spillways during the incident is thought to have been about 10.1 m^3/sec (6.1 m^3/sec down the old masonry channel and 4.0 m^3/sec down the 1943 spillway). The table below suggests that the flood had a return period of about 200 years.

The following table is based on calculations by Jeremy Benn Associates of Skipton presented in their report of November 2006. Adjustments have, however, been made to the assumed discharge coefficients and the results related to the datum used for the 2006 survey.

Return Period (Years)	Peak Water Level (m AOD)	Peak Outflow (m3/sec)	Wave freeboard (m)
FEH 150 year	53.25	7.00	1.95
FEH 1,000 year	53.61	14.44	1.59
FSR 1,000 year	53.47	11.24	1.73
FSR 10,000 year	53.98	26.32	1.22
FEH PMF Summer	54.95	71.56	0.25
FEH PMF winter	54.69	57.99	0.51

The above figures show that the peak flow in the incident was only about 14 % of the PMF and 38 % of the 10,000 year outflow.

Figure 3 - Scour hole photographed on Tuesday 26 June, 2007

There is a low level outlet with its intake near the right abutment of the dam. The control for this is now at the downstream end of the tunnel. The capacity of the low level outlet is given as 18,000 m^3/day (0.21 m^3/sec) in the Prescribed Form of Record.

LESSONS LEARNED FROM THE INCIDENT

It would appear that the erosion of the downstream embankment shell was initiated by the collapse of the masonry chute running down the left mitre. Mechanisms proposed for this collapse have included:-

- high turbulence flows in the chute 'plucking out' the masonry blocks or pushing them out by water pressure from behind.

- overtopping of the chute walls eroding fill behind the walls as well as, perhaps, causing some backpressure on them.

Figure 4 - Turbulent Flow in Spillway channel

The chute in question was not simply stepped but comprised a series of separate plunge basins formed by low walls across the chute at each step. Not only was the hydraulic action in these basins extremely turbulent, as evidenced by photographs, but there would have been a mix of turbulent high pressure and low pressure zones, the former caused by flow impact from the basin above.

In conjunction with this the wall facing blocks were, in effect, open textured. That is to say although there were remnants of mortar pointing, the facing blocks did not seem well bedded in mortar. Furthermore these outer facing blocks were backed only by rubble mortar with no keying between the two parts. There would have been little to prevent individual blocks being removed by pressures from high turbulent zones feeding as back-pressure behind facing blocks in lower pressure zones.

Had the chute been sited across natural ground any collapse and erosion of adjacent soil would simply have been a matter of rebuilding. The incident was compounded in this case by the chute being sited at the toe of the embankment. It is also clear from photographs taken at the time that debris from the initial collapse caused enhanced turbulence on the chute and deflected flows in such a way as to further enhance embankment erosion.

Figure 5 - Local hydrodynamic stone removal. Note also general condition of wall.

Calculations carried out since the incident suggest that 'skimming flow' would have set in and that the peak flow velocity probably exceeded 8 m/sec. The high velocities and turbulent nature of the flow in the channel would appear to have been quite sufficient to generate internal pressures within the masonry sufficient for stones to be extracted. Near the upper end of the chute a number of carefully shaped rectangular stones were seen to have been plucked out of the wall. This was probably by high water pressure behind the stones although suction from in front may also have played a part.

The hydrodynamics of flows in stepped spillways are further discussed in the paper by Mason and Hinks in the March 2008 issue of Dams and Reservoirs

EMERGENCY RESPONSE

The response to the emergency at Ulley was impressive and illustrates a number of key lessons for reservoirs in general. Once the crisis was identified the Rotherham Metropolitan Borough Council (RMBC) Emergency Planning Procedures were invoked. The Council already had such procedures in place following the requirements of the 2004 Civil Contingencies Act and the duty of County Councils to establish Regional Resilience Forums.

Under the emergency procedures already existing, a Bronze (Operational) Command centre was established at Ulley reservoir. This comprised appropriate on-site expertise able to identify immediate needs and resource requirements, such as pumps, plant and ballast. These requirements were passed upwards to Silver (Tactical) Command.

Silver Command was sited at a local district police centre with the purpose of resourcing any requests from Bronze and arranging, for example, required plant and material to be taken to the affected site.

Gold (Strategic) Command sits above this and operates at county level, monitoring crises in different areas, adjudicating the relative seriousness of conflicting requirements and deciding strategically on orders of precedence in terms of response. For a number of reasons, the resources needed to mitigate the potential failure of Ulley reservoir were given precedence by Gold Command over other flooding elsewhere in Rotherham.

Figure 6 - Repaired scour hole after incident

RESERVOIR DRAW-DOWN CAPABILITY

It is not normally practicable retrospectively to provide permanent low-level outlet facilities capable of drawing down reservoir levels in the event of major flood events. In fact the draw-down facilities provided at Ulley had served to lower reservoir levels on previous occasions sufficiently to allow maintenance work to take place. Nevertheless, their sizing would appear to be rather low. Their capacity is quoted in the Prescribed Form of Record as being 18,000 m³ per day. Given the area of the reservoir this constitutes a draw-down rate at maximum reservoir level in the order of 150 mm per day, which is less than is desirable for a dam where human life is at risk (a figure of the order of 500 mm/day being desirable in our opinion).

The 2007 report under Section 10 of the Reservoirs Act recommended replacement of the scour pipe with a larger diameter pipe discharging to the channel downstream of the dam.

ACTIONS TAKEN SINCE INCIDENT

Within a very short time of the incident being declared the erosion hole in the downstream shoulder of the dam was filled with stone and the old south spillway was closed off temporarily with a loaded skip. Subsequently it was permanently closed off with a concrete wall. Water level in the reservoir was lowered using temporary pumps.

The Section 10 Inspection which had been called for by the Supervising Engineer made several recommendations in the Interests of Safety. These included, *inter alia*:

- Changing the Dam from Category C to Category A, a review of the flood hydrology and construction of spillway capacity to pass the PMF.
- A site investigation, which has shown that the core raising carried out in 1967 using plastic concrete was brittle and crumbling.
- Filling in the Morthern bywash channel with stone to support the toe of the embankment.
- Increasing the size of the draw-off pipework in the tunnel

Discontinuance of the reservoir was considered and rejected by the Undertaker and at the time of writing options for repair are being considered.

POSSIBLE RESEARCH ISSUES

The incident at Boltby in June 2005 and the Ulley incident both point to a need for research into the hydraulic conditions that can be withstood by masonry walls of various types. Superficially the masonry walls beside the southern bywash spillway at Ulley appeared to be in reasonable condition. However, events showed that they were unable to withstand the hydraulic conditions to which they were exposed.

Figure 7 - Damaged spillway at Boltby in 2005

Peak flow in the southern bywash spillway at Ulley has been estimated at 6.1 m^3/sec. The width of channel is given as 1.83 m so the discharge was 3.33 m^3/sec/m width. Obviously this was too high although the form of the spillway, comprising a series of stepped plunge pools, probably exacerbated the situation.

With UK dams averaging about 110 years in age there are very many masonry channel walls at UK dams which play a vital role in ensuring the stability of the structures. It is thought that the deployment of resources to study this problem, and provide advice to Inspecting Engineers, would be well justified.

CONCLUSIONS

The incident was most unfortunate and many people were severely inconvenienced including those evacuated from their homes and those delayed as a result of the 40 hour closure of the M1 motorway. However the situation was very well controlled by the prompt and efficient action of Rotherham MBC, the emergency services and J.N.Bentley Ltd. who were a framework contractor to Yorkshire Water. The threat to the security of the dam was posed by the collapse of the masonry spillway walls. This, in turn, was caused by high velocity 'skimming' flow down the old spillway whose sill was set 1.23 m lower than the sill of the more modern concrete spillway built in 1943.

Similar failures are known to have occurred at not less than nine other masonry spillways including Toddbrook in 1985 and Boltby in 2005. None of these have led to the failure of a dam but they highlight the risk particularly where the spillways run close to the toe of the dam. The authors of this paper believe that Inspecting Engineers should be made aware of the risks and that research should be promoted into possible failure mechanisms. There were two high pressure (37 bar) gas mains of 1.22 m diameter just downstream of Ulley Dam of which the last Inspecting Engineer was not aware. Such infrastructure should be identified and taken into account when assigning a category to a reservoir.

REFERENCES

CIRIA Report 33, *The Hydraulic Design of Stepped Spillways.* 1978.

CIRIA Technical Note 134, *Guide to Analysis of Open-Channel Spillway Flow,* 1989.

Hinks J.L. and Mason P.J. *Ulley Dam: Post-Incident Review.* Report prepared by Halcrow Group Ltd. for The Environment Agency, November 2007.

Mason P.J. and Hinks J.L. *The Security of Stepped Masonry Spillways – Lessons from Ulley Dam,* Dams and Reservoirs, March 2008.

Sanchez-Juny M, Blade E and Dolz J. *Pressures on a Stepped Spillway,* Journal of Hydraulic Research, Vol 45, No 4, pp 505 – 511, 2007.

Thorwarth J., *Self-Induced Unsteady Flow on Pooled Stepped Chutes,* 14th German Dam Symposium, Freising, September 2007.

ACKNOWLEDGEMENTS

The authors of this paper gratefully acknowledge the support and assistance given by the officers of Rotherham MBC, The Environment Agency, Arup Consulting Engineers and Mr. Nick Reilly, Independent Consulting Engineer. A number of the photographs were kindly supplied by Mr. Mark Chapman of the Environment Agency.

Supervising Engineer's Account and Reflections on a Major Reservoir Safety Incident

D. M. CROOK, Arup, Leeds, UK
P. KELHAM, Arup, Leeds, UK
D PHILLIPS, Rotherham Metropolitan Borough Council, UK
I MACHIN, Formerly Rotherham Metropolitan Borough Council, UK

SYNOPSIS. This paper details the first response to an emergency incident that took place at Ulley Reservoir in June 2007, and provides a first hand account of the events that took place during the initial twelve hours and looks at the actions taken by the Undertaker, Rotherham Metropolitan Borough Council (RMBC), and the Supervising Engineer during that period. The paper also provides a commentary on the actions taken by the AR Panel Engineer, the emergency services and Undertaker's and contractor's staff in carrying out further emergency works to stabilise the dam during the following 48 hours. It provides a detailed account of an emergency plan being put into action and also how the emergency plan can be flexible enough to allow improvisation to deal with particular problems as they are identified. The paper concludes by providing a summary of the main lessons learnt during the emergency response and comments on general housekeeping matters such as the availability of reservoir records, communications and welfare facilities.

The problem at Ulley was the disintegration of one of the spillways that led to the rapid erosion of the toe and downstream shoulder of the embankment. The erosion process was out of control and a solution had to be found to reduce the flow along the damaged spillway. In the absence of penstocks, stop logs or other similar equipment that could be used to control the flow of water into the spillway, an improvised plug was installed, which comprised of an 8 tonne skip packed with large sandbags wedged into the upstream end of the spillway. The Undertaker's emergency plan allowed for plant and materials to be obtained at any time of the day. Emergency pumping to lower the reservoir water level was initially provided by the fire service, but these were later supplemented by the addition of hired in pumps. By the time the AR Panel Engineer arrived, the situation, though still critical was under control.

INTRODUCTION

This is a Supervising Engineer's account, supplemented by information from RMBC's Forward Liaison Officer (FLO), about the issues raised and the lessons learnt in attending a serious reservoir safety incident. First hand experience provides a good example of the application of emergency planning to avert a potential disaster and gives a valuable insight to those professionals who may be involved in the first response to a reservoir safety incident in the future and help them to be ready.

DESCRIPTION OF THE RESERVOIR

The principal data for the reservoir is given in Table 1 below.

Table 1 – Principal Reservoir Data

Capacity	580 000 m^3
Top water level	51.75 maOD
Surface area	0.12 km^2
Catchment area	11.87 km^2
Embankment height above natural ground	16 m
Embankment length	205 m
Upstream slope angle	1 in 3
Downstream slope angle	1 in 2 for 9 m with berm
Crest level	55.5 maOD

DESCRIPTION OF THE OVERFLOW ARRANGEMENT

The overflow arrangement is not straightforward. Each spillway at the reservoir has its own historical name and these are used in the description of the incident. The spillway names, locations and details are shown below in Figure 1 and Table 2.

Prior to the incident, the reservoir top water level was controlled by the Ulley spillway. When the reservoir level rose by 1.23m, then the 1943 spillway would come into operation. A further level rise of 0.58m would be required to bring the Morthen spillway into operation.

Table 2 – Weir Details

Spillway Name	Weir Length	Weir Level
Ulley	1.8 m	51.75 maOD
1943	12.2 m	52.98 maOD
Morthen	6.1 m	53.56 maOD

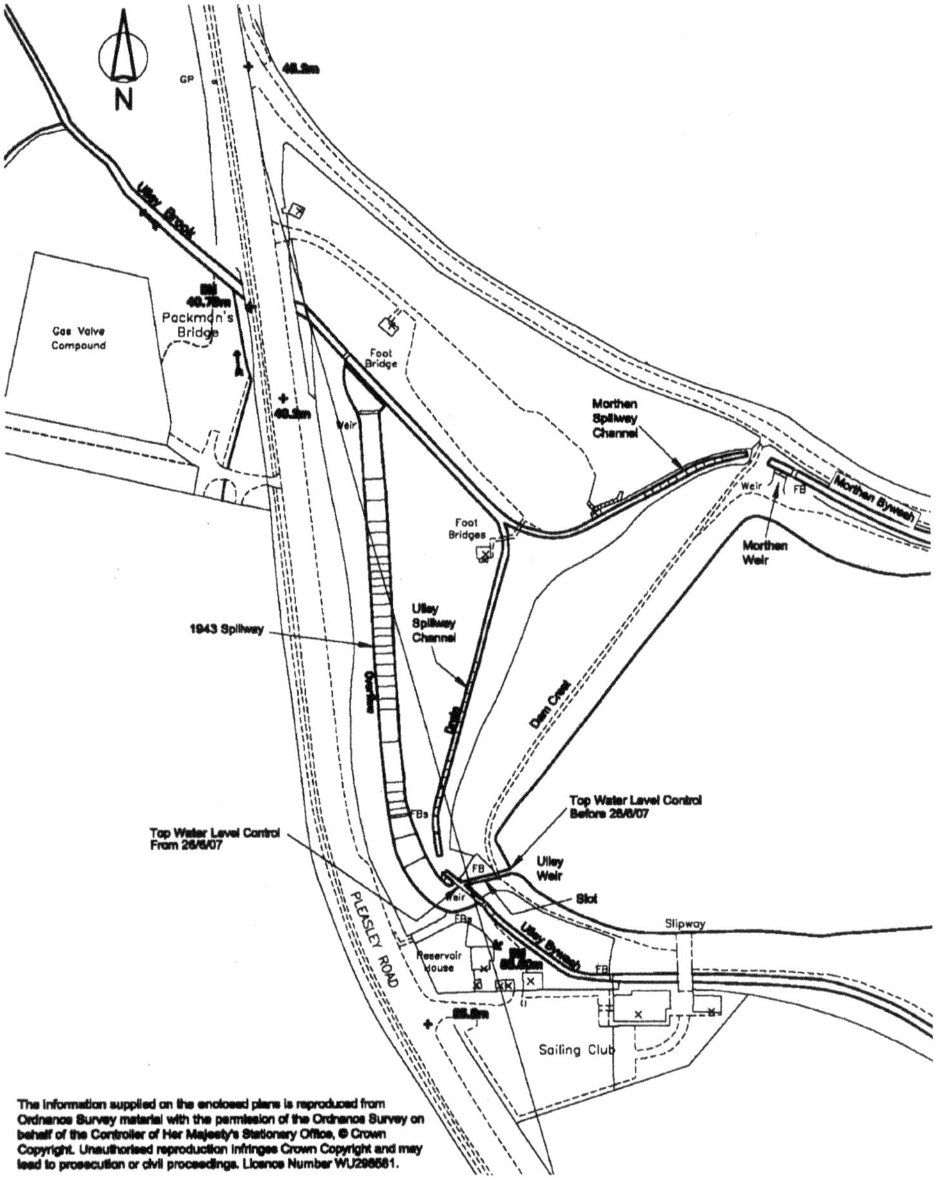

Figure 1 – Overflow Layout

DESCRIPTION OF THE ULLEY INCIDENT

The incident followed a day of heavy rain in the area around and upstream of Rotherham. Rangers based at Ulley were asked to check the reservoir periodically during the afternoon and early evening of 25th June. At about 8pm, one of the rangers reported damage occurring to the embankment. The Country Park Manager, who was also one of the FLO's for RMBC's emergency planning unit, went to site. In consultation with the local police, a major incident was declared. This allowed the emergency management systems to be established. This included establishing a bronze command at the reservoir and closing the perimeter roads around the reservoir including the main A618.

An engineer from RMBC arrived on site at 10pm and recognised that the situation was serious. The Supervising Engineer for the reservoir under the 1975 Reservoirs Act was notified of the problem and went immediately to site. He arrived on site at about midnight and went onto the embankment to look at the damage. Speaking to the RMBC staff on site, it was clear that the problem was getting worse but there was insufficient light to make a clear assessment of the problem. He was asked by the police if the dam would survive until the morning. Based on the damage that was visible and the rate of erosion, he was not sure that the dam would survive until dawn. This answer was passed to Gold Command who decided to evacuate residents from areas believed to be at risk and to close the M1. It is worth noting that at this early stage, decisions about which premises should be evacuated were based around the brief study of maps at both Bronze and Silver control as no evacuation plans for premises downstream of Ulley existed.

The immediate priority was to try and get a clearer idea of the amount of deterioration of the dam. RMBC were asked to obtain lights to allow a clearer view. Fairly rapidly, the Supervising Engineer realised the problem was surface erosion caused by the water flow down the Ulley spillway and a solution was to try to stop that flow. There was no mechanical means of closing the upstream end of the channel. The idea of sand bags was dismissed because they were two small and would be washed away. The next idea was to use intermediate bulk containers (IBC) filled with gravel. It was realised that to place the bags a crane would be required. RMBC, as part of the emergency plan, have arrangements to collect materials during out of hours for emergency purposes.15 IBC's were obtained. A crane was obtained but due to the flooding in the middle of Rotherham took some time to arrive. The first attempt to place an IBC was unsuccessful; the flow of water caused it to virtually disintegrate. It was then realised that something more substantial was needed to block the channel. The answer was an 8T skip. RMBC obtained two of these as a resilience measure in case one failed, however only one was required. Once the skip was lowered into the channel, the flow was considerably reduced. The IBC's were then lowered into and packed around the skip. With this reduced flow, it became possible to look more closely at the damage to the embankment. At that time, the water was flowing over exposed rock foundation and no longer eroding the embankment material.

In parallel to the blocking exercise, it was realised that there was a need to lower the reservoir since water could no longer be allowed along the damaged spillway. RMBC requested fire service support at about 2:30 am. The first pumps arrived at about 6 am and there was a steady stream of arrivals thereafter. The first pumps became operational at about 6:30 am.

The following are photographs taken during the incident between 2:45 pm on 25th June and 11:17 am on 26th June 2007.

PHOTOGRAPH 1
Taken at 14:55 on 25th June

This is the last photograph of the Ulley spillway before its failure. As can be seen, there is considerable turbulence in the flow and it appears to be overtopping slightly at the bottom of the steep slope.

PHOTOGRAPH 2
Taken at 20:26 on 25th June

The water level has continued to rise. It is now above the lower rail on the footbridge.

PHOTOGRAPH 3
Taken at 20:26 on 25th June

The flood is now passing down the 1943 spillway.

PHOTOGRAPH 4
Taken at 07:45 on 26th June

This shows the flow down the Ulley spillway and the water eroding the embankment.

PHOTOGRAPH 5
Taken at 08:32 on 26th June

This shows the skip being placed to act as an improvised plug. The flow into the Ulley spillway can be seen below the skip.

PHOTOGRAPH 6
Taken at 10:32 on 26th June

This shows how the plug had reduced the flow considerably so it was no longer eroding the embankment.

PHOTOGRAPH 7
Taken at 11:17 on 26th June

This shows the damage to the embankment and the spillway.

SPECIFIC ISSUES RELATING TO THE ROLE OF THE SUPERVISING ENGINEER

In being presented with an incident at a reservoir, there are a number of matters to consider. These are as follows: -

- Assess the situation. On arrival at site, the first question to be asked is 'is it safe?' It is only possible to start to answer such a question by looking at the damage and trying to make an assessment of the rate of deterioration. Poor light made this task very difficult.
- Decide if AR Panel Engineer required. As soon as the report was received, it was clear that something serious was taking place and the first reaction, before leaving home, was to try and contact an AR Panel Engineer.
- Identify possible mitigation measures. There is a room full of people who are worried about their dam. Can the cause of deterioration be identified? Can anything be done to reduce the rate of deterioration? In this case, it was clear that there was an alternative route for the water. This led to the suggestion to simultaneously block the channel and start emergency pumping operations.
- Implement. Once a decision to do something was made and a means of achieving it identified, then the actual logistics were passed via the on site FLO to the emergency control room. On site, it was a case of waiting to be advised that the materials/equipment were on their way and then for their actual arrival.
- Response to questions: -
    - Third party. There is not just the Undertaker asking questions but also the police and other bodies. At Ulley this included Transco who have a high pressure main close to the toe of the embankment. A lot of pressure was applied by the Transco representative, quite rightly, to determine the risk to their infrastructure to protect regional gas supplies. Being prepared to say 'I do not know' required some will power.
    - Own thoughts. As mentioned above, there are a lot of worried people looking for guidance from a reservoir engineer. Seeking clarification about the actual incident provides some thinking time. There is a possibility of saying something rashly that in the cold light of day does not make sense. The most important thing is not to panic.
- Maintain communication. There are a number of people who want information. The information has to be accurate. In the early hours of the morning, it can be difficult to think clearly and it is of great benefit to bounce ideas off another reservoir engineer.
- Keep a record of events using notes and photographs. There are bound to be questions after the event to try and determine the approximate return period of the flood and similar. Events are moving fairly quickly

that it is difficult to take measurements and remember to write full notes.

IMPLEMENTATION OF THE EMERGENCY PLAN

Under the Civil Contingencies Act 2004, a framework was established to allow multi-agency emergency plans to be prepared. The Borough Emergency Plan is a framework within which the emergency, or emergencies, can be managed. The framework allows risks to be established, incidents to be prioritised and resources to be allocated and delivered to sites.

- Gold/Silver/Bronze Command. The three layers have different roles. Gold is responsible for strategic decision making over a wide area. Bronze is generally established at the site of the actual incident and is responsible for operational response at the incident site. There may be a number of incidents each with their own Bronze commander. Silver acts as a buffer between Gold and the various Bronze sites. Silver is responsible tactical decision making and arranges for resources such as required plant and materials to available and be taken to affected sites.
- Local Authority duty Forward Liaison Officer (FLO). The FLO provides a single point of contact through which requests for information, materials and services from the council or its contractors can be made.
- RMBC emergency control room. The control room controls and co-ordinates the response to the incident(s) by such means as responding to requests made by the FLO and will arrange actual delivery of such. The control room keeps a log of information and events so that after the incident it should be possible to recreate a timeline as part of the feedback on the incident, and to provide detailed information for any subsequent enquiries.

LESSONS LEARNED

The incident showed a number of matters that apply not just to this incident but could be of benefit to other Supervising Engineers facing an emergency. The principal lessons are: -

Supervising Engineer's equipment/documents/knowledge

When the Supervising Engineer is advised of an incident, this will be accompanied by a request to go immediately to site. Rather than trying to improvise in what is quite a stressful situation, it is better to have something prepared in advance. There should be a checklist of equipment and documents stored in a known location. The following is a suggestion: -

- Supervising Engineer's own file on the reservoir including the last Inspecting Engineer's report, selected drawings, the last annual statement and Section 11 records.

- Other documents include a contact list of All Reservoir panel engineers which has not just mobile telephone numbers but home telephone numbers as well. A possible alternative is for the Environment Agency to keep such a list in their control room for emergencies.
- The event needs to be recorded to capture lessons that could apply to other reservoirs. The type of information is notes and measurements but it is difficult to do this in the dark. Also, people are asking for information so thinking time is at a premium. Digital cameras are useful especially if high resolution. Even at night, if the flash does not appear to work, data can still be collected which can be used for later back analysis.
- Communications are vital. A mobile telephone is essential. However, during a regional emergency, the mobile telephone networks can be overwhelmed and crash (see emergency management below).
- Is transport available? There was a problem that the car was low on petrol and finding fuel late at night can be a slow process. As a matter of course, should the fuel tank be kept half full to allow enough fuel to get to the reservoir? It is worth noting that in RMBC, there is a policy that designated emergency vehicles are never allowed to have less than half a tank of fuel. The issue that this raises is the comfortable distance that a Supervising Engineer would have to drive to reach an incident. Is an hour a sensible maximum or would longer be acceptable? The decision has to lie with the individual Supervising Engineers. How many reservoirs are the responsibility of an individual and how would multiple incidents be prioritised?
- Knowledge of the reservoir and its behaviour is important. There is also knowledge of land use downstream of the dam and what emergency procedures that can be activated. As part of the normal duties, scenarios should be considered for each reservoir being supervised by each Supervising Engineer. Under the Water Act 2003, Supervising Engineers need to be aware of the emergency plans and to assist the undertakers in their preparation. Should this role be more pro-active and consider the unthinkable?
- Thought should be given by Supervising Engineers likely to be called to such incidents of the need for a 'grab bag' containing PPE, torches, water, food/snacks, phone, charger, flask and other items that may make life easier at an incident site. RMBC have such a system for their FLO's.

<u>Emergency Management</u>
There is a requirement for emergency planning under the Civil Contingencies Act 2004 augmented by elements from the Water Act 2003. By their nature, emergency plans cannot cover every eventuality and

flexibility must be built in. In formulating an emergency plan, the following should be considered: -

- Access – theoretically there may be very good access to a site. During an emergency, this access might no longer exist. At Ulley, the M1 was shut as well as numerous roads in the Rotherham area. This is difficult to predict though there are bound to be situations where a road closure is more probable. Do alternative routes exist even if they are 'off-road'. Consideration should be given to requesting the police to provide an escort. At this incident, the police provided 'blue light' escorts for specialist staff and/or resources to be delivered to site.
- Welfare – a large number of people arrive on site. They need food, drinks, somewhere to dry out/ warm up and toilets. Ulley was lucky because there is a visitor's centre which provides the basics. How would a more remote site function? This could be provided by the local authority or at least sourced by them if planned. RMBC hold about 1000 ready meals for use in an emergency.
- Communications – as mentioned previously, the mobile telephone network was occasionally overwhelmed and stopped working. RMBC, as part of their emergency plan, are part of the ACCOLC (ACCess OverLoad Control) for mobile telephones whereby the telephone system is restricted to those people who have a recognised essential requirement to use the telephone. Hence communications between site and the emergency control room are maintained. The site also has good mobile telephone network coverage.
- Working hours – the incident does not conform to a normal working day. Long hours are inevitable. Are staff fit to drive home at the end of a stint or should they be booked into local hotels? It is very difficult to get rest on a busy site.
- Multiple incidents – on the night of 25th June, there were a number of incidents in Sheffield and Rotherham related to the flooding. There were locations where lives were more at risk to events not related to Ulley and resources were diverted to these incidents. Before dawn on 26th June, it would be reasonable to state that Ulley was not the top priority but second or third. A thought is what could happen in a reservoir cascade and which reservoir would be allocated the resources. The allocation will be made off-site at Gold Command.
- It is essential that the local authority emergency planning unit are involved in preparing reservoir emergency plans and discussing the need for resources to respond to an emergency. The integration of reservoir emergency plans within the local authority plans and the full support of the emergency planning officers is integral to the success of dealing with an emergency.

Reservoir Management

The undertaker can take steps to improve emergency responses if the normal management of the reservoir is established in the right manner, for example:

- Monitoring – the reservoir is well monitored. There are rangers based at the reservoir whose main role is the running of the country park but who can also provide a rapid 'eyes-and-ears' service if there is a possibility of an incident. This level of surveillance is not universal and there would be no certainty that other undertakers will identify the early initiation of an incident.
- Records – the Undertaker was able to bring drawings to the reservoir. However, there is the issue of keeping records secure and where to have the secure place. The obvious choice for Rotherham is at the main council offices. However, these were cut off when the river Don rose. Even if the records can be taken to site, there is no guarantee that they will be safe.
- Management of supervision process – the effectiveness of the Supervising Engineer depends on a long association with a reservoir and an understanding of how it works and any idiosyncrasies. There is an argument against frequent changes in Supervising Engineers and a preference for longer contracts of say three to five years. There is also a duty on Supervising Engineers that if they know they are looking to retire that they do not take on new reservoirs. Obviously there will be instances where changes are necessary but they should not be on purely commercial reasons. The current appointment as Supervising Engineer for the RMBC reservoirs commenced in January 2007.
- Alternative Supervising Engineer's – Arup have a network and provide a list to clients so if the named Supervising Engineer is unavailable there are alternatives. Undertakers need to take this into consideration when making appointments.

Matters that might have been dealt with differently during the incident.

Hindsight is a wonderful thing. There were aspects that when thought about later might have been dealt with in a different manner.

- Record of event. Under the Supervising Engineer's heading above, was a reminder to record things.
- Location of pipework on the spillway. The pipelines from the fire service pumps were laid over the 1943 weir and on the spillway. The effect of doing this is to obstruct the spillway so if there was a further flood then there could be problems letting the water get away from the reservoir.
- Benefits of dambreak plan. Although the reservoir was classed as a Category C reservoir, it meant that a dambreak analysis would not necessarily need to be carried out under current proposals for Reservoir

Flood Plans as part of the Water Act 2003. When faced with the possibility of failure, it is very difficult to formulate a considered strategy. In fact the approach used was to look at the contour from the toe of the embankment and any properties, even if not in the direct line of the flow from a breach, were designated for evacuation. If a dambreak had been previously done, it is possible that fewer people would have needed to be evacuated.

- The effects of mining subsidence probably need to be looked at more carefully. There was a situation where the spillway at the bottom of the mitre was more or less level.
- When asking for resources, the importance of prompt delivery and an explanation of the urgency of the particular request are very important otherwise delays may occur. An example was the request for the skips which could have been given a low priority unless the FLO had explained the crucial role in which they were to be employed.

<u>General reservoir safety</u>

This is more the territory of the Inspecting Engineer though the Supervising Engineer has a role though a longer involvement with a reservoir. There will be issues that cause more concern and they might not be obvious.

- Consideration of spillway capacity not just spillweir capacity. Traditionally the approach has been to ensure the water can be removed safely over the weir and possibly miss the point that significant flows cannot be safely discharged from the embankment.
- Integrity of masonry spillway. The masonry appeared to be in good condition comprising ashlar masonry. In fact the units were tapered to give the appearance of ashlar and were only in contact at the exposed edges. Water was able to get behind the exposed face and there are a number of holes where individual stones have been pulled/pushed out of the wall. There is an implication for Supervising Engineers where the reservoir has another eight or so years to the next inspection whether intervention by an AR Panel Engineer is needed sooner or even a full Section 10 Inspection.
- Safe access onto an embankment during an event is important. The main access from the visitor's centre was across a flooded footbridge upstream of the weir or across the 1943 spillway.
- Masonry lined channels. Where a channel is lined using masonry walls, what is the integrity of those walls if the water gets out of the channel? Will the walls collapse leading to further erosion or will they stay upright and limit any further erosion? The walls need to be able to act monolithically which means the different skins need to be tied together. Any voids need to be filled to stop water entering the masonry and pressurising the voids.

- Where there is an abrupt change in gradient from steep to slack, the spillway walls need to be high enough to contain the hydraulic jump and direct the water flow away from the embankment toe.
- The last two items in 'things done differently' also apply here as well.

IMMEDIATE POST INCIDENT ACTIVITIES

The Supervising Engineer's role on the site did not finish with the arrival on site of an AR Panel Engineer. The role changed to a support role for the AR Panel Engineer and involved monitoring of conditions around the embankment and reporting changes to the AR Panel Engineer. The AR Panel Engineer had on his way to the site started to make arrangements for the attendance of an experienced contractor to carry out the emergency stabilisation works. This work was carried out very efficiently. The Supervising Engineer's site involvement was full time for two weeks after the incident and then reduced to weekly visits as confidence in the stability of the embankment improved.

Given the seriousness of the incident, the Supervising Engineer recommended that a Section 10 Inspection under Section 11(3) of the 1975 Reservoirs Act. Jim Claydon was appointed by RMBC to undertake the inspection and the Supervising Engineer provided assistance to the Inspecting Engineer in collecting site information and inspecting the tunnels and shafts. This led to a small number of tasks that could be overseen by the Supervising Engineer in liaison with the Inspecting Engineer.

Finally, the Supervising Engineer prepared a Post Incident Report which was submitted to the Environment Agency.

CONCLUSION

The incident could have been much worse. The problem was identified by RMBC at an early stage which allowed the appropriate personnel to be assembled on site. The emergency was managed without a major collapse, there were no fatalities and no accidents. This is very much due to the efforts put in by all involved and the emergency procedures put in place by RMBC and the multi agency approach taken. Once launched along the path of dealing with the emergency, there is very little that would be changed.

It is also important that when preparing emergency plans that advice must be obtained from the relevant local authority emergency planning team.

REFERENCES

Hinks J.L. Mason P.J. and Claydon J. (2008). *Ulley Reservoir and high velocity spillway flows,* in Ensuring Reservoir Safety, Proceedings of the BDS Warwick Conference, Thomas Telford, London.

The Safety of Small British Reservoirs

C A GOFF, Halcrow Group Ltd
A L WARREN, Halcrow Group Ltd

SYNOPSIS. This paper discusses the risk posed by small, non-statutory reservoirs in Great Britain and the benefits that arise from carrying out simple risk assessments of such reservoirs. By drawing on two major studies carried out on behalf of the Environment Agency involving visits to, and assessment of, approximately 400 non-statutory reservoirs in England and Wales, the paper aims to describe and discuss:

- Appropriate methods of qualitative risk assessment for large portfolios of reservoirs;
- The risk posed to life and property by non-statutory reservoirs relative to statutory reservoirs;
- The use of reservoir condition assessments to improve flood risk management; and
- The implications for possible amendments to reservoir safety legislation in Great Britain.

INTRODUCTION

Since the introduction of reservoir safety legislation in 1930, no dam failures causing loss of life have occurred in Great Britain. The introduction of the Reservoirs Act 1975 in the 1980's introduced the role of the Supervising Engineer to provide continuous supervision of reservoir safety matters. This has been beneficial in identifying and resolving reservoir safety issues as well as providing a means of promoting safety legislation compliance by reservoir owners or users. However, two recent studies carried out by Halcrow Group for the Environment Agency suggest that there remains much scope for improvement in the current legislation with respect to small reservoirs that endanger lives or have the potential to cause significant damage. The purpose of this paper is to describe the scope and findings of these studies and to discuss the results in light of the current reservoir safety legislation.

SOUTH WESSEX FLOOD RISK ASSESSMENTS OF RESERVOIRS

Project Aim

This study was carried out on 104 reservoirs in the South Wessex Area (mostly in Dorset and South Wiltshire) between 2005 and 2007. The breakdown of these reservoirs by type is shown below.

- 85 non-statutory reservoirs (less than 25,000m^3); 19 statutory reservoirs (more than 25,000m^3)
- 97 impounding reservoirs, 5 non-impounding reservoirs, 2 service reservoirs.

The main aim of the study was to evaluate and rank the reservoirs in terms of likelihood of failure. This information was then used to inform the Environment Agency's emergency and contingency planning provisions. In a small number of very poor condition non-statutory reservoirs, the Environment Agency wrote to the dam owners to inform them of the concerns. However the driver for the project was the need to understand the flood risk posed by reservoirs in order to inform flood risk management. The Environment Agency is a Category 1 responder under the Civil Contingencies Act 2004 and lead on flood risk issues for Local Resilience Forums (LRF's). The Act places a general duty on the Environment Agency to maintain arrangements to warn the public and provide advice and information about flooding. The Environment Agency's experience from dealing with the failure of a non-statutory reservoir in the New Forest in 2002 was a further driver for the project. Under the Water Act 2003, the Environment Agency became the enforcement authority for the Reservoirs Act 1975. However, to fulfill its wider functions and responsibilities, knowledge of the level of risk posed by all reservoirs, whether statutory or not, was deemed necessary to:

- Reduce flood risk where possible;
- Make informed decisions based on good data;
- Offer informed advice to others such as utilities, planners, local authorities, LRF's and the public.

It was recognized that there would be additional potential benefits, such as:

- Ensuring that any review of the current reservoir safety legislation is informed by reservoir flood risk studies;
- By identifying non-statutory reservoirs owners and making contact with them to discuss reservoir safety assessments, the owners are more likely to take interest and responsibility for the condition of their structures

- Where non-statutory reservoirs are believed to fall within the ambit of the Act, bathymetric surveys would be carried out and the national reservoir safety team notified where appropriate.

Methodology

An appropriate methodology was developed to assess the risk of dam failure and to enable the regional population of reservoirs to be ranked. This methodology was broadly based on a method previously used internationally for portfolio risk assessments (Hinks and Dedja, 2002). This method is a qualitative assessment whereby it is not necessary to estimate the annual probability of failure. Hydrological risk was assessed by estimating the ratio of desirable freeboard to actual freeboard. Geotechnical risk was assessed against a wide range of indicators covering both the intrinsic design condition and the current condition. Hydrological and geotechnical risk indices were then combined to allow the reservoirs to be ranked.

Results

Of the 104 reservoirs assessed, only 19 were statutory reservoirs and 95% were impounding reservoirs. The distribution of the 19 statutory reservoirs within the overall qualitative ranking by risk of failure is shown in table 1.

Table 1. South Wessex study: distribution of statutory reservoirs by rank

Reservoir Ranking (1 = highest risk of failure)	Number of Non-Statutory Reservoirs by band	Number of Statutory Reservoirs by band
1 – 15	15	0
16 - 30	14	1
31 – 45	14	1
46 – 60	14	1
61 – 75	13	2
76 – 90	9	6
91 – 104	6	8
Total	**85**	**19**

Table 1 demonstrates that in the South Wessex area there is a very clear indication that statutory reservoirs have a lower risk of failure than non-statutory reservoirs.

The highest risk rated reservoirs were commonly found to have inadequate spillway and freeboard provisions and some showed indications of having been overtopped. Other common problems were:

- well developed seepage problems;
- dense vegetation preventing any effective surveillance and poor tree management;
- uneven dam crest levels;
- wave erosion to upstream face;
- poor levels of maintenance to structures;
- lack of monitoring;
- lack of routine surveillance.

This might not be of great concern if non-statutory reservoirs did not pose a risk to life. Unfortunately, many of them do. An analysis of the 85 non-statutory reservoirs by dam category is provided in table 2 below.

Table 2. South Wessex non-statutory reservoirs by dam category (ICE, 1996)

Dam category	Percentage
A	11
B	24
C	35
D	31

INVESTIGATION OF POTENTIAL RESERVOIRS PROJECT

Project Aim

The 'Potentials' project commenced in 2006 and is being carried out by Halcrow Group Ltd on behalf of the Environment Agency. The aim of the project is to identify and assess small reservoirs in England and Wales that are currently not on the National Register of Large Raised Reservoirs but may fall within the ambit of the Reservoirs Act 1975. To date the project has assessed 319 reservoirs of which 141 are now considered not to fall within the ambit of the Act (i.e. hold less than 25,000m^3). The project is described in detail in this journal by Goff and Hope (2008). The project is of relevance to this paper as a large number of small reservoirs have been visited and assessed according not only to their capacity but also their condition and the apparent downstream hazard. Although detailed condition assessments, risk assessments and dambreak assessments were beyond the scope of the project, the condition score and flood category assigned to the reservoirs is of relevance when considering the question of the safety of small British reservoirs.

Methodology

A pilot study at the commencement of this project confirmed that site visits were essential to provide reasonable confidence in volume estimates: GIS techniques were insufficient in themselves to meet the aim of the project. Through visiting each of the 'candidate' reservoirs, it was then also possible to assign a condition score to each dam. Scores were assigned according to the conditions described in Table 3.

Table 3. Guidance on dam condition score.

1	Excellent condition, no significant cause for concern.
2	Good condition, some minor maintenance issues.
3	Some cause for concern. Problems evident that may lead to a dam failure if not addressed.
4	Problems causing immediate concern for safety. Urgent works required.
5	Poor condition. Imminent dam failure possible.

Results

Summary tables of results from an analysis of the potential reservoirs are provided below. The results are given for all of the reservoirs assessed rather than just those with a volume less than 25,000m^3 as this provides better insight to the condition of dams not benefiting from current reservoir safety legislation.

Table 4 - Results by condition score

Condition Score	No. Reservoirs	Proportion of total
1	40	13%
2	117	37%
3	80	25%
4	17	5%
5	6	2%
-*	59	18%
Total	319	100%

Table 5 - Results by dam category

	Number of Reservoirs				
Estimated Volume (,000m^3)	**Dam Category A**	**Dam Category B**	**Dam Category C**	**Dam Category D**	**Total**
0	n/a	n/a	n/a	n/a	57
1-10	0	5	9	4	18
10-20	8	13	23	12	56
20-25	11	15	33	8	67
25-30	2	6	19	0	27
30-40	1	10	13	6	30
40-50	3	6	9	3	21
50+	10	13	14	6	43
Total	**35**	**68**	**120**	**39**	**319**

Table 6a - Condition and dam category by number (non-statutory)

Dam Category	No. of non-statutory reservoirs by dam condition score						Total
	1	2	3	4	5	-	
A	3	8	4	2	2	0	19
B	3	15	12	1	2	0	33
C	10	29	20	3	1	2	65
D	2	10	9	3	0	0	24
-*	0	0	0	0	0	0	0
Total	18	62	45	9	5	2	141

Table 6b - Condition and dam category by percentage (non-statutory)

Dam Category	% of non-statutory reservoirs by dam condition score						Total
	1	2	3	4	5	-	
A	2.1%	5.7%	2.8%	1.4%	1.4%	0.0%	13.5%
B	2.1%	10.6%	8.5%	0.7%	1.4%	0.0%	23.4%
C	7.1%	20.6%	14.2%	2.1%	0.7%	1.4%	46.1%
D	1.4%	7.1%	6.4%	2.1%	0.0%	0.0%	17.0%
-*	0.0%	0.0%	0.0%	0.0%	0.0%	0.0%	0.0%
Total	12.8%	44.0%	31.9%	6.4%	3.6%	1.4%	100.0%

* In some cases a dam condition score and dam category was not assigned. This typically arose where the water body was found not to be a raised reservoir.

POST-INCIDENT REPORTING SYSTEM

The new Post-Incident Reporting System (PIR) is being administered by the Environment Agency. The system was trialed by Halcrow Group Ltd through 2006 and the finalized system has been in operation since January 2007 (Hamilton-King et al, 2008). Historically, reservoir incidents were recorded in a database administered by the Building Research Establishment. This data is now held in the new database administered by the Environment Agency. Prior to 2006, incidents were only added to the database if they became public knowledge through, for example, details being provided in technical papers or by word of mouth between reservoir safety specialists and the BRE. In many cases these were 'high profile' incidents. The breakdown of these incidents before and after the start of the new system is as follows:

Table 7. Reported incidents at non-statutory reservoirs

Period	Number of reportable incidents at non-statutory reservoirs
1985 - 2005	0
2006 - 2007	5

It is evident from Table 7 that although there has been an incident database available for many years to record reservoir incidents in the UK, the mechanism for reporting had been poor prior to the introduction of the new system. The percentage of incidents reported for non-statutory reservoirs is very small compared with those for statutory reservoirs. However, four out of the eleven reportable incidents arising in 2007 were for non-statutory reservoirs.

Incidents at statutory reservoirs are commonly reported by a panel engineer or an informed undertaker. Incidents at non-statutory reservoirs generally only come to light when there has been concern for the flooding effects of dambreak and the Environment Agency is well placed to identify such incidents.

The PIR cannot currently be used to assess the risk posed by non-statutory reservoirs. This is due to the means by which incidents have historically been identified. The new system has however improved the situation for non-statutory reservoirs in England and Wales with many incidents at non-statutory reservoirs now being identified and investigated where appropriate. As a group, the owners of non-statutory reservoirs are less well informed than owners of statutory reservoirs regarding both the mechanism and collective benefits of openly sharing information on incidents. As a result, incidents at non-statutory reservoirs generally only come to light in serious cases where the Environment Agency becomes involved to manage the flood risk, or where the owner contacts a Panel Engineer for urgent advice.

HISTORICAL FAILURES OF SMALL RESERVOIRS

There are numerous references on historical dam failures in the UK that led to loss of life (for example, Charles, 2002). It is interesting to note that the Skelmorlie disaster of 1925 in south-west Scotland that killed five people and contributed to the introduction of the Reservoirs (Safety Provisions) Act, 1930, was of 24,000m^3. In addition, the Darwen disaster of 1848 in Lancashire that killed twelve people arose from the failure of a 5m high embankment impounding about 20,000m^3. Both of these reservoirs would be regarded as non-statutory reservoirs today. Binnie (1987) reported that there was only 2m of water behind the subsidiary dam at Diggle Moss when it failed in 1810 causing five deaths in the Huddersfield area.

It is clear that even quite small reservoirs can pose a significant hazard, especially if they are not adequately inspected and supervised. Recent experience from investigating incidents at non-statutory reservoirs under the

PIR suggests that the failure of non-statutory reservoirs might be common today with perhaps at least one small reservoir in the UK failing in most years. It is simply very fortunate that none of the failures in modern times have led to loss of life.

SUMMARY OF FINDINGS

The South Wessex project confirmed that within a population of statutory and non-statutory British reservoirs, one can expect that there is a very marked difference in terms of safety with statutory reservoirs generally being much safer than non-statutory reservoirs. In this particular population, 35% of the non-statutory reservoirs were considered to pose a significant threat to life in the event of a failure.

The investigation of potential reservoirs project assessed a large sample of 319 reservoirs that were not subject to reservoir safety legislation. 141 of these reservoirs were found to be too small to fall within the ambit of the Reservoirs Act 1975 and would therefore be termed 'non-statutory'. However, in nearly 42% of these non-statutory cases, maintenance works on the dam were considered either desirable or urgent, and some 37% had the potential to cause loss of life.

The post-incident reporting system cannot currently be used to assess the risk posed by small reservoirs based on the statistics of historical events. However, the new system is better able to capture incidents at small reservoirs and has served to underline the fact that many serious incidents at non-statutory reservoirs can be expected in a typical year. The other studies provide evidence that a small percentage of such reservoirs can be expected to pose a risk to life were they to fail.

There are no statistics on the number of small reservoirs in Britain but studies to date indicate that there is likely to be many that both pose a risk to life and are in need of urgent safety works.

POSSIBLE IMPLICATIONS FOR LEGISLATIVE CHANGE

There is evidence to suggest that each introduction of change to reservoir safety legislation to date has been progressively more effective in reducing the risk of reservoir failure. A summary of British reservoir safety legislation is provided by Charles (2002). It can be argued that the use of reservoir volume alone to define a statutory reservoir has many advantages: it is a simple criterion that in most cases is relatively easy to apply, but it does not always reflect the risk posed by reservoirs. In the view of the authors, future legislation should introduce sufficient flexibility to allow the application of effective safety provisions to any raised reservoir irrespective of reservoir volume where it is judged that lives are clearly at risk. To any homeowner living immediately downstream of a small but very poorly maintained reservoir, such legislation would seem appropriate. One could equally argue that there are a small number of statutory reservoirs where the consequences of failure are not life threatening and where the statutory provisions could be reduced where certified by a panel engineer.

REFERENCES

Binnie, G.M., (1987) *Early dam builders in Britain,* Thomas Telford, London.

Charles, J.A. (2002) *A historical perspective on reservoir safety legislation in the United Kingdom* in Reservoirs in a Changing World, the 12th Conference of the British Dam Society, Trinity College, Dublin. Thomas Telford.

Goff, C.A. and Hope, I.M. (2008) *Investigation of potential reservoirs* in Ensuring Reservoir Safety, the proceedings of the BDS Conference, Warwick.

Hamilton-King, LJ., Hope, I.M. and Warren, A.L. (2008) *Post-Incident Reporting: Learning from Experience to Promote Reservoir Safety,* in Ensuring Reservoir Safety, the proceedings of the BDS Conference, Warwick.

Hinks, J.L. and Dedja, Y. (2002) *Rehabilitation of Irrigation Dams in Albania* in Reservoirs in a Changing World, the 12th Conference of the British Dam Society, Trinity College, Dublin. Thomas Telford.

HMSO (1975) *Reservoirs Act*

HMSO (2003) *Water Act*

HMSO (2004) *Civil Contingencies Act*

Institution of Civil Engineers (1996) *Floods and Reservoir Safety, 3rd edition.* Thomas Telford.

Investigation of potential reservoirs

C. A. GOFF, Halcrow Group Ltd, Swindon, UK
I. M. HOPE, Environment Agency, Exeter, UK

SYNOPSIS. Following the introduction of the Water Act 2003, the Environment Agency took over the role of enforcement authority for England and Wales for the Reservoirs Act 1975 from 136 local authorities in October 2004.

As part of their responsibility to maintain a national register of large raised reservoirs, the Environment Agency commissioned Halcrow Group Ltd. to undertake a project to find currently unregistered reservoirs that should be subject to the Reservoirs Act 1975.

The project identified 121 large raised reservoirs that should have been registered. The potential risk to the general public from an uncontrolled release of water has been reduced by identifying and registering these reservoirs.

This paper provides an overview of the project to identify these 'potential' reservoirs, by explaining the legal background, the approach used, the results and future considerations.

INTRODUCTION

Currently the Reservoirs Act 1975 applies to some 2100 reservoirs in England and Wales, details of which were transferred from local authorities to the Environment Agency, when it took over its new role of enforcement authority in October 2004 (Hope & Hughes, 2004).

The Environment Agency introduced a comprehensive quality assurance process to make sure data held on the register was accurate. To make sure that this register was complete, they also commissioned a study to investigate 'potential' reservoirs with an escapable volume of more than 25,000 cubic metres above natural ground level that weren't already registered.

LEGAL BACKGROUND

The original Reservoirs (Safety Provisions) Act of 1930 (HMSO 1930) was introduced following the failures of Eigiau, Coedty and Skelmorie dams, which resulted in a number of deaths. Due to the size of these reservoirs, the minimum volume of a 'large reservoir' within the law was set at five million gallons. (Hughes, 2006)

This legislation was later replaced by the Reservoirs Act 1975 (HMSO 1975) and the minimum volume of a 'large reservoir' increased slightly to that containing more than 25,000m^3 of water above the lowest point of the surrounding natural land.

The 1975 Act also introduced the role of the enforcement authority. 136 separate local authorities in England and Wales held this role until the Water Act 2003 (HMSO 2003) was introduced. As a result, this led to the Act being inconsistently applied and a wide difference in the quality of information recorded.

As part of the Water Act 2003, the role of enforcement authority for the Reservoirs Act 1975 was transferred to the Environment Agency (for England and Wales) with effect from 1 October 2004. It is managed by the Reservoir Safety Team based in Exeter as a 'national once only service'. One of the enforcement authority's main roles is to maintain a public register and make this information available to the public. This is defined by Section 2 (2) of the Act:

"It shall be the duty of each relevant authority to establish and maintain for their area a register showing the large raised reservoirs situated wholly or partly in the area, and giving the prescribed information about each of them" (Reservoirs Act 1975, Section 2(2))

As part of this responsibility, the Environment Agency compiled a national register of large raised reservoirs based on the existing records held by previous enforcement authorities. After a quality assurance exercise was undertaken, the register comprised some 1924 reservoirs (Environment Agency 2007). They then set about finding unrecorded reservoirs that could 'potentially' be large raised reservoirs under the Act.

LOCATING POTENTIAL RESERVOIRS

In 2004 the Environment Agency commissioned ESRI (UK) Ltd to carry out a geographical information system (GIS) based search of Ordnance Survey MasterMap™ data for England and Wales to identify all inland bodies of water. (Sampson 2004)

In considering the minimum volume limit of the Act, ESRI was asked to report back only on those bodies of water with a surface area greater than 10,000m^2. Flood storage reservoirs were excluded from the project scope.

ESRI searched all the map data for suitably sized polygons on the 'inland water' layer and also within the 'text' layer for the word "reservoirs" and its derivatives. The grid references of the results were compared with those of reservoirs already on the national register, and where these matched, the water bodies were removed from the list. After the checking and validation of the data, this study produced a list of 205 potential reservoirs to investigate further.

As well as the ESRI study, the Environment Agency reviewed the incomplete data for some 600 reservoirs handed over from the local authorities.

Previous enforcement authorities, All Reservoirs Panel Engineers and Environment Agency area enforcement officers were also contacted for details of any suggested potential reservoirs leading to the identification of a further 100 sites.

Once duplications (e.g. different names for the same reservoir) and previously registered reservoirs had been accounted for, there was an initial total of 302 'potential' reservoirs to investigate.

Throughout the project, assessors, owners, panel engineers and Environment Agency staff identified a further 78 ‘potential’ reservoirs. By the end of the project, a total of 380 'potential' reservoirs had been considered by the Environment Agency and of these, 345 required further assessments.

ASSESSING POTENTIAL RESERVOIRS

Initial approach

Following evaluation of competitive tenders in 2005, Halcrow Group Ltd was commissioned to carry out the physical assessment and investigation of these ‘potential’ reservoirs.

As the role of enforcement authority is wholly regulatory and administrative, it was important that the assessments were overseen by an independent qualified civil engineer. To this end, the Halcrow Project Director was also an All Reservoirs Panel Engineer. Police and Criminal Evidence Act (PACE)(HMSO 1984) training was provided by the Environment Agency in the event that evidence and site assessment would be required in subsequent legal proceedings.

The Environment Agency wrote to all owners of 'potential' reservoirs asking for any information (such as construction drawings) that would help estimate the escapable volume. Communications with reservoir owners and operators were sent in order to seek their permission for an assessor to enter their land and carry out an assessment.

Locally based Environment Agency staff were involved in approaching reservoir owners, and arranging and attending site visits.

It was proposed that a pilot study of a small sample of the potential reservoirs should be carried out to develop an approach that would give an acceptable balance between the cost and accuracy of the volume estimate.

Pilot study

The Environment Agency selected a representative sample of 20 'potential' reservoirs in the south of England. These sites ranged from Cornwall to East Anglia and included non-impounding reservoirs and impounding reservoirs.

LIDAR topography data for each site was provided so that a digital terrain model (DTM) could be produced using geographical information systems (GIS) and an estimation of volume made from this.

Each site was visited to measure key parameters, such as embankment height and the amount of freeboard at the spillway. The results of these site assessments were used to estimate the volume to compare with the GIS method.

Coincidentally, bathymetric survey data was available for five of the sites from an earlier project Halcrow Group Ltd carried out on behalf of the Environment Agency's Wessex (South) Flood Risk Management Team. The results of these surveys were useful in comparing the estimates from the two methods to establish how accurate the desk study was and whether the site visits were necessary.

The pilot study concluded that:

- 40% of the reservoirs contained more than 25,000m^3 of water and therefore should have been registered.
- The site visit results (as opposed to the desk study results) offered the best compromise between economy and accuracy - especially for non-impounding reservoirs where the reservoir bed was more likely to be flat and access generally better due to lack of vegetation.
- There was a higher proportion of non-impounding reservoirs than expected among the potential reservoirs to assess. It is suspected that this occurred because the ESRI search identified discrete polygons,

whereas impounding reservoirs, which had long stream lines attached, were sometimes identified as part of a river and not counted.

- The horizontal grid size (5m x 5m) and vertical accuracy (+/-0.5m) some of the LIDAR data provided at the time meant that the estimates of volume using this method were inaccurate.
- The bathymetric survey data most accurately represented the volume of the reservoirs. However, this method was too expensive for assessing over 300 'potential' reservoirs.

As a result of this appraisal, the approach selected for the main project was to visit and assess all the 'potential' reservoirs to enable a 'first pass' volume estimation.

Main project

The project was programmed to run over two years between 2006 and 2008 and included the remaining 325 'potential' reservoirs, 319 of which have been assessed at the time of writing.

Faced with visiting a large number of sites spread over a wide area, it was important to take a cost-effective, risk-based approach.

Initially, an engineer examined the 1:50,000 scale map of each site and gave the reservoir a dam category A-D (subsequently reassessed on site) in accordance with Table 1 of Floods and Reservoir Safety (ICE 1996). This categorisation was adopted for all types of reservoir. It was adopted to ensure that a risk based approach was taken to the deployment of resources to resolve any non-compliance that arose. All the potential reservoir sites were then plotted on a map of England and Wales and assigned to some 40 regional groups across the country.

The number of reservoirs in each category for each group was analysed and the groups were ranked according to highest risk. For example, the highest risk group contained nine category A, two category B, one category C and one category D reservoir. The lowest risk group contained only one, remote, category C reservoir.

Grouping the reservoirs by geographical area meant the reservoir assessors could visit several sites in one trip, thus optimising cost. Visiting the groups in order of the highest risk made sure that the risk of dam failure at a site not yet visited was reduced as far as possible.

Written permission was sought from the owner of the reservoir for an engineer (the assessor) to visit the site. The assessor then visited each reservoir and sought to measure the properties of the reservoir to estimate the volume retained above ground level. In most cases this included measuring:

- maximum embankment/dam height above the level of the lowest point of the surrounding natural ground;
- the minimum freeboard provided above a fixed spillway crest;
- the lengths of the sides of the reservoir;
- the angles of the internal slopes of each side of the reservoir.

Typical problems assessors encountered on site included dense vegetation that prevented accurate measurement of embankment height, no access to some parts of the reservoir, offshore spillways that made freeboard difficult to estimate when the reservoir was drawn down, and estimation of an average internal slope where the slope varies or is hidden by turbid water. Engineering judgement was used in these cases.

Whilst on site, the assessor would note the condition of the dam on a scale of 1 to 5 as shown in Table 1 below. The principles are adapted from the Environment Agency condition assessment manual (Environment Agency 2006). If a dam was considered to be condition 4 or 5, the owner was advised to seek engineering advice, in order to relieve Halcrow or the Environment Agency of any potential liability that could follow.

Table 1 - Dam condition score guidance

Score	Typical attributes
1	Excellent condition, no cause for concern
2	Good condition, some minor maintenance issues or early indication of minor seepage.(e.g. unusual reed growth, wet patches on downstream face, a few small animal burrows)
3	Some cause for concern. Wear problems that may get worse if unattended. (e.g. erosion of crest, blocked spillway or major animal burrows)
4	Problems causing immediate concern for safety (e.g. Seepage through downstream face, unstable large trees on dam, inoperable spillway or bottom outlet pipes/valves)
5	Poor condition. Imminent failure possible. (e.g. Major seepage, recent large slips on crest or downstream face, major erosion of dam crest or downstream face)

As the majority of the potential reservoirs were expected to be non-impounding reservoirs, the default volume calculation methodology was to multiply the surface area by the maximum retainable water depth above the surrounding land, and then subtract the volume of any islands and the internal slopes into the reservoir.

For some cases, including triangular shaped impounding reservoirs, this method was not suitable and so, in these cases, the assessor would carry out separate calculations, taking into account the slope of the valley on the reservoir bed.

For every reservoir visited, the assessor produced a one-page data sheet. This contained a summary of:

- the properties of the reservoir, including owner contact details, location, shape, orientation, dam category, and condition score;
- the dimensions measured on site, including embankment height, freeboard, slope angles, and side lengths;
- comments on the limitations of the site visit, assumptions made during the calculations, justification for the dam category and condition score assigned;
- the current best estimation of the escapable volume and a confidence score in that figure.

The Environment Agency established a monthly panel process to review these data sheets together with any accompanying calculations and to decide whether or not to register the reservoir or to seek further information or clarification. This process ensured that a consistent approach was adopted in deciding whether or not to register a reservoir.

By using the assessed dam category, capacity, confidence score and condition a risk based approach was adopted to prioritise any enforcement activities.

Where a decision to register the reservoir was taken, the Environment Agency contacted the owner to invite them to either discuss this decision within 28 days or to ask them to appoint both a supervising engineer and an inspecting engineer. If the owner did not agree with the decision to register the reservoir within 28 days, they were contacted again and reminded of their responsibilities under the Act. In some cases where no response was received, enforcement action was initiated.

If owners did not agree with a decision, they were encouraged to provide further information (such as construction drawings, or depth measurements) so that volume estimates could be reviewed and updated as appropriate.

Difficult decisions

Throughout the course of the project many difficult issues had to be confronted and decisions taken. Most of these were due to the wording of the Reservoirs Act 1975.

For example, the following questions were addressed:

- What size of spillway sill or pipe is required to define top water level in a non-impounding reservoir?
- On a flood storage reservoir with no overflow sill, where should top water level be taken?
- In order for the "designed to hold" statement in the Act (Cl 1(1)a) to apply, what level of proof of design is required?
- Should a reservoir that has been "designed to hold" more than 25,000m^3, but that is now so silted up that it now holds less than 25,000m^3, be within the Act?
- Or the converse argument: should a reservoir designed to hold less than 25,000m^3, but that is actually "capable of holding" more than this amount, be within the Act?

Although the seven reservoir assessors from Halcrow were not qualified civil engineers (QCEs) within the meaning of the Act, the project was undertaken under the direction of an All Reservoirs Panel Engineer. For complex or difficult cases, his review and written opinion was sought. This opinion was provided in an agreed format to enable it to be used in any subsequent litigation. In these difficult cases, the Environment Agency based its decision on whether to register the reservoir, or to carry out further studies, on this advice.

Results

In total, 319 reservoirs were assessed between May 2006 and February 2008. Table 2 shows the proportion of potential reservoirs with estimated escapable volumes above and below the 25,000m^3 threshold of the Act.

Table 2 - Results of the assessment of potential reservoirs

Estimated escapable volume (m^3)	No. of 'potential' reservoirs	Proportion of the total
>25,000	121	38%
<25,000	141	44%
0	57	18%
Total	319	100%

As a result of the project, 121 previously unregistered and unregulated reservoirs were found to fall within the Act. The relatively high percentage (38%) of this group validates the search criteria of 10,000m^2 surface area, used for the initial GIS search.

The reservoirs in Table 2 that have an escapable volume of zero are those that were found to be not raised above ground level. The initial GIS search identified these as having surface areas greater than 10,000m^2, but the lack of embankment above ground level only became apparent when the site was visited.

Table 3 shows a breakdown of the dam category of the reservoirs now considered to be within the Act.

Table 3 - The risk posed by unregistered reservoirs

Dam category	No. of reservoirs	Proportion of the total
A	16	13%
B	35	29%
C	55	46%
D	15	12%
Total	121	100%

42% of the reservoirs now considered to be within the Act fall within dam categories A (posing risk to life in a community) and B (posing risk to life not in a community). By registering these newly discovered reservoirs and requiring the undertakers to comply with the Act, the Environment Agency has helped to reduce the risk of damage or injury to the general public caused by the uncontrolled release of water from these reservoirs.

Figure 1 shows the distribution of dam category by volume for all 319 reservoirs assessed.

Table 3 and Figure 1 both demonstrate that the majority (46%) of previously unregistered statutory reservoirs fall within category C. This means that a breach in these dams would pose a negligible risk to life and cause limited damage. It could also account for the fact that they had remained undiscovered.

Figure 1 also demonstrates that, of the reservoirs with volumes of between 20-25,000m^3 (that is, just below the threshold of the Act), there is a relatively high proportion (39%) of dam category A and B reservoirs. Goff & Warren 2008 address this issue further.

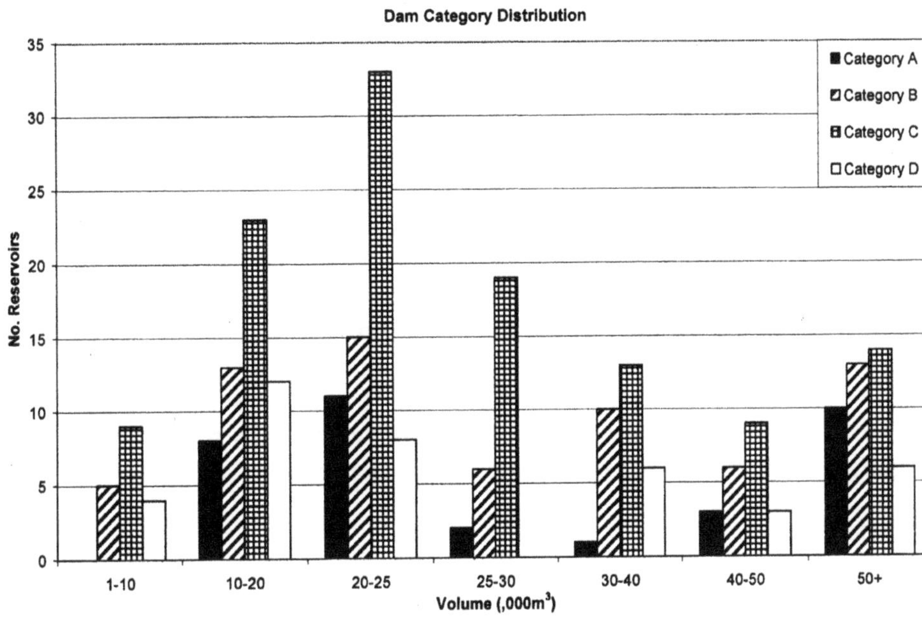

Figure 1 - Distribution of dam category by volume

FUTURE CONSIDERATIONS

During the study, several areas for further investigation were noted.

Firstly, due to limitations in the methods used for estimating the escapable volume, it is recognised that in some borderline cases more detailed analyses may result in a volume immediately above or below the 25,000m^3 threshold. More detailed analyses may include, for example, a detailed bathymetric or topographical survey or the presentation of previously undisclosed construction drawings.

Secondly, the higher than expected proportion of non-impounding reservoirs located by the original GIS search suggested that some impounding potential reservoirs may have gone undetected. The limitations of this original GIS search have been overcome by recent advances in technology and innovative research carried out by Halcrow Group Ltd. A further GIS search of England and Wales is now possible and it is expected this would find further previously unidentified impounding potential reservoirs.

CONCLUSIONS

As a direct result of the Investigation of Potential Reservoirs Project public safety has been improved by the registration of 121 previously unidentified and unknown reservoirs, 42% of which are dam category A or B.

The methodology, risk based approach and proportionate enforcement action that was adopted, ensured the most efficient use of public money. By transparent and open communication with reservoir owners, the project has raised public awareness of the Reservoirs Act 1975, its purpose, and the legal responsibilities on Undertakers imposed by it.

The project has enabled the Environment Agency to carry out its statutory responsibility as enforcement authority under Section 2(2) of the Reservoirs Act 1975. The project has also increased and improved the enforcement authority's understanding and awareness of possible 'grey areas' within the Reservoirs Act 1975, and informed their recommendations for improving the law including the call for the adoption of a risk based approach. The dam condition score and dam category assigned to each reservoir has also highlighted the risk posed by non-statutory reservoirs.

ACKNOWLEDGEMENTS
The authors wish to thank the Environment Agency for permission to publish this paper.

REFERENCES

Environment Agency, 2006. *Managing Flood Risk: Condition Assessment Manual,* Document Ref 166_03_SD01, Environment Agency, Bristol

Environment Agency, 2007. *Improving safety protecting lives:* Biennial report on reservoir safety 2005 - 2007, Environment Agency, Bristol

Goff C. A. & Warren A. L. 2008. *The Safety of Small British Reservoirs,* Ensuring Reservoir Safety, Thomas Telford, London

HMSO. 1930. *Reservoirs (Safety Provisions) Act*

HMSO. 1975. *Reservoirs Act*

HMSO. 1984. *Police and Criminal Evidence Act*

HMSO. 2003. *Water Act*

Hope, I. M. & Hughes, A. K. 2004. *Reservoirs Act 1975 – progress on the implementation of the Environment Agency as enforcement authority*

Hughes, A. K. 2006. *Reservoir Safety - Are things improving?* Improvements in reservoir construction, operation and maintenance, Thomas Telford, London, pp3-12

ICE 1996. *Floods and Reservoir Safety,* Thomas Telford Ltd. London

Sampson A. 2004. *Reservoir Identification Query Results,* ESRI (UK) Ltd., Aylesbury, Not published.

Post–incident reporting: learning from experience to promote reservoir safety

L J HAMILTON-KING, Environment Agency, Exeter, UK
I M HOPE, Environment Agency, Exeter, UK
A L WARREN, Halcrow Group Ltd, Swindon, UK

SYNOPSIS. This paper explains how the new role of post-incident reporting for UK dams is working and considers the challenges for the future.

The paper will:

- explain how the post-incident reporting system works;
- summarise the type of incidents reported to the Environment Agency;
- highlight findings from post-incident investigations;
- summarise how the industry has been kept informed (bulletins, first annual report etc.) and expand on communications issues (freedom of information requests etc);
- expand on how similar systems are being engaged (e.g. The Standing Committee on Structural Safety (SCOSS) and Confidential Reporting on Structural Safety (CROSS);
- consider the need for mandatory post-incident reporting in the future;
- explain how the system can inform the reservoir research and development programme.

INTRODUCTION

The average age of dams in the UK is more than 110 years. Whilst some new reservoirs are still being built, the expectation is that the existing stock of dams will continue to serve future generations.

As well as planned improvements and repairs, the reservoir industry needs to have more accurate information on the likely lifespan of these critical structures and the likelihood of potential modes of failure. This will be achieved by promoting and adding to the new post-incident reporting database. Having a formal process in place to alert undertakers and panel engineers to critical events and issues will further help to improve reservoir safety.

The new post-incident reporting system for UK reservoirs (Warren and Hope, 2006) began on 1 January 2007. The system is administered by the Environment Agency on behalf of the reservoir industry and is overseen by an independent inspecting engineer. It draws upon previous work carried out by the Building Research Establishment (Charles 2005).

The system is operated on a voluntary basis: there is currently no legislation in place to require a reservoir owner to share details of an incident when dam safety is threatened or even when dams fail. Other industries have long recognised the need to continually improve safety by sharing knowledge when incidents happen. Learned societies such as the British Dam Society play a valuable role in helping to share experiences to benefit everyone.

The system provides a consistent way of capturing, analysing and sharing information about incidents, which is a valuable tool for the industry. The information can also be used to inform research and development priorities.

This paper aims to summarise the incidents and investigations carried out in the first year of administering the system and considers future developments.

A SUMMARY OF THE NEW SYSTEM

Incident reporting procedure

Anyone can report an incident at a reservoir and the procedure for doing so is summarised in Figure 1. The procedure distinguishes between dealing with the initial emergency and reporting the incident to the Reservoir Safety team once the incident has been brought under control. Hence the term 'post-incident reporting'.

UK-wide reporting

The database allows post-incident data to be entered for any reservoir within the UK, both statutory and non-statutory. This is because lessons can be learned from incidents at any dam, irrespective of its size or location.

The database contains information on all statutory reservoirs in England and Wales. Information on non-statutory reservoirs and reservoirs in Scotland and Northern Ireland is entered as and when information becomes available. The Scottish Government and the Northern Ireland Rivers Authority support this process.

Figure 1. Procedure for reporting an incident

<u>Incident level definitions</u>

Incidents are entered on the new database if we consider that they are reportable. Assigning a level to each incident is important for managing the system and for reporting on incidents according to how severe they are. Table 1 below defines what a reportable incident is.

Table 1. Reportable incidents

Incident level	Definition
One	Failure (uncontrolled sudden large release of retained water)
Two	Serious incident involving any of the following: o emergency drawdown o emergency works o serious operational failure in an emergency
Three	Any incident leading to: o an unscheduled visit by an inspecting engineer o a precautionary drawdown o unplanned physical works o human error leading to a major (adverse) change in operating procedures

How incidents are reported

Incidents are reported using a post-incident report form which can be downloaded from the Environment Agency website or requested by e-mail. When the system administrator receives a completed post-incident report form, the information is added to the database.

Database

A new database was developed from the existing Building Research Establishment (BRE) database of incidents. The database contains information on historical incidents recorded by BRE as well as information on more recent incidents. The database can hold information on dam characteristics as well as incidents.

The database can be used to produce reports on the nature of incidents, the lessons learned and, over time, it will also help to identify trends.

Dam characteristics

A detailed database of the characteristics of UK dams is important, as it will allow the reservoir industry to make best use of the post-incident data and provide a focus for future research.

Information on dam characteristics is gathered via a reservoir data sheet in one of two ways. For each reported incident, the undertaker, supervising engineer or investigating engineer is asked to complete a reservoir data sheet as well as a post-incident report form. The data sheet comprises a simple spreadsheet, which typically takes around 30 minutes to complete. This is the only way data is collected for non-statutory reservoirs and reservoirs in Scotland and Northern Ireland. For statutory reservoirs in England and Wales a reservoir data sheet is sent to the inspecting engineer when he is appointed to do the next statutory inspection of a reservoir. The inspecting engineer is asked to complete and return the data sheet as part of his/her inspection.

System reporting to the reservoir industry

An annual report is prepared for the reservoir industry, which aims to set a structured and consistent approach to reporting that can be referred to in the future. The following information was presented in the 2007 annual report:

- incidents in 2007 and over the past four years by types of lesson learned;
- incidents in 2007 and over the past four years by the main threat posed to the reservoir (internal or external);

- incident severity against:
    - the mechanism of deterioration
    - flood category

- a summary of the main lessons learned from reported incidents.

Apart from high profile incidents, such as Ulley reservoir in 2007, the reservoir names, locations and owners are not disclosed. This is to encourage voluntary reporting.

Bulletins are prepared, when appropriate, to provide information about an incident or group of incidents where there are particular lessons that should be shared with the reservoir industry. The aim of the bulletins is to alert undertakers and panel engineers to the issues that have arisen and the actions they should take.

Anyone wanting further information from the database can contact the Environment Agency's Reservoir Safety team.

RESERVOIR SAFETY INCIDENTS

Before January 2007 there was no formal system for reporting reservoir incidents. In the past, incident data was gathered on an ad-hoc basis from a variety of sources, including technical papers (Warren and Hope, 2006). The new post-incident reporting database currently holds data on 218 reportable incidents dating from 1800 to 2007. To date, 25 level 1, 85 level 2 and 108 level 3 incidents have been recorded. Some additional historical incidents have yet to be processed and added to the database.

Reservoir safety incidents since 2004

There have been 24 reportable incidents recorded since 2004. Figure 2 shows the breakdown of these incidents in each year by severity.

Dams in the UK can be categorised by the hazard they pose to life and property (ICE, 1996). The categories range from category A for reservoirs that pose a significant threat to life, to category D where no risk to life is likely in the event of dam failure. The majority of reported incidents between 2004 and 2007 occurred at category A dams as shown in Figure 3. It is unclear if this truly represents a greater number of incidents at category A dams or merely a greater level of reporting of incidents at category A dams. It might be that incidents at category A dams are more high profile, therefore more likely to be reported. The dam category is not known in some cases and this reflects incidents reported at non-statutory reservoirs or those still under construction.

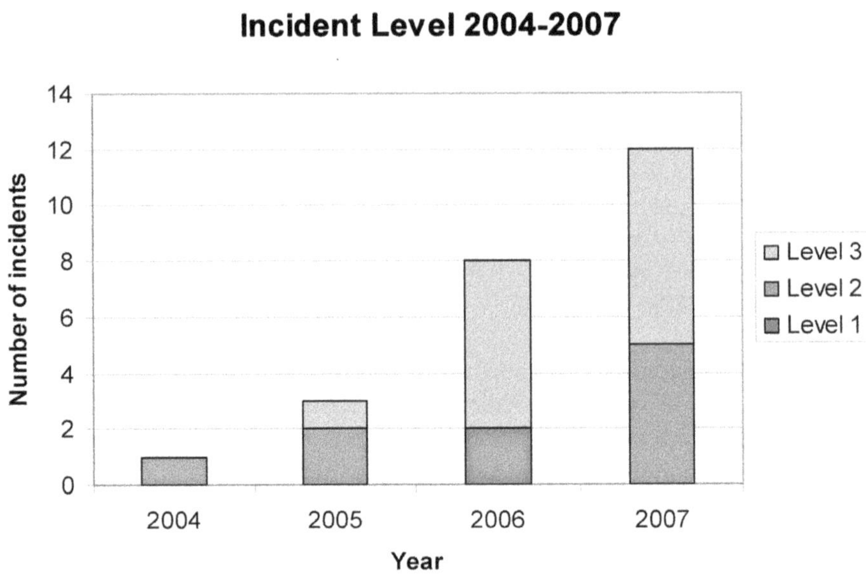

Figure 2. UK reservoir incidents by severity, 2004-2007

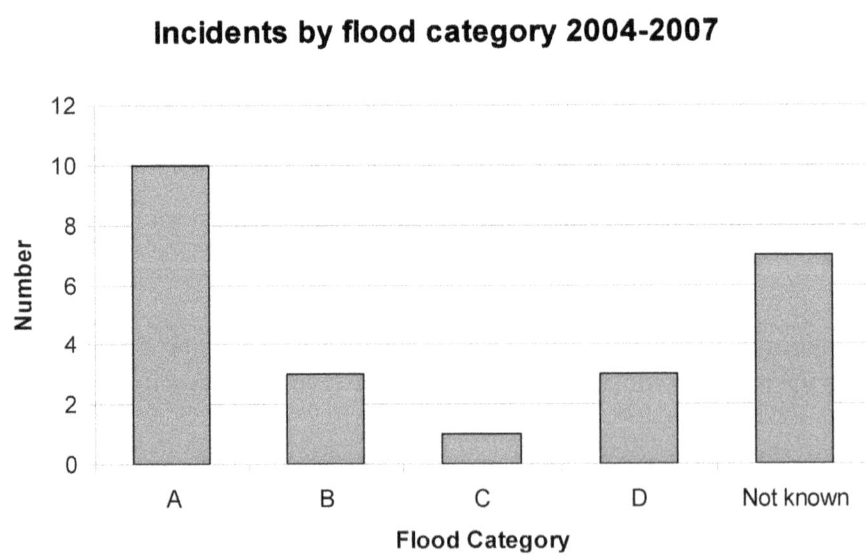

Figure 3. Reported incidents by flood category, 2004-2007

The great majority of incidents occurred at earthfill embankments as shown in Figure 4. This is not surprising given that 85% of dams in England and Wales are of the earthfill type.

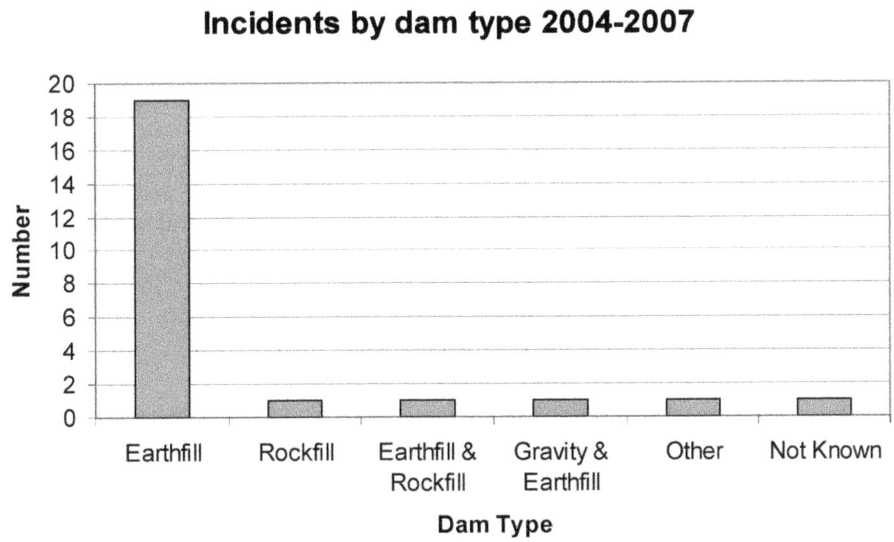

Figure 4. Reported incidents by dam type, 2004-2007

Incidents recorded in the database are classified by the type of lesson learned. The lessons learned are split into eight categories. The five main types are explained in Table 2. The other three are 'none', 'other' and 'not known'. Incidents reported by lesson type are illustrated in Figure 5.

The database allows for more than one of the lesson types to be assigned to any particular incident where appropriate, but only the main lesson type is reported here. Categorising the lessons learned in this way will make it easier to highlight trends in the sort of incident arising.

Table 2. Types of lesson that can be learned from reservoir incidents

Type	Examples	Possible implications
Surveillance	Inadequate surveillance or processing of instrument observations.	Reservoirs require more or better monitoring and surveillance.
Operation	Malfunction or mis-use of reservoir control facilities.	Reservoirs require more or better trained staff or security against mis-use.
Physical (current condition)	Inadequate performance due to deterioration of a design element by erosion, wear, weathering, corrosion, vandalism, poor maintenance, etc.	Reservoir components are inadequately maintained.
Physical features (intrinsic)	Inadequate performance due to the original design and/or construction of a structure, or through changes in the loading (structural or hydraulic) experienced.	Reservoir components are inadequately designed or built to meet current physical conditions.
Emergency planning	Incidents relating to the application of emergency planning provisions (alarms, evacuations, etc).	There is a need for more effective use of emergency planning provisions at reservoirs.

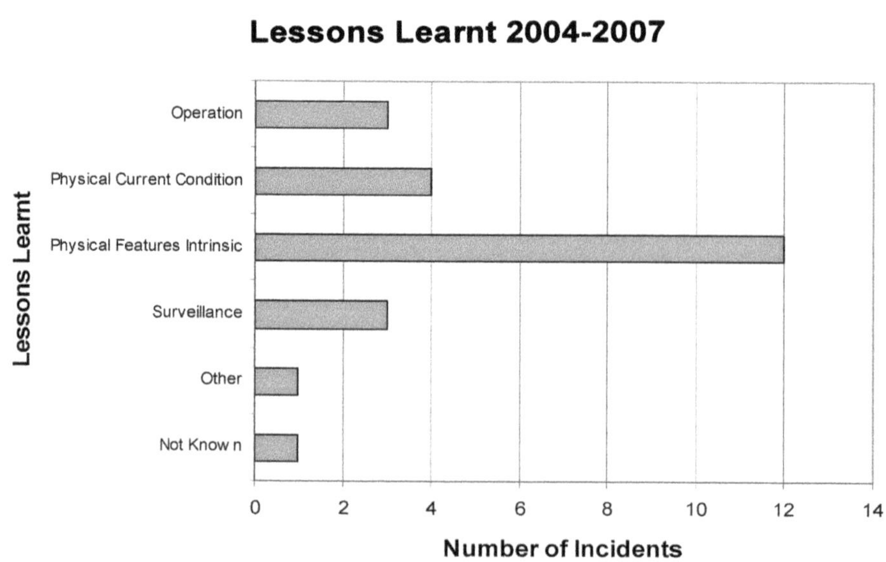

Figure 5. Reported incidents by type of lesson learnt, 2004-2007

INCIDENT INVESTIGATIONS

Following review with a panel engineer, post-incident investigations are carried out for the most serious or complex incidents where permission has been gained from the dam owner. A post-incident investigation has been carried out for four reservoir incidents to date. Each investigation was carried out by a qualified civil engineer to explore the root cause of the incident without apportioning blame. Investigations are also commissioned for significant incidents at non-statutory reservoirs where, generally, there is no supervising engineer or other person with enough technical knowledge about the reservoir to gather information for the database.

Investigations will often be commissioned by the Environment Agency where there may be important lessons to share with the industry through a bulletin or technical paper. A good example is the investigation into the causes of the Ulley reservoir incident in 2007, where the investigation:

- made sure that the information held on the database was well-informed and accurate;
- provided detailed information to the reservoir industry on the key findings of the investigation in a bulletin;
- provided valuable information for future research;
- allowed the reservoir industry to demonstrate that it responded in an effective and responsible way, taking account of the severity of the incident, to try to understand what contributed to the incident.

LIMITATIONS AND CONCERNS FOR VOLUNTARY REPORTING

Completeness of reporting

The main difficulty with a voluntary incident reporting system is that accurate information on how often incidents occur cannot be gathered, due to considerable uncertainty about the completeness of reporting. This limits the value of using the current system for quantitative risk assessments which rely on information relating to incident frequency (probability) by incident type.

Voluntary reporting relies on individuals giving up a small amount of their time to complete incidents forms and reservoir data sheets. Some have had difficulty in making this time available or have simply not reported incidents for personal, corporate or other reasons. When deciding whether to take the time to report an incident it is worth considering that:

- the Environment Agency is administering the system on behalf of the reservoir industry as a whole; it does not use the information for

any secondary purpose in particular prosecution under the Reservoirs Act 1975;
- the effectiveness of voluntary reporting has influenced the call for a mandatory reporting system as part of the proposed changes to the current reservoir legislation.

The completeness of reporting for incidents at statutory reservoirs is low and might be below 50%, but it is impossible to estimate from the information available. However, the new system has been effective in capturing more incident data for non-statutory reservoirs, and raising our awareness of the threats that small reservoirs can pose (Goff and Warren, 2008).

Under reporting – a loss to the industry

There is anecdotal information from the industry that around the time of the Ulley incident there were problems at a number of other masonry spillway channels but these were not reported. In particular, there was a case of high velocity flows removing blockwork on a spillway channel which was only 15 years old.

The Environment Agency's first bulletin to the reservoir industry was based on incidents involving Ashlar block spillways commonly used during the Victorian era. Arguably some engineers and undertakers whose spillway(s) are of the more recent concrete block variety will read that bulletin and consider that it does not apply to them.

Taum Sauk – a case study

The publication of the cause of the failure of the upper reservoir at the Taum Sauk pumped-storage hydroelectric plant provides an excellent example of good practice. Detailed information on the cause of the incident was publicised ahead of any lawsuit.

Immediately following the incident, which occurred in December 2005, the regulator FERC (Federal Energy Regulatory Commission) promptly established a dedicated webpage, provided briefings and updates, and appointed a panel of experts to conduct a detailed forensic investigation. Their report was available for public comment within 5 months, and in addition to an independent detailed analysis of the causes of the incident, the report provides a concise history of plant operation and issues that arose.

We have cited this incident at previous BDS presentations and asked the audience if they were aware of the incident and its causes. Encouragingly, all those involved in the power generating industry knew about the case because it has been so openly reported, and had taken action accordingly.

A number of reports on the incident are available on the FERC website at http://www.ferc.gov/industries/hydropower/safety/projects/taum-sauk.asp

Figure 6. Taum Sauk Dam

Regulatory change – mandatory incident reporting

As the post-incident reporting system relies on voluntary reporting, a strategy for communicating with the reservoir industry was developed (Hope 2007). The aim of the strategy was to encourage undertakers and panel engineers to contribute to the new system. However, as discussed, we believe there has been under-reporting of incidents under the voluntary system.

The ICOLD bulletin 59 highlights the need for regulators to continually review 'safety regulations and procedures'. The Environment Agency has previously highlighted a number of shortfalls in the Reservoirs Act 1975 (Hope 2007) and has alerted the Pitt Review to its concerns about under-reporting of incidents. One of the Environment Agency's proposals to the Pitt Review is to establish a mandatory reporting system.

Confidentiality

Some concern has been expressed that information provided to the Environment Agency about incidents could enter the public domain. This concern arises because, as a public body, the Environment Agency is subject to the Freedom of Information Act 2004 and the associated Environmental Information Regulations 2004.

Following the incident at Ulley in June 2007, the Environment Agency received requests from the media for post-incident report forms for a number of incidents. However, the information requested was not released, as disclosing it would have adversely affected public safety and national security. This decision was reached after referring to the Ministry of Justice and was upheld following a subsequent appeal.

Fear of prosecution

Concern has been expressed that the Environment Agency's role as the enforcement authority for the Reservoirs Act 1975 conflicts with the post-incident reporting role. However, the aim of post-incident reporting is only to improve reservoir safety. The Environment Agency does not intend to use information acquired through post-incident reporting to retrospectively initiate enforcement action under the Reservoirs Act 1975.

As the enforcement authority, the Environment Agency is well placed to administer the post-incident reporting system as it already holds reservoir details and acts as a single point of contact for dam owners and engineers on a number of issues. In addition, due to its wider flood risk management role, the Environment Agency can pick up reports of incidents at non-statutory reservoirs through its incident hotline (0800 80 70 60) and Floodline (0845 988 1188), as happened during the 2007 summer floods.

OVERVIEW OF SIMILAR SYSTEMS

The Institution of Civil Engineers, the Institution of Structural Engineers and the Health & Safety Executive have established the Standing Committee on Structural Safety (SCOSS). The role of SCOSS is to monitor and investigate trends or practices that might cause structural failure. They have also established Confidential Reporting on Structural Safety (CROSS), which is a voluntary system.

Since the Environment Agency became the enforcement authority for the Reservoirs Act 1975, it has held annual liaison meetings with the Health & Safety Executive to make sure there is a coordinated approach to all reservoir safety matters. A productive working relationship has also been established with SCOSS who publish annual reports and bulletins which can be viewed on their website. (http://www.scoss.org.uk/).

ICOLD are planning to introduce a system of incident reporting 'Dam Failures and Incidents Database' which will draw on the data and information held by member countries. The University of Stanford in the United States has established a web-based reporting system (http://npdp.stanford.edu/index.html).

Issues for developing a mandatory system

In September 2007 the Environment Agency provided evidence to Sir Michael Pitt as part of his review into the summer 2007 floods. The Environment Agency called for a number of changes to the Reservoirs Act 1975, including the requirement for mandatory incident reporting.

The proposal is in its early stages and many detailed policy decisions will need to be resolved. These include who should be responsible for incident reporting, which reservoirs should it apply to (statutory/non-statutory) and should there be an incentive to encourage reporting or a fine for not reporting? The definition of a reportable incident may also need to be more clearly defined.

INFORMING RESEARCH AND DEVELOPMENT

Historical reservoir safety incidents and the original BRE database have helped to inform many reservoir safety guides commonly used by the UK reservoir industry today. A national strategy for research and development can only be effective if it is informed by good quality information on the many safety issues that arise every year. The system can inform legislative change as well as research and has served to raise the awareness of incidents at small reservoirs. In time, it is anticipated that the system will inform quantitative risk assessments by providing information that can help determine the frequency of incidents. Mandatory incident reporting would be important in this respect to promote completeness of reporting.

CONCLUSION

2007 might be regarded as a pivotal year for reservoir incident reporting in the UK. It was the first year of operating the new post-incident reporting system. The summer floods resulted in a large number of incidents in England. The Ulley incident was a high profile event and the new system was in place to investigate the causes and to share points of learning. Many incidents were reported to the Environment Agency but there were also some that were not reported.

The aim of the post-incident reporting system is to improve reservoir safety by sharing lessons learned with the reservoir industry and informing research and development. The Environment Agency will publish annual reports and special bulletins on incidents that occur each year and administer the continuing development of the national database. The value of these documents and the database to the UK reservoir industry will depend in part on the level of support provided by reservoir owners and panel engineers.

ACKNOWLEDGEMENTS

The authors wish to thank the Environment Agency for permission to publish this paper.

REFERENCES

Charles, J.A., 2005. *Use of incident reporting and data collection in enhancing reservoir safety*. Dams and Reservoirs Vol.13 No.2 November 2005.

Environment Agency, 2007. *Learning from Experience: Post-incident reporting for UK dams.*

Goff, C.A. and Warren, A.L. (2008) "The safety of small British reservoirs" in *Ensuring reservoir safety*, proceedings of conference of the British Dam Society, Warwick. Thomas Telford, London.

Halcrow Group Ltd, 2007. *Post-incident reporting system for UK dams – Final Report.*

Hope, I.M., 2007. *Reservoir Safety in England and Wales – A Time of Change*. Dams and Reservoirs Vol.17 No.1 April 2007.

ICOLD Bulletin 59, 1987. Dam Safety Guidelines, ICOLD.

ICE 1996. *Floods and Reservoir Safety*. Thomas Telford, London. Institution of Civil Engineers, 1996, *Floods and Reservoir Safety*, 3rd edition, Thomas Telford.

Warren, A.L. and Hope, I.M., 2006. *A new incident reporting system for UK dams*. Proceedings of the 14th conference of the British Dam Society, Durham. Thomas Telford, London.

www.environment-agency.gov.uk/reservoirsafety

http://www.scoss.org.uk/

http://npdp.stanford.edu/index.html.

The application of the Mines and Quarries (Tips) and the Reservoirs Acts

M. CAMBRIDGE, Cantab Consulting Ltd, Ashford, Kent, UK

SYNOPSIS. Correspondence in the technical press over the last thirty years has shown repeated concerns on the part of the reservoir engineering community that tailings dams do not comply with appropriate design and construction standards. This has been accompanied by repeated calls for mine waste disposal facilities to be included in the Register of Large Raised Reservoirs, and thus be covered by the Reservoirs Act rather than fall under the remit of the Mines and Quarries (Tips) Act. However, these concerns have been accompanied by a certain amount of confusion amongst academics and Panel Engineers alike as to the application of the UK legislation for tailings management facilities.

This paper presents a personal interpretation of the application of the UK legislation based on more than thirty years' experience of the inspection of reservoirs and tailings dams under both Reservoirs and Mines and Quarries (Tips) legislation. The paper concludes that much of the implied criticism of the Mines and Quarries (Tips) legislation, and of the perceived lower engineering standards applied to mine tips, is ill-founded. It is further evident that any criticism stems both from a lack of knowledge of, and from poor interpretation of, the legislation and that any recent failings in design standards relate more to the approach adopted by individual experts than to any significant deficiencies in the Acts themselves.

INTRODUCTION

In the year in which the Reservoirs Act 1975 came into force there was a major and catastrophic failure of the Stava tailings dam in Italy which resulted in the death of 269 people. In the aftermath of this failure, numerous articles appeared in the technical press, particularly the NCE, calling for an immediate revision of the Reservoirs Act and for mine tailings dams to be included under the definition of large raised reservoirs. The justification for the proposed revision was the perceived lack of appropriate design standards for tailings dams and the belief that, in the case of the Stava failure, this had been a contributory factor. Various correspondents implied that there was a similar risk within the UK and concluded that, by incorporating mine waste facilities under the Reservoirs Act, the design and construction of these would automatically be improved and failures prevented. In subsequent years there has remained an undercurrent of criticism at BDS meetings, and in the technical press, of the design standards for mine waste sites. This included an authoritative prediction, post the Baia Mare event in Romania, that without major changes in legislation there would be a year-on-year increase in the failure rate of tailings disposal facilities. This assessment has proved to be wholly unfounded.

There is no dispute that the untoward occurrences at Stava and Baia Mare were preventable, and it is clear that there was a lack of appropriate engineering design in both cases. However, the assumptions made in the published post-event articles were unfounded, appearing to be based not on any knowledge of tailings management facilities currently under construction in the UK but showing, rather, an inadequate understanding of the application of the Mines and Quarries (Tips) Act and of the design and construction standards applied to UK mine waste facilities since 1970. Rational analysis of both water dam and tailings management facility failures shows that reservoir legislation is not a universal panacea for resolving the design and operational problems of tailings lagoons.

The confusion concerning UK legislation applicable to mine and quarry lagoons continues to the present day. This is demonstrated by an abstract from an EU-funded report recently prepared by academic teams from around Europe, including one from a leading UK university minerals department. The report reviewed current EU legislation for mine waste facilities and included the erroneous statement that: "In the United Kingdom tailings dam safety laws fall under the Reservoirs Act 1975 (chapter 23 which enters into force on December 1 1991). UK Legislation provides a mandatory safety regime for all reservoirs which are capable of holding 25,000m^3 of water above natural ground level." It is evident, therefore, that there is confusion in academic circles, amongst reservoir engineers and even in major mining consultancies, as witnessed by a recent due diligence study which incorrectly reported on the statutory inspection procedures for a current UK tailings management facility. This clearly demonstrates the poor understanding of the legislation by the industry in general.

Rather than confirming the worst expectations of the critics, a practitioner's review of the similarities and differences between the two Acts in the context of industry practice indicates that tailings management facilities in the UK constructed since 1970 have generally been designed to the highest standards.

UK LEGISLATION

Chapter 23, Section 1 of the Reservoirs Act 1975 includes the following definition:

"For purposes of this Act "reservoir" means a reservoir for water as such (and accordingly does not include a mine or quarry lagoon which is a tip within the meaning of the Mines and Quarries (Tips) Act 1969); and -
(a) a reservoir is a "raised reservoir" if it is designed to hold, or capable of holding, water above the natural level of any part of the land adjoining the reservoir; and
(b) a raised reservoir is a "large raised reservoir" if it is designed to hold, or capable of holding, more than 25,000 cubic metres of water above that level."

Even those advocating incorporation of tailings management facilities under the aegis of this Act acknowledge that Section 1 specifically excludes mine tips, and a major revision of reservoir legislation would be required to bring these within its remit. It is therefore worth examining the existing mine waste tips' legislation to assess whether such a revision would be justified.

The Mines and Quarries (Tips) Act 1969, Part I, Section 2 defines a tip as "an accumulation or deposit of refuse from a mine or quarry (whether in a solid state or in solution or suspension) other than an accumulation or deposit situated underground, and where any wall or other structure retains or confines a tip then, whether or not that wall or structure is itself composed of refuse, it shall be deemed to form part of the tip for the purposes of this Act". The Act addresses, in addition, both operating and closed tips, defining these as "active classified tips" and "closed classified tips" respectively. The subsequent Mines and Quarries (Tips) Regulations 1971, Part I, Regulation 2 further defines the physical nature of tips registered under the Act as follows:

"*Interpretation*

2. – (1) In these regulations…"classified tip" means a tip to which Part I of the 1969 Act applies, being a tip of any of the following classes-

   (b) the tip consists of refuse accumulated or deposited wholly or mainly in solution or suspension and-
      (i) any part of the tip (other than any wall or other structure retaining or confining it but including any liquid in it) is more than 4 metres above the level of any part of the neighbouring land within 50 metres of the perimeter of the tip; or
      (ii) the volume of the tip (other than any wall or other structure retaining or confining it but including any liquid in it) exceeds 10,000 cubic metres:"

These regulations also impose a duty on the Owner and on the Manager of every mine to ensure that the provisions of the Act are strictly applied and that the nature and scope of the responsibilities of all statutory appointments are written and fully understood. It is also a requirement that tipping rules are prepared and that written records of all activities pertaining to the tip are kept at the site, are up to date, are regularly reviewed by the Mine Manager and made available to the Mines Inspectorate.

The Quarries Regulations 1999 endorsed the 1969 legislation, and further strengthened it by ensuring that the, generally smaller, quarry silt lagoons were also included. These Regulations detail the approach to be taken in the design and execution of tips, and define the nature of geotechnical hazards which require a specified level of competence in assessment, design, operation and monitoring. The definition of a significant geotechnical hazard is as follows:

"Clause 300 The hazard should be treated as significant and the tip subject to a geotechnical assessment if it is, or will be:

- (b) a lagoon containing any liquid or material wholly or mainly in solution or suspension (i.e. likely to flow if not contained); and
  - (i) the contents of the lagoon are more than 4m above the level of any land which is within 50m of its perimeter; or
  - (ii) the contents of the lagoon exceed 10 000 cubic metres; or
- (c) irrespective of the size of the tip, other factors, for example the geology, location or proximity to an excavation, mean that there is a significant hazard....".

Any facility containing a volume of more than 10,000m^3 of material in liquid or potentially liquid form therefore needs to meet certain standards of design and operation under tips legislation. This compares with a volume of 25,000m^3 of such material under the Reservoirs Act. The Quarry Regulations extend the definition to include any at-risk facility, acknowledging that a simple volumetric definition does not necessarily address the problems posed by high-risk structures, i.e. a 500m^3 lagoon above a village could pose greater risk to life than a 50,000m^3 facility on remote moorland. Thus sub-clause (c) above requires the employment of a competent geotechnical specialist to assess the risk, regardless of the size of the facility, and to define suitable mitigation methods in design, construction and monitoring. The statutory appointments under the relevant legislation are shown in Table 1 and illustrate the often more prescriptive monitoring and recording requirements for mine waste tips.

Table 1: Summary of Statutory Appointments under UK Legislation

	Reservoirs Act	Mines & Quarries (Tips) Regulations (i)
Regular supervision	Supervising Engineer at minimum of annual inspection	Regulation 5 Appointee – weekly inspections
Independent inspection	10-yearly reporting	2-yearly reporting (ii) 5-yearly for closed classified tips
Statutory records	Prescribed form of record	Weekly record of waste tipped Recording of all defects and remedial actions

Notes

(i) The frequency of inspections and record keeping under the Mines and Quarries (Tips) Regulations are endorsed and strengthened by the Quarries Regulations 1999.

(ii) The Mines and Quarries(Tips) Regulations 1971, Part III, Regulation 11 states that "it shall not be lawful for tipping operations to be carried out at that tip unless a report has been obtained in the last two preceding years…."

Even the hydrological concerns that have been raised by the proponents of an all-embracing Reservoirs Act do not necessarily withstand scrutiny. Mine and quarry lagoons (tailings management facilities) are, as indicated above, specifically excluded from the ambit of the UK Reservoirs Act 1975. However, those in the industry recognise that good practice requires tailings storage facilities to be designed, constructed and operated to the same standards and in accordance with similar risk categories as UK reservoirs. The Reservoirs Act requires that the flood design standards to be adopted for each reservoir accord with the risk categories defined in the ICE publication "Floods and Reservoir Safety". In the UK, most tailings dams would be placed in the highest risk category, i.e. Category A, due to the implications of an untoward release for both life and the environment in the downstream catchment. Any tailings storage facility defined as a "classified tip" within the meaning of the Mines and Quarries (Tips) Act 1969 which has the ability to store more than 25,000m^3 of water, and thus otherwise defined as a "large raised reservoir" under the Reservoirs Act 1975, would therefore require special consideration for flood provision. By applying equivalent standards, therefore, a Mine Owner would be required to ensure that a suitably qualified civil engineer be engaged to undertake any necessary hydrological assessment for such tips and be responsible for defining the necessary flood standards to be applied. Further, the definitions under the Mines and Quarries (Tips) Act and the subsequent Quarries Regulations require that similar risk assessments be undertaken for smaller facilities, i.e. for those with storage volumes of between 10,000m^3 and 25,000m^3 or where specific hazards have been identified. It could therefore be inferred that the hydrological standards required for a mine or quarry facility are higher and more rational than those which apply to a large raised reservoir.

There are other lagoons which are either not, or not directly, associated with a mine, and for which it is evident that the requisite standards of design and construction may not be fully appreciated. The Health and Safety Executive is unequivocal concerning the definition of ash lagoons, for example, which, though not directly associated with a mine, are defined as tips and fall under the Mines and Quarries (Tips) Act. Other tips and lagoons which fall outside both the Mines and Quarries (Tips) and the Reservoirs Acts, and which hold liquid/fluid waste or "waste with the potential to flow" from manufacturing industry, are covered by the Health and Safety at Work Act. In both cases the HSE has a responsibility for enforcement and applies the nearest equivalent standards principle, i.e. application of the Mines and Quarries (Tips) Regulations as described above.

PERFORMANCE OF UK TAILINGS MANAGEMENT FACILITIES

In 1970 the UK had in excess of 2 billion tonnes of mine waste on surface and, prior to 1969, was regulated only by the Town and Country Planning Act and by specific mine-related Health and Safety Legislation. Following the Aberfan disaster, and the Mines and Quarries (Tips) Act 1969 receiving royal assent, tailings management facilities within the UK became subject not only to general planning, environmental and health and safety legislation but, more importantly, to strict control by the subsequent Mines and Quarries (Tips) Regulations (1971) issued pursuant to the 1969 Act. Design, construction, operation and closure of facilities classified under the Act became strictly regulated by these Regulations. These Regulations impose a strict duty on the Owner of the mine or quarry to ensure that all necessary engineering studies and plans, including tipping plans, be prepared in advance of commencement of tipping and that these documents be continually updated during the development of the facility.

Though no detailed guidance is provided, the Health and Safety Executive requires that the design, construction, operation and closure of the confining structures be undertaken in accordance with current UK design standards and, as appropriate, with ICOLD guidelines. Further, the Regulations require the Owner to make a number of statutory appointments for the day-to-day management of the tip and to appoint a Competent Person to undertake independent inspection and to report on "every matter which might affect the security of each tip". Without such reporting "it shall not be lawful for tipping operations to be carried out at that tip". The 1971 Regulations were further strengthened by the Quarries Regulations 1999. In addition to requiring an operator to ensure that excavations and tips are designed, constructed, operated and monitored to ensure safety and stability, and that suitable and sufficient rules are made, they also require the Owner to make suitable written appointments within the context of an overall health and safety plan for the site. Documentation and recording are also strengthened, requiring the preparation and regular updating of tips rules governing the operation of a lagoon or tailings dam, with inter alia, respect to maximum storage levels, freeboard, inspection, monitoring and instrumentation.

Though the majority of coal slurry tips in 1970 were non-impounding, with small storage capacity, a significant number of larger, and often impounding, metalliferous and industrial mineral waste facilities were either designed and constructed or permitted in the immediate aftermath of Aberfan. A list of UK facilities is shown in Table 2 and, though not considered to be fully comprehensive, is a reflection of the author's direct experience over the last thirty years or so. It is noted that all the facilities listed involved a suitably qualified civil engineer at relevant stages of design, construction and operation.

The application of both Acts is exemplified by the design and construction standards adopted for the Clemows Valley Tailings Dam, one of the first facilities to be influenced by changes in legislation following the Aberfan disaster. The facility was permitted on a "greenfield site" in 1969 whilst the Regulations were still in draft. The basis for the design was influenced by both international practice and by current and impending UK legislation, including the Reservoirs (Safety Provisions) Act 1930. The design and construction responsibilities were clearly separated into hydrological and geotechnical disciplines, the project being coordinated by a senior geotechnical partner with independent overview and statutory inspection responsibility under the direction of an All Reservoirs Panel Engineer. This was undertaken in accordance with the then extant 1930 Act, and thus the flood design and spillway capacity met with best UK practice for reservoirs. The hydrological design was based on the PMF, spillway design being updated on a regular basis to meet the constraint of a stage-raised facility. Annual inspection was carried out by a Panel Engineer, and Certificates issued permitting the reservoir and tailings storage levels to be raised commensurate with the new spillway and embankment crest level. Once mining operations commenced in 1970 and the 1971 Regulations were in force, modifications were instigated to meet the demands of the Mines and Quarries (Tips) Act. Statutory appointments under the Regulations were made by the Owner, and the inspection routines modified to meet the more stringent inspection and monitoring routines imposed by this Act. In addition, the reservoir continued to receive inspections under the 1930 Reservoirs Act, and Certificates were issued for each spillway raise until the final upstream wall was commissioned in 1977 and the facility became non-impounding. With the change in flood design criteria, the spillway design became reliant upon PMP, process water balance and groundwater control. Continued supervision by a reservoir engineer has, however, continued since this date. The current Competent Person for the facility under the Mines and Quarries (Tips) Act is also a suitably qualified civil engineer. Through its life the Clemows Valley Tailings Dam has met the highest standards set by both the Mines and Quarries (Tips) and the Reservoirs Acts.

EU LEGISLATION

A review of the details of the operational or permitted mine sites in Europe indicates that most tailings management facilities are covered by some form of legislation. The majority of EU countries have national regulatory bodies for water dams which require designs to be in full accordance with internationally accepted criteria for embankment dams, both for stability and hydrological control. These dams comply with international criteria such as those produced by ICOLD, and are subject to ongoing national embankment inspection routines. Many tailings-confining embankments, whether designed and constructed in the recent past, i.e. during the last twenty years or earlier, have been governed by local legislation – historically that which relates to water dams. This legislation, though not entirely appropriate, results at least in an ongoing programme of controls at design and construction and, in some instances, during the operational phase.

Specific regulations exist in a number of countries for regular monitoring of tailings management facilities. In some countries such as Spain and Portugal the guidance remains within the water sector, which can unfortunately lead to anomalies, particularly with the legal requirement for an emergency outlet, essential for water dams but totally inappropriate for tailings management facilities. The result across the EU as a whole is a large number of tailings management facilities which meet international standards of design and construction despite the varied legislative provisions, and an almost similarly sized group whose day-to-day operations receive at least some level of independent scrutiny. Moreover, an increasing number of tailings management facilities are regularly monitored. However, the type of monitoring regime adopted at many sites remains less than ideal, and the inspection routines would benefit from better regulation. Though most confining embankments are considered to be robust at the initial design stage, neither operational nor closure phases have to date met the same standards.

Table 2: Examples of recent UK Mines and Quarries Tips

Facility	Location	Mine	Mineral	Type of Facility	Height (m)	Storage Volume ($x10^6 m^3$)	Involvement of suitably qualified civil engineer
Clemows Valley	Cornwall	Wheal Jane	Tin/copper zinc	Valley dam	50	6	(iii)
Middleton Moor	Derbyshire	Glebe Mines	Fluorspar/barytes	Paddock dam	22	1.25	(iii)
Blakedon Hollow	Derbyshire	Glebe Mines	Fluorspar/barytes	Paddock dam	29	3.5	(iii)
Wheal Maid	Cornwall	Wheal Maid	Tin/copper zinc	Valley dam	30	4	(iii)
TD 1&2	Derbyshire	Glebe Mines	Fluorspar/barytes	Paddock dam	15	0.9	(iii)
Carnon Valley	Cornwall	Wheal Maid	Tin	5 Paddock dams	7	1	(iii)
Cononish (i)	Perthshire	Cononish Mine	Gold	Valley dam	40	0.9	(iv)
Cavanacaw (i)	Tyrone	Omagh Mine	Gold	Valley dam	12	1	(iv)
Warren North	Norfolk	Kings Lynn Quarry	Industrial sand	Paddock dam	16	1	(iv)
Mintlyn Wood	Norfolk	Kings Lynn Quarry	Industrial sand	Paddock dam	4	1	(iv)
Redmoor (ii)	Cornwall	Redmoor	Tin	Valley dam	54	2.7	(iv)
Acre Nook	Cheshire	Chelford	Industrial sand	Paddock dam	17	1.3	(iv)
Bent Farm	Cheshire	Congleton	Industrial sand	Paddock dam	7	0.5	(iv)
Portworthy	Devon	Lee Moor	Kaolin	Valley dam	55	(v)	(iv)
Kernick	Cornwall	St Austell	Kaolin	Valley dam	90	17	(iv)
Norton Bog	Staffordshire	Cannock Chase	Coal	Paddock dam	6	0.09	(iii)

Notes

i) Permitted but not constructed
ii) Feasibility study completed
iii) Certification under Reservoirs Act
iv) Involvement of suitably qualified civil engineer during design
v) Not known

In May 2008 the Directive on waste from the extractive industries will come into force and will affect the legislative approach across the EU. This legislation is intended to ensure that suitable regulations exist in all member states to prevent a repeat of untoward events such as that which occurred at Baia Mare, and to ensure that these relate specifically to mine waste and NOT to water dams. Though increasing the scope of UK legislation relating to tailings dams, the Directive recognises that the current regulations under the Mines and Quarries (Tips) Act are both appropriate and in accordance with its central tenets. There will not, therefore, be a need for the UK to modify current legislation, which is considered to be competent and has indeed been used as a model for some of the Annexes currently in draft. The Directive will not affect current design and operation practice, though it is likely to reinforce aspects of closure and financial guarantees.

CONCLUSIONS

Correspondence in the technical press over the last thirty years has repeatedly expressed concerns that tailings dams do not comply with appropriate design and construction standards. These concerns have been accompanied by repeated calls for such mine waste disposal facilities to be included in the Register of Large Raised Reservoirs, and thus be covered by the Reservoirs Act rather than fall under the remit of the Mines and Quarries (Tips) Act. However, the proponents of this approach appear to have limited experience of the content and application of this post-Aberfan legislation, demonstrated most recently by a number of misleading papers and reports.

Though mine and quarry lagoons are specifically excluded from the ambit of the UK Reservoirs Act 1975, it is accepted that good practice requires tailings storage facilities to be designed, constructed and operated to the same standards and in accordance with the same risk categories as UK reservoirs. In the UK, most tailings dams would be placed in the highest risk category, i.e. Category A, due to the implications of an untoward release for both life and the environment in the downstream catchment. Any tailings storage facility defined as a "classified tip" within the meaning of the Mines and Quarries (Tips) Act 1969 which has the ability to store more than 25,000m^3 of water would therefore require special consideration for flood provision. By applying equivalent standards, therefore, a Mine Owner would be required to ensure that a suitably qualified civil engineer be engaged to undertake any necessary hydrological assessment for such tips and to be responsible for defining the necessary flood standards to be applied. However, both the Mines and Quarries (Tips) and the Quarries Regulations require that any lagoon intended for the storage of solids in liquid or suspended form which has a capacity in excess of 10,000m^3, or indeed which poses a significant risk regardless of size, would need to meet the same design standard. This implies that a tailings management facility requires higher rather than lesser standards, not only of design and construction but also of operation, since the Regulations clearly prescribe statutory appointments for the day-to-day management and supervision of the facility. The UK legislation for tailings management facilities is likely to be further strengthened with the implementation of the "extractive waste Directive" on 1st May 2008.

Evidence from more than thirty years' personal experience of the inspection of reservoirs and tailings dams under both Reservoirs and Mines and Quarries (Tips) legislation concludes that much of the implied criticism and the perceived lower engineering standards applied to mine tips is ill-founded. It is evident that this stems both from a lack of knowledge and from poor interpretation of the legislation. The evidence shows that a significant number of mine and industrial mineral tips have been designed to a standard which UK reservoir engineers would recognise as being consistent with good practice. Any failings there may have been in these tips arise from the approach adopted by individuals rather than to any significant deficiencies in the Acts themselves. As with any comparable legislation, its ultimate effectiveness relies not only on the integrity of the independent inspections and audits but also on the degree of enforcement applied by the Regulator which is, of course, a function of political will.

Acknowledgements
The author would like to thank Donald Lamont of the HSE for his helpful comments on the final draft.

BIBLIOGRAPHY

Cambridge, M., The *importance of failure in the design process;* International Workshop in Geoenvironment and Geotechnics, September 2005

Cambridge, M., *Instrumentation and Monitoring to Prevent Failure;* International Workshop in Geoenvironment and Geotechnics, September 2005

Cambridge, M., *Tailings Disposal in Cornwall – Past and Present;* Professor Kontopoulos Memorial Volume, April 2004

Cambridge, M., The Future of Tailings Disposal in Europe, *Minerals and Energy,* Vol 18, No. 4, 2003

Cambridge, M., A Review of Tailings Dam Failures; *International Water Power & Dam Construction,* May 2001

Cyanide Spill at Baia Mare, Romania; UNEP/OCHA Mission, March 2000

Directive 2006/21/EC of the European Parliament and of the Council of 15th March 2006 on the management of waste from the extractive industries.

Health & Safety Commission, *Health and safety at quarries,* Quarries Regulations 1999, Approved Code of Practice, 1999

HMSO Mines and Quarries (Tips), Regulations 1971

HMSO Mines and Quarries (Tips) Act 1969

ICE, *A Guide to the Reservoirs Act 1975,* Thomas Telford, 2000

ICE, *Floods and reservoir safety,* Thomas Telford 1996

ICOLD *Tailings Dams, Risks of Dangerous Occurrences;* Bulletin 121, 2001

ICOLD *Dam Failures; Statistical Analysis;* Bulletin 99, 1995

5. Construction, remedial works and discontinuance

Bradford Area Reservoirs' Group - Flood Routing Remedial Works at Leeming, Doe Park and Hewenden Reservoirs

M.AIREY, Mott MacDonald Ltd, Cambridge, UK
G.CARRUTHERS, J N Bentley Ltd, Skipton, UK

SYNOPSIS. Following detailed flood study investigations and extensive physical model testing, it was demonstrated that the overflow works at three existing Yorkshire Water reservoirs situated to the west of Bradford were inadequate to handle the PMF design flood. A programme of design and construction of remedial works was therefore carried out to overcome the deficiencies.

In each case, remedial works were required to increase the capacity of the spillway and to provide adequate embankment freeboard for both still water flood rise and wave surcharge in order to prevent overtopping. However, whilst the three projects had certain similarities, there were also differences in the nature of the works and the approach that was adopted. This paper describes the remedial works carried out at each of the sites as delivered by the MMB joint venture team. For each project the emphasis was on the development of innovative solutions in order to reduce costs and minimize construction time, whilst providing a viable technical solution which was in keeping with the local environment.

At Leeming reservoir the solution involved deepening and widening of the tumble bay and the enlargement of the spillway using conventional reinforced concrete and masonry cladding construction. At Doe Park there were extensive works to raise the dam crest and a pre-cast wave wall solution was adopted to facilitate the construction and to maintain access for local residents during the works. At Hewenden the spillway walls were heightened in in-situ concrete using molded formwork and concrete staining to give the appearance of stone. In the lower section of the spillway there was a certain amount of out of channel flow that was tolerable and appropriate protection works were included at the embankment toe.

INTRODUCTION

The Bradford area group of reservoirs are located approximately 5 miles to the west of Bradford and consist of six dams of a traditional pennine type embankment construction. All of the dams were constructed in the latter half of the 19th century to supply Bradford's ever increasing demand for water.

The reservoirs are Leeshaw and Leeming which are located to the South of Haworth and which discharge ultimately into the River Worth, and Thornton Moor, Stubden, Doe Park and Hewenden which are in cascade and are located in and around the village of Denholme, and discharge into the River Aire at Bingley.

Following statutory inspections that were carried out in 2004 by the late J.Beaver, it was recommended that a flood study should be carried out to determine if the reservoirs had sufficient freeboard and spillway capacity to pass the design flood event (PMF). The flood study confirmed that there were significant deficiencies and that capital works were required to rectify the problems.

The works at Thornton Moor and Stubden proved to be minor and were carried out straight away. Works at Leeshaw were more significant and were completed as part of Yorkshire Water's AMP3 capital programme in 2004. The works at Leeming, Doe Park and Hewenden were scheduled for the current AMP4 programme and are the subject of this paper. The key parameters for the three reservoirs that determine the design criteria for the improvement works are summarised below.

Reservoir	TWL (mAOD)	Wave Surcharge (m)	Permissible Flood rise (m)	Min Crest Level (mAOD)	Min Wave Wall level (mAOD)	Peak outflow (m^3/s)
Leeming IRE	254.97	0.60	1.32	256.29	257.53	89
Doe Park IRE	244.91	0.65	1.76	246.67	247.54	120
Hewenden IRE	205.74	0.65	1.65	207.39	207.39	79

LEEMING RESERVOIR

Physical Hydraulic Model

A physical hydraulic model was constructed to assess the performance of the existing spillway arrangement. The model consisted of the weir and its approach channel, the tumble bay arrangement including part of the by-wash channel and the spillway chute. The testing confirmed that the restricted exit to the tumble-bay caused the weir to drown under PMF conditions, with the result that the flood rise exceeded the permissible value. To prevent the weir from drowning it was concluded that a widening by 1m and deepening by 2m of the downstream end of the tumble bay was required. The model also showed that over a large part of the chute the flow depths were greater than the height of the side walls. The problem was exacerbated by small up-stands at the downstream end of each cascade step which were shown to cause a localised increase in flow depth. It was decided therefore that the up-stands should be in-filled so as to induce skimming flow and reduce depths.

Preferred Solution

The remedial works project comprised the following components:-

- Deepening and widening of the tumble bay
- Enlargement of the spillway chute
- New access bridge across the spillway
- Reinforcement of the embankment wave wall

Optioneering concentrated on the construction of these works in order to determine the most practical and cost effective solutions. Several alternatives were considered for the wave wall including pressure grouting to reinforce the lower part of the wall (up to still water flood level), demolition and re-building of a like for like wall, or construction of a completely new wave wall with a central concrete section that would be linked to the clay core. For the lower end of the spillway the possibility of limited out of channel flow with appropriate protection works such as gabion mattresses or concrete slabs was also examined.

However, following the Boltby incident it was decided that in this case all of the flow should be contained and hence a new, larger, reinforced concrete channel was required. Given the nature of the existing spillway and it's proximity to the embankment mitre, it was determined that the most practical solution was to construct a new enlarged reinforced concrete channel as a lining to the existing masonry chute.

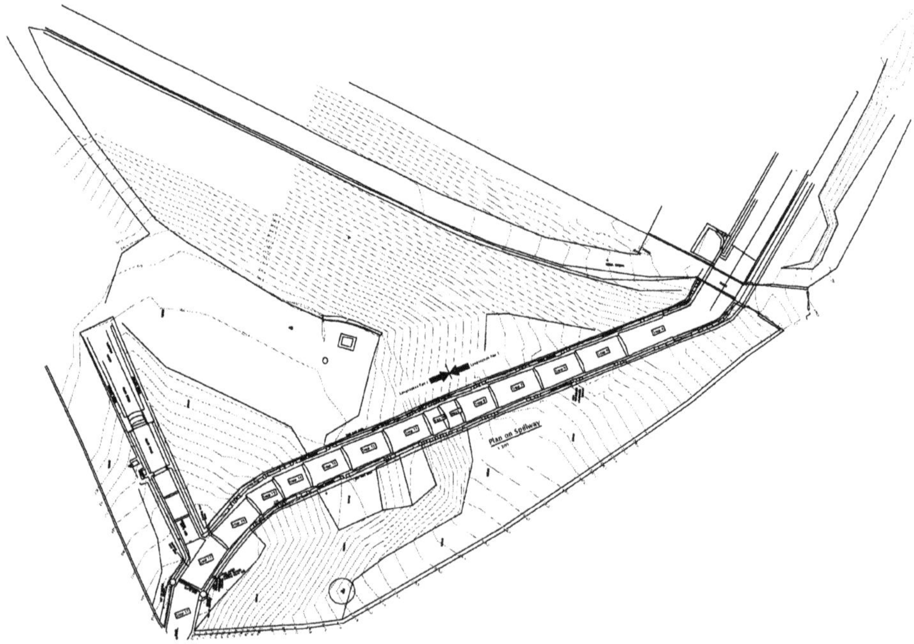

Figure 1: Leeming Reservoir Embankment and Spillway Arrangement

Construction and flood risk protection

Reservoir safety works of this type require a detailed contingency and temporary drawdown plan for the construction phase. The design criteria demanded that the 1:100yr flood should be accommodated without causing inundation of the partially completed works or overtopping of the embankment. The storage buffer that was needed resulted in a temporary drawdown of 4.5m at Leeming, together with a 1m drawdown of Thornton Moor Reservoir which is located upstream. In addition a weir on the incoming catch-water system was temporarily removed at Thornton Moor to prevent overflow into Leeming catchwater, this meant that as much as possible of the incoming flow could be diverted away from the reservoir.

The spillway was constructed in three distinct phases. First the base of the spillway was constructed, then the existing spillway structure was lined using a single face shuttering system and finally the top section was cast using a two faced shutter system. Overall, the depth of the new spillway cascade averages 6m. The structure is predominantly buried with only the top 1.1m visible from the embankment. All of the external faces have been clad using local stone from re-claimed sources.

The modifications to the tumble bay were the most complex parts of the project in that the new works had to be incorporated within the existing structure. The excavation to deepen the tumble bay was about 8m deep, and Larssen LX16 piles 10m long were pre-driven and braces were installed prior to demolition and excavation of the existing arrangement. The bracing arrangement was complex as there was only one side of an excavation to push-off, so that raking props had to be used to provide the required thrust.

As part of the contingency plan all excavated surfaces were to be blinded with concrete at the end of every shift to provide some erosion protection should the reservoir have filled and overflowed. The tumble bay was constructed using reinforced concrete with a masonry clad face on exposed vertical surfaces. To increase the length of any potential leakage path a key was cast around the outside of the structure where it passed through the clay core. Upon completion of the new reinforced concrete works the majority of the sheet piles were removed. However, those which had been driven into the clay core were left in place and were cut down to crest level. The interface between the new structure and the piles was also a potential leakage path, and to overcome this problem a puddle clay plug was constructed between the key and the face of the sheet piles.

The access bridge across the spillway provides the only access for residents to the west of the dam, and access had to be maintained for all vehicles at all times throughout the duration of the scheme. It was necessary to provide a temporary access bridge and diversion route in order that works on the tumble bay could proceed.

Raising of the clay core proved to be particularly interesting. The sourcing of good puddle clay in Yorkshire proved to be difficult: there are many sources of clay but testing showed that several of these contained traces of arsenic and for obvious reasons could not be used. Upon finding a suitable source of good clay it was necessary to re-work the clay to the specified parameters. This operation of re-working the clay was carried out at the source site as there was insufficient room on-site for this re-working process. A concrete lined and covered bin was provided on site to enable the deliveries of clay to be kept free of contamination and at the required level of workability. The raising of the clay core was carried out using 2 excavators and 2 dumpers each of them handling either new or old clay to prevent contamination of the new clay. Compaction of the clay was carried out using a vibrating plate attachment on the clean excavator, compacting the clay in 150mm layers.

At the outset of the works it was intended that the wave wall should be strengthened by pressure grouting. This approach was found to be impractical as the central cavity of the wall had, over time, filled with debris. It was not possible to grout the wall with the debris present so the decision was taken to re-build the wall utilizing the existing walling stone with a reinforced concrete core.

Figure 2: Tumble bay under construction

DOE PARK RESERVOIR

Investigations

The flood study and subsequent testing of the spillway indicated that there was a significant shortfall in both embankment freeboard and spillway discharge capacity. Remedial works were essential to overcome these deficiencies and again it was determined that the outflow discharge (PMF) should be fully contained along the length of the spillway channel. In addition the existing wave wall along the crest of the embankment was in a poor state of repair and needed to be replaced.

Preferred Solution

The main components of the remedial works project were:

- Increased wall heights for the spillway channel
- New wall to upstream side of crest to retain wave and flood rise
- New boundary wall to downstream side of crest
- New road bridge across the spillway
- Raising of the valve shaft to accommodate the predicted flood rise.

Various value engineering options were examined to facilitate construction and to minimise costs. For the crest walls the following alternatives were considered:

(a) Re-construct with wet-walling to current standards
(b) In-situ concrete wall with stone cladding
(c) In-situ concrete with form liner and stain finish
(d) Pre-cast concrete with form liner and stain finish

In the case of the new 17m span spillway bridge the focus was on the arrangement and construction method that would cause minimum disruption. These included:-

(a) Re-construct bridge pier with knife edge and strengthen bridge
(b) Construct a new double span bridge with knife edge pier
(c) Construct a new single span bridge in pre-cast concrete
(d) Construct a new single span bridge in steel.

For the spillway channel the alternatives considered involved the choice between masonry cladding or a form liner finish to the concrete. The possibility of battered walls was also compared with vertical walls, where the increased channel capacity and hence reduced height had to be measured against the more complicated formwork that would be required.

For the valve shaft modifications the alternative of raising the outer walls to a level that was above the flood rise level was compared with the option of capping and sealing the top of the shaft which in any case was no longer to be used as an overflow.

In the preferred solution the wave and crest retaining walls were re-built using pre-cast concrete wall units that were linked to the clay core using an in-situ cut-off trench. Essentially this widened the crest by 450mm as the reinforced concrete wall units were thinner than the existing wall units. The bridge across the spillway was built of pre-cast bridge beams with an in-situ deck. This solution dispensed with the risk of blockage beneath the bridge. The spillway was re-constructed in reinforced concrete with a formliner and stain finish to the visible external surfaces. The shaft was raised in concrete with an ashlar formliner and staining to the external surface. The concrete stain that was selected was 'ebony' which best matched the concrete and masonry of the surrounding environment.

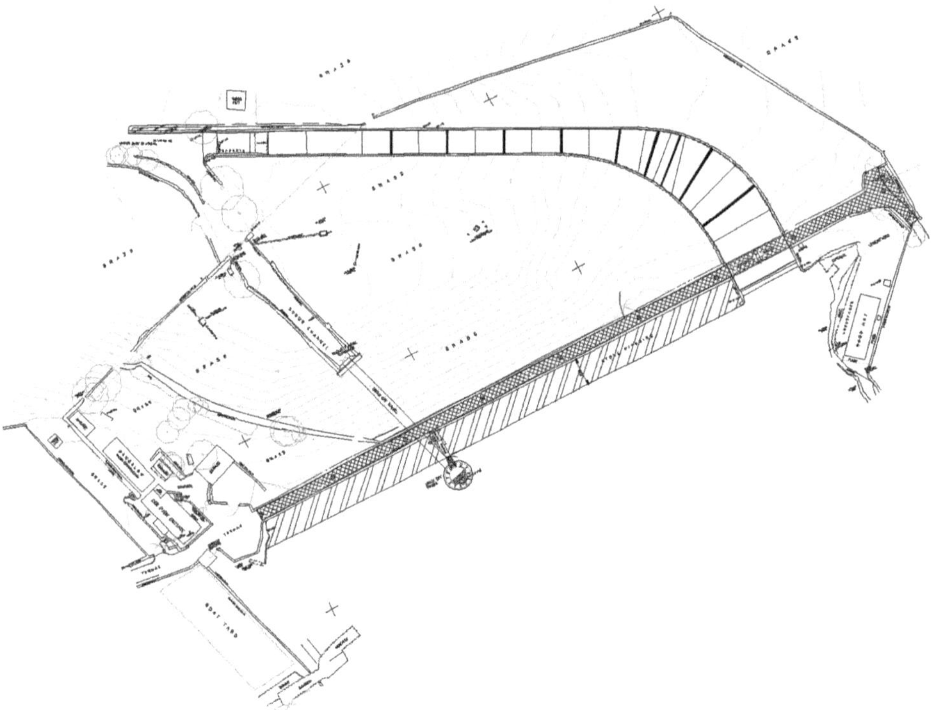

Figure 3: Doe Park Reservoir - General Arrangement

Construction and flood risk protection

The remedial works at Doe Park were to be constructed concurrently with similar works at Hewenden reservoir, which is located directly downstream. Thus the contingency plan was formulated to give adequate protection to both of the projects. It required a temporary drawdown of Doe Park reservoir by 4.5m, together with draw downs of 1m and 2m at the upstream reservoirs of Thornton Moor and Stubden respectively.

The first phase of spillway construction was to form temporary mass concrete protection bunds on either side of the existing channel. This was done prior to the demolition of the existing walls. The provision of the temporary channel proved its worth in December 2006 when after a period of prolonged and heavy rainfall all of the reservoirs in the cascade filled and overflowed. The contingency plan that was in operation at the time meant that the incident was well managed. All parties involved knew what to do and who should be contacted. Whilst spilling from the reservoir was unavoidable, the temporary works proved to be very effective in that the discharge was restricted to the middle portion of the partially completed channel. There was some disruption to construction activities but no damage was sustained to either temporary or permanent works.

Figure 4: Doe Park Reservoir - Overview post construction

HEWENDEN RESERVOIR

Investigations

The model testing of the Hewenden spillway showed that the horseshoe shaped weir operated under free discharge conditions, but even so the predicted flood rise under PMF conditions was above the embankment crest. Hence works were required to raise the core of the embankment and to provide a substantial wave wall that could withstand both the still water flood rise and the wave surcharge. It was also shown that overtopping of the spillway walls would occur over a large part of the channel and to address this deficiency, modifications were needed to enhance the channel capacity and contain the flow.

Preferred Solution

The main components of the remedial works project were

- A reinforced concrete wave wall along the embankment crest with a concrete nib beneath its base slab that extended into the clay core.

- A new concrete slab invert and concrete side wall to part of the overflow channel; this replaced the original masonry that was in poor condition.
- Wall heightening to both sides of the spillway channel over most of its length.
- Protection works in the downstream area between the end of the spillway and the outlet channel from the draw-off tunnel. This area would be inundated under extreme flood conditions but the movement of flow was shown to be away from the embankment toe.
- Modifications to the top of the valve shaft to contain the maximum surcharge level.

In a similar way to the other schemes a number of alternatives were considered during the planning and design of the works in order to improve constructability, minimize the visual impact of the new works and reduce the overall construction period, thereby reducing costs. In the case of the wave wall, conventional masonry construction with a concrete core was compared with a reinforced concrete wall with either masonry cladding or form liner finish. It was the concrete wall with a stained form liner finish that proved to be the most economic, whilst providing a final appearance that was acceptable to all parties. Similar comparisons were made for the raising of the spillway walls where it was very important that the appearance of the new walls should be in keeping with the existing masonry.

<u>Construction and flood risk protection</u>

Works at Hewenden reservoir were timed to coincide with works at Doe Park reservoir in order to reduce the impact on water stocks. A combined flood contingency plan based upon protection for floods up to the 1 in 100 year event was therefore developed. This required a 4m temporary drawdown at Hewenden along with the 4.5m of flood storage that was available at the upper reservoir. The two sites were managed as a single operation and there was close liaison with YW Operations staff to ensure that the plan was implemented successfully.

Works to the spillway were split into three phases; re-construct the 60m long upper portion of the channel in reinforced concrete, raising of the remainder of the channel walls and construction of the scour protection slab to the downstream area beyond the embankment toe. For the building of the wave wall, the use of pre-fabricated reinforcement cages greatly reduced the construction time, with the average rate of construction of 10m/day being achieved. Upon completion of the concreting works the structure was left to cure for a period of thirty days before the final stain was added. In tests on site it had become apparent that the stain reacted differently to fresh and aged concrete and that this inevitably affected the final appearance.

Figure 5: Hewenden Reservoir – Completed Wave Wall

Conclusions

The three projects were similar in nature as each involved the provision of a new wave wall to an existing embankment dam and significant modifications to the overflow works in order to contain the PMF discharge. However the special characteristics of each site meant that different approaches were required for both the design and construction of the remedial measures. In addition new techniques were tested that facilitated the construction and reduced costs, whilst providing a solution that was technically viable and comparable with conventional construction methods. In particular the following conclusions can be taken from the execution of these projects:-

- At Leeming the continuous access requirements through the site and over the spillway channel were a significant challenge. A ‘first time’ permanent bridge solution in preference to a temporary diversion bridge and the permanent bridge on a different alignment would have resulted in less disruption.
- The use of concrete with form liner and staining has been shown to be a real alternative to masonry cladding. Construction times are greatly reduced and the cost is significantly less. In order to achieve a uniform staining and final appearance of the concrete there needs to be a consistent concrete mix combined with tight control of curing and stain application.

- At Doe Park the use of pre-cast units for the new wall gave a significant reduction in the construction period with the main length of wall (approx. 150m) being completed in just 3 weeks. This had a positive impact in minimizing the disruption to access along the crest.
- Pre-fabricated reinforcement as used at Hewenden was found to be cost neutral on installation, but as the construction process is speeded up there were considerable savings in site preliminary costs.
- The final appearance of the completed works at all three locations is very pleasing. The new works are in sympathy with those parts of the existing works that have been retained, and in general they blend well with the surrounding environment.
- The contract strategy that Yorkshire Water employs for the delivery of its projects can work to great advantage. In each case the turnaround from the issue of project contract to the beneficial completion of the works was approximately 18 months meaning that compliance dates were met in all cases. Issues arising were dealt with quickly and without any contractual wrangling.
- There was close cooperation between the designers, the construction team and the operations staff. This was demonstrated in the successful development and implementation of flood contingency and drawdown plans which ensured the safety of the reservoirs during the construction of the improvement works.

REFERENCES

Halcrow. 2004. *Bradford Area Reservoirs Flood Study.*

Hydrotec Consultants. 2004. *Leeming Reservoir.* Physical Hydraulic Model Report.

Hydrotec Consultants. 2004. *Hewenden Reservoir.* Physical Hydraulic Model Report.

Hydrotec Consultants. 2005. *Doe Park Reservoir.* Physical Hydraulic Model Report.

ICE. 1996. *Floods And Reservoir Safety.* 3rd Edition.

MMB. 2005. *Leeming Reservoir. Spillway Works.* Water level Control and Contingency Plan.

MMB. 2006. *Hewenden and Doe Park Reservoirs.* Water level Control and Contingency Plan.

Remedial works at Sutton Bingham Reservoir

J WELBANK, Wessex Water
R SOLOMAN, Wessex Water
J L HINKS, Halcrow Group Limited
G GREEN, Halcrow Group Limited
R PHILLIPS, Halcrow Group Limited

SYNOPSIS. Sutton Bingham Dam, near Yeovil in Somerset, was completed in 1955 and is possibly the last embankment dam in the UK to have been constructed with a puddle clay core. The dam behaved satisfactorily until October 2006 when the annual settlement readings along the crest showed a sudden increase of up to 80 mm over a length of about 45 metres. There was further settlement of 73 mm during the period from October 2006 to January 2008. The settlement together with the observed distortion of the concrete slabs was indicative of an incipient slope failure in the upstream shoulder of the dam.

As soon as the problem was drawn to the attention of the Panel Engineer in October 2006 he asked for the water level to be held down 2 m below Top Water Level to mitigate the consequences of any future slip. Meanwhile intensive monitoring and site investigations were put in hand. Considerable difficulty was encountered maintaining the target water level when significant inflows occurred during the early months of 2007 and this highlighted the need to increase the drawdown capacity of the reservoir.

The reasons for the slope failure are discussed in this paper and include the effects of the annual operational reservoir drawdown during the summer months. The major component of the remedial works was the flattening of the upstream slope of the dam to 1(V):5(H) using rock fill to give a satisfactory factor of safety.

The timing of the construction of the remedial works required detailed consideration to optimise construction activities without jeopardizing abstraction from the reservoir and operation of the water treatment works.

EMERGENCY DRAW-DOWN FACILITIES

The original bottom outlet from the reservoir takes the form of a 450 mm diameter pipe and valve discharging into the bottom of the bell mouth spillway 11.8 m below Top Water Level. With the 57.5 ha reservoir full this has a capacity of about 1.74 m^3/sec which is sufficient to lower water level by 262 mm/day in the absence of inflow to the reservoir. The average inflow is estimated to be 0.1m^3/s. However the reservoir has a catchment area of 30km^2 and typically overflows for several months each winter. If only periods when the reservoir is overflowing are considered, the Q50 and Q10 inflows are estimated to be 0.52 m^3/sec and 1.46 m^3/sec respectively. At the higher of these inflows there was only sufficient capacity to lower water level by 42 mm/day.

The Inspecting Engineer asked for the bottom outlet capacity to be increased by at least 1.7 m^3/s but the Owner elected to provide a total of 4.33 m^3/s new capacity, bringing the bottom outlet capacity to 6.07 m^3/s. This was done by installing two 600 mm diameter pipes through the wall of the bellmouth spillway 5 m below TWL. Each pipe was fitted with a control valve and a guard valve. Operating in conjunction with the existing scour pipe, the final arrangement allows the water level to be lowered by 838 mm/day from top water level against an inflow of 0.5m^3/s and to draw the reservoir down to 75% level within approx. 3 days. The reasons for choosing a higher capacity than that originally requested were that, due to the cost of the temporary works, the additional cost of two pipes was relatively modest and also to provide some additional control over water levels during construction of the rock fill stabilisation work.

Twin 700 mm diameter holes were first drilled through the 3 m thick concrete wall of the bell mouth into a temporary steel limpet type cofferdam which had been installed on the outside of the spillway wall. The 600 mm diameter pipes were then inserted and grouted in. The installation of the valves was completed later with the reservoir level drawn down.

Figure 1- Limpet type cofferdam

The 600 mm diameter valves each weighed 0.77 tonnes so there was a need for a steel platform bolted to the spillway to support the four valves. This structure was provided with impressed current cathodic protection.

Figure 2 - New bottom outlet valves

The valves are operated by electric actuators installed on a new steel bridge leading from the original valve house.

The north-west side of the spillway shaft has settled since construction and, in 2006, it was 38 mm lower than the south-east side. It is therefore unfortunate that the weight of the valves, platform and access bridge had to be carried on the north-west side of the shaft. There was no obvious alternative. However analyses suggested that there will be negligible acceleration of the tilt of the shaft. This will need to be kept under observation in the future.

Figure 3 - New bridge providing access to valve spindles and actuators

UPSTREAM SLOPE STABILITY

Sutton Bingham Dam has a maximum height of 15.2 m and impounds a reservoir with a capacity of 2.6 Mm^3. The reservoir area is underlain by Middle and Upper Jurassic formations consisting mainly of clays and limestones overlain, in the valley bottom, by a thin cover of recent alluvium. Water in the reservoir submerges successively the Oxford clay, Cornbrash limestone, Forest Marble (clays, mudstones and shelly limestone) and, at the shallow end, Fullers' Earth while the dam itself is founded on Forest Marble with its ends in the overlying Cornbrash limestone and Oxford clay.

Figure 4 - Wave wall and concrete slabbing prior to the start of the works

The Cornbrash limestone is very permeable and a cut-off trench was excavated to a depth of more than 70 feet (21.3 metres) beneath the surface on the left bank. The grout curtain extends beneath the dam and for some distance at either end.

The upstream face of the dam, with which this paper is primarily concerned, slopes at 1(V): 3(H) at the top slackening to 1(V): 4 (H) and ultimately 1(V): 6(H) further down. A paper in 1957 by R.C.S.Walters and R.J.C.Walton stated that the shoulders of the dam were composed of Cornbrash limestone and Forest Marble Clay. The site investigations in 2006/2007 however discovered very little of the Cornbrash limestone fill suggesting that, at least in the area exhibiting movement, the upstream shoulder was composed mostly of the Forest Marble Clay.

Discussion on the above paper in March 1958 drew the comment by A.C.Penman that piezometers in the upstream shoulder of the dam were particularly slow to respond to changes of water level in the reservoir. In the light of the 2006 movements this remark now seems particularly significant.

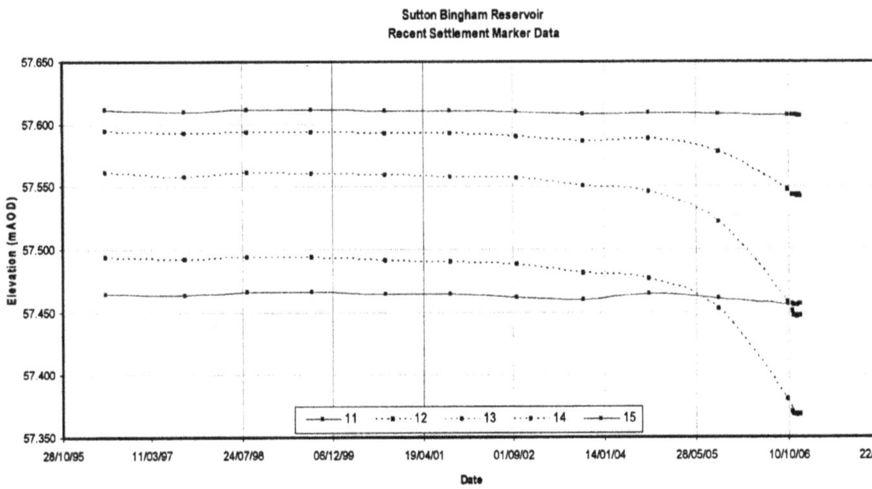

Figure 5 – Settlement at selected monitoring points

In 1974 Binnie and Partners had been commissioned by the Wessex Water Authority 'To investigate and report on the feasibility and cost of enlarging Sutton Bingham Reservoir as a means of augmenting water resources within the Parrett catchment'. As part of this investigation there were five boreholes and three trial pits on the dam. Working from these data a cohesion of 10 kPa and Ø' of 25 degrees were assumed for the Forest Marble in calculations performed at the end of 2006. With rapid drawdown assumed to 4.1 m below TWL the Factor of Safety falls to 1.01.

The subsequent installation of inclinometers showed the depth of the slip surface to be close to that predicted by these early analyses. However the analyses do not explain why the concrete slabbing on the upstream face of the dam had apparently slipped about 100 mm down the face overturning the toe beam. It would appear that this was an independent phenomenon although probably also related to the low shear strength of the Forest Marble.

Early in the investigation cracks were noted in the fill immediately downstream of the wave wall. When these were explored with trial pits softening of the sides of the cracks was observed prompting speculation that the cracks had opened during the dry summer of 2006 and that water had entered them during the autumn. This may have played a part, but the principal mechanism of failure was probably the annual 'rapid' drawdown which occurs during the summer months. During the periods May to September 2005 and 2006 levels were drawn down by about 4 m with the upstream piezometers showing very little response.

Recent settlement data is shown on Figure 5. An unanswered question is why significant settlement was only observed in 2005 (when it caused little concern) and in 2006. The reason that there was negligible settlement in previous years is still unexplained although the appearance of cracks immediately downstream of the wave wall in 2006 (and possibly also in 2005) might possibly offer an explanation.

OPTIONS FOR REMEDIAL WORKS

The following four options were considered for the remedial works:

Option 1 Remove the concrete slabs and slacken the upstream face of the dam to 1(V): 5(H) by placing rock fill.

Option 2 Construct a granular toe berm with its crest about 3.4m below TWL and then place rockfill at a gradient of 1(V): 5(H) above it.

Option 3 Construct a sheet pile wall along the upstream face of the dam and use it to retain rock fill above it at a slope of 1(V): 5(H).

Option 4 Dig out the slipped Forest Marble in the upstream shoulder of the dam and replace it with granular material.

Option 1 was eventually chosen for the following reasons:

- Conventional and robust solution.
- Least sensitive solution to variations in ground model and geotechnical parameters.
- Construction works generally of one form – "earthworks".
- Construction works are relatively unaffected by changes in reservoir level and/or adverse weather conditions.

The main disadvantages with the option were that it would require the deployment of waterborne plant and require the greatest volume of rock fill – with associated logistical and environmental issues.

DESIGN OF UPSTREAM SLOPE

Factors of safety for the existing and proposed slopes were calculated as follows:

Design Case	Required	Existing	Remedial Option
Operational Drawdown	1.3	1.0	1.42
Operational Drawdown plus Seismic	1.05		1.13
Reservoir Full	1.5	1.29	1.67
Reservoir Full + Seismic	1.2		1.40
Emergency Drawdown	1.1		1.32

The 10,000 year Peak Ground Acceleration was 0.08 g. The analyses considered 2/3 of this figure as pseudostatic methods were used.

The above factors of safety were considered satisfactory.

The design required the breaking up and removal of the existing 125mm thick concrete slabs. There was then a fine filter (Filter 2) overlain by a coarse filter (Filter 1) on the uncovered surface. Filter 1 and 2 layers both have a minimum thickness of 150mm. The rip rap wave protection layer is 360mm (2-rocks layer) thick. The wave condition used in the design of the rip rap sizes was hind-cast using Jonswap spectrum for a fetch length of 1,000m and 1 in 200 years return period +50% wind speed.

The rip rap was designed using Van der Meer (1998) for Initial (minimal) Damage Level, which corresponds to no damage level (0 to 5%) in Hudson's formula. The design rock size, W50 is 15kg, based on a 1 in 5 slope and a permeability factor of 0.1 for the impermeable embankment material.

Achieving the correct grading for the rip-rap proved extremely demanding. First the quarry sent material that was much too small and when this was rejected they supplied material that was far too large. Many loads had to be sent back to the quarry and even when the stones were in place some had to be dug out and replaced.

Figure 6 - Recently placed rip-rapUnderwater Works

Fill at the upstream toe of the dam and part way up the slope had to be placed under water. This required the establishment of a temporary wharf on the left bank of the reservoir and the importation of barges and a pontoon.

Figure 7 - Plant and barges for underwater work

PROGRAMME AND CONSTRUCTION MANAGEMENT

The objectives set by the owner were to complete the measures in the interests of safety as quickly as possible whilst not jeopardising the water supply to its customers. The water treatment works fed from the reservoir supplies around 55,000 customers. Due to the extensive placement of rock fill underwater it was expected that the water quality in the reservoir would deteriorate such that it was not treatable (this proved to be the case when work started in August 2007). Additional constraints included the need to avoid winter working on the clay embankment, allowance for the fact that the existing scour may not be able to fully control water levels, and inclusion of sufficient time after completion of the work for refilling of the reservoir before the following summer.

After detailed consideration of all these factors the agreed approach was to commence work in late summer (August 2007) after the first summer peak demands, with a target completion date of January 2008, allowing 2 months for the reservoir to refill before 1 April. The first activity was the installation of the additional scour pipes so that water levels could be controlled but this would have to commence when the water level was 2m below TWL, hence the need for the limpet dam referred to above. As the underwater rock fill level came up the water level in the reservoir was drawn down. During this particularly sensitive period the monitoring frequency for the settlement markers and inclinometers was reduced to every other day.

As there remained the possibility of high water demands during the early part of the construction programme and the risk of loss of supply from the water treatment plant, Wessex Water also implemented a resource contingency plan involving maximisation of the output of other treatment plants, increasing the transmission capacity into the supply area and bringing into service standby sources. Although primarily acceleration of future projects, it should be noted that the cost of these works exceeded the cost of the remedial works to the dam.

The project was constructed by Wessex Engineering and Construction Services (WECS), Wessex Water's in house delivery arm, with the design undertaken by Halcrow. Both parties worked closely together during the design development to ensure optimum constructability, within the project constraints. The only major sub contract was for the construction of the temporary wharf and provision of all the water borne construction plant. All materials including the rock were procured directly and the rock placing above water carried out directly by WECS.

The time constraints required the options appraisal to be carried out in parallel with the ground investigation. In addition an independent peer review by an All Reservoirs Panel Engineer was integrated into both the optioneering and detail design stages. These activities enabled the design concept to be agreed in March 2007 and the project budget set. Detail design and procurement then progressed in parallel, with site establishment and stockpiling of rock commencing in July 2007.

COST

The cost of the remedial work is approx. £ 3.4 million. A simplified breakdown is given below:

Description	**Cost £m**
Site investigations, monitoring	0.253
Environmental mitigation, archaeological surveys etc	0.036
Slope stabilisation works	
Enabling works and temporary access road	0.164
Rock fill (23,500 t) and rip-rap (6,500 t)	0.402
Plant for under water placing (10,000 t)	1.651
Plant for above water placing (20,000 t)	0.181
Miscellaneous repairs	0.010
Emergency Draw-down Facilities	
Drilling, 2 No. 600 mm diameter pipes, valves, bridge etc.	0.192
Design, project management, preliminaries, supervision etc	0.515
Total	3.404

CONCLUSIONS

The slope movements at Sutton Bingham were not particularly dramatic but might well have become so. Very careful monitoring was carried out during the period of the remedial works to give early warning of any acceleration of the movement. Careful control was also maintained over changes in water level during this period.

The time taken from the first site visit by the Panel Engineer (19 October 2006) to the completion of the £ 3.4 million remedial works (February 2008) was about 16 months. During this period the initial appraisal was made followed by site investigation, detailed design and construction. The compressed timescale was made possible by the use of Wessex Water's in-house construction arm.

A very obvious lesson from the incident was the value of the annual level survey even where there had been 50 uneventful years since the original completion of the dam.

Another useful lesson learned from the incident was the value of a bottom outlet with sufficient capacity to control water levels in an emergency. This was provided relatively easily at Sutton Bingham – in many cases it would be more difficult.

REFERENCES

Van de Meer (1988), *Rock slopes and gravel beaches under wave attack*, Doctoral thesis, Delft University of Technology

R.C.S.Walters and R.J.C.Walton (1957) *Proceedings of the Institution of Civil Engineers,* Water Supply for the Yeovil District (Sutton Bingham Scheme) and Discussion on the above paper in March 1958

Binnie and Partners (1974) *Report for Wessex Water Authority,* 'To investigate and report on the feasibility and cost of enlarging Sutton Bingham Reservoir as a means of augmenting water resources within the Parrett catchment'

ACKNOWLEDGEMENTS

The authors acknowledge the kind permission of Wessex Water Services Ltd. to publish this paper

Llwyn-On Reservoir Scour Valve Refurbishment

J.M.LEWTHWAITE, Technical Director, Blackhall Engineering Ltd.
D.RICHMOND, Site Operations Director, Blackhall Engineering Ltd.
A.R.BEEVER, Senior Project Manager, Blackhall Engineering Ltd.
AK HUGHES, Director of Dams & Water Resources, Atkins Ltd

SYNOPSIS. This paper has been prepared as a case study looking at the refurbishment of critical valves that had previously been declared to be beyond economical repair. It describes the inspection and repair of two 900mm diameter Larner Johnson valves installed some 80 years ago. There were considerable challenges in gaining safe access to enable the valves to be removed from the outlet tunnel and taken back to the factory for refurbishment. Once the valves had been removed the gears and plunger were refurbished, bronze liners were cast and fitted, all iron surfaces were coated with epoxy paint and all bolts were replaced with stainless steel ones. The valves were successfully reinstalled and commissioned.

SITE INFORMATION AND HISTORY

Llwyn-on reservoir is retained by an earth embankment with clay core some 23.85m high and 400m long. It was built between 1910 and 1926 on the Taf Fawr and is the lowest of three dams in the locality. The embankment has a puddle clay core set on a deep concrete filled cut-off trench. The upper part of the slope is protected by powled pitching 450mm thick whilst the lower slopes are protected by shingling 300mm thick.

Fig 1: Extract from OS Map

Fig 2: View of tunnel exit and stilling pool

Ensuring reservoir safety into the future 2008, Thomas Telford, London

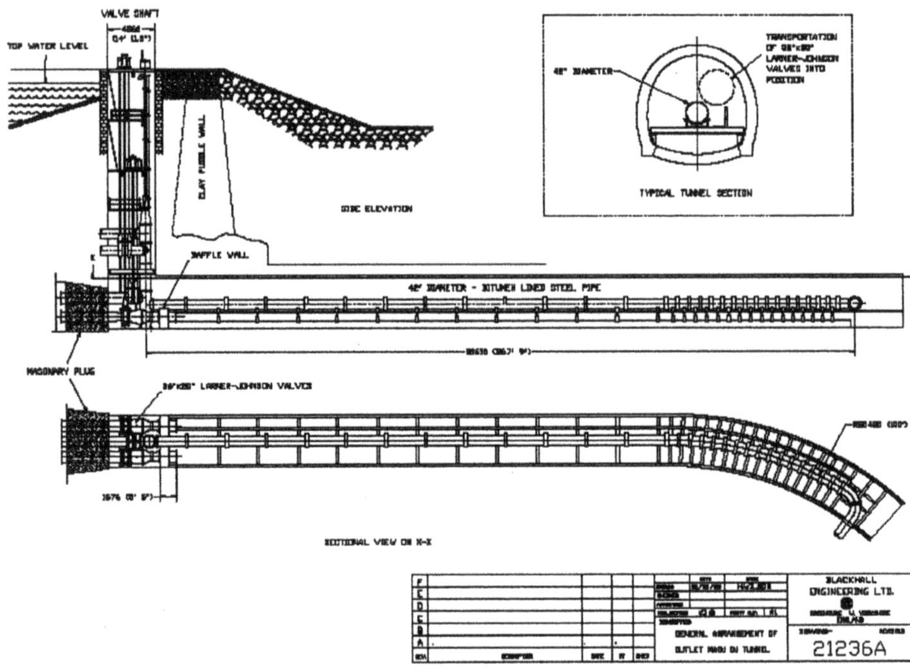

Fig 3: Cross sectional drawing ot embankment, valve tower and discharge tunnel.

The crest is 8.5m wide and carries a road and a wave wall on the upstream side. The reservoir has a capacity of 5,504,550 cubic metres and a surface area of 0.59 square metres when full to the top water level of 260m above sea level. The valve shaft is 4.6m in diameter in the upper part and is situated such that its centre is 9.2m upstream of the centreline of the cutoff. As a result the downstream tunnel is 86m long whilst the upstream tunnel is 68m long.

The dam is equipped with three 900mm diameter drawoff pipes at various levels each with valves in tandem and connected to a 1066mm diameter main standpipe. There are two 1066/900mm scour pipes with 900mm valves (1 sluice valve and 1 Larner Johnson) on each pipe discharging onto the floor of the tunnel at the base of the shaft.

The reservoir was inspected under Section 10 of the Reservoirs Act 1975 by Dr A K Hughes of Atkins. A recommendation in the interests of safety was made which stated "the inoperable Larner Johnson valve be made to operate. I would expect this work to be complete by the end of 2007." In fact the owner decided to repair both valves once the condition of the inoperable valve had been seen.

History of Valves at the Dam

The valves have had a mixed and varied history since they were commissioned in 1926.

- During the 1939-45 war the bottom of the valve tower was filled with gravel to minimise possible damage from falling high explosives.
- The chambers were not cleared until 1964 at which point the valves were pronounced inoperable.
- The original manufacturer was consulted and came to the conclusion that there had been a build up of manganese causing the valves to seize. This is now puzzling as no recent signs of manganese have been observed. The treatment was to fill and flush with acetic acid.
- An attempt was made in 1965 to remove one of the valves. The west baffle wall was 90% removed, but the contractor panicked when movement of valve and supporting pipework was thought to have taken place. Studying the site records now it is hard to determine if there had been some movement or a recording error. The uneasiness was compounded as it was thought movement had also taken place in the masonry plug and chamber walls, detected by cracked glass tell tales. The result was that the baffle wall was hastily replaced with concrete and work on the valve abandoned.
- In 1967 the west valve was removed after the gate valves were restrained with large strong back beams complete with timber thrust pads. The beams were let 12 inches into the chamber walls. Any sign of movement was carefully monitored as the valve was removed. Repair work was done to one valve without removing it from the chamber. Some movement was restored to one valve but no steps were taken to prevent further degradation.
- No further action was taken until 1982, although at some unrecorded time attempts were made, no doubt with good intention, to operate the east valve. This was done by the tried and tested method of extending the diameter of the handwheel with levers, resulting in a bent handwheel and, as would be discovered later, broken valve internals.
- An inspection was made and a report written in 1982 and again in 1990 by Welsh Water's designated consultants. Their findings and recommendations were:
    - That according to the manufacturer the original valves were irreparable.
    - The valves should be replaced with butterfly valves.

- The valve chamber was "a congested and inaccessible place"

Fig 4: View of bottom of valve tower

- The recommendation of butterfly valves was based on cost, available chamber space and access. We now believe this would have been a disastrous choice: a quick check of the predicted flow velocity through the system would have shown that it well exceeded the 5 m/s recommended level for a standard butterfly valve.
- No further action was taken until Blackhall's involvement in 2005.

Acknowledgment is here given to United Utilities who where able to provide much historical information on previous reports, site records and correspondence on these valves.

Valve Type & History

The Larner Johnson valve design originates back to 1910 in the USA. The valves were made exclusively under licence by Blakeborough in the UK for all worldwide markets outside of USA from 1920 until 1989, and thereafter by Blackhall Engineering.

The valve is an inline needle control valve with a very streamlined flow path even at partly open positions. The flow path splits symmetrically around the central controlling plunger, the position of which can be altered to control the annular area forming the valve throat. As the flow merges again after the throat the flow impinges upon itself helping to dissipate energy from the flow.

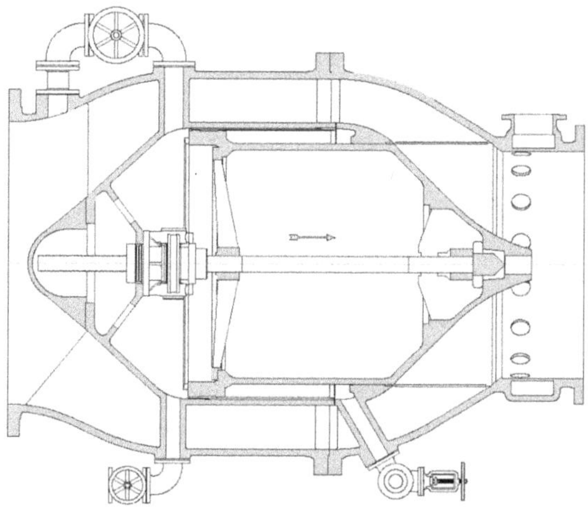

Fig 5: Cross section of a Larner Johnson Valve

The continuous annular throttling cross section at any opening position ensures linear regulation characteristics. These are the key features of the valve that enable it to handle high flow velocities of up to around 12m/s. Operating forces are kept low in service by utilising the available pressure differential to assist in moving the control plunger. Any type of valve operating with high pressure differentials will always be in danger of some cavitation. The fundamental design layout of the Larner Johnson valve easily lends itself to cavitation control where other designs cannot. This is achieved by the addition of an air belt to introduce air into the flow stream immediately after the valve throat, the point of highest flow velocity.

The typical applications for this type of valve are in water handling and treatment installations, anywhere that high pressure differentials result in high velocities. Reservoir and dam scouring and emergency drawdown are classic examples of this. When the discharge is into a tunnel or aqueduct then the controlled and focused jet issuing from a Larner Johnson is ideal. Very early valves were purely pressure operated, however all subsequent designs have some form of mechanical control operating an internal pilot valve. There are many variations of operating mechanisms: worm and wheel, lead screw and nut, lever and linkage, rack and pinion and in this case a double rack and pinion. These operate as "pressure assisted" valves and are inherently safer.

VALVE REFURBISHMENT SCHEME

The problem was quite clear when work began on the reservoir in 1964. There was no effective scour draw down capacity to drop water levels quickly. The west valve was in very poor condition with a limited internal movement and an excessive seat leakage. The east valve was seized solid and had been since before 1964. The accessibility to the base of the valve tower is very poor and there was a history of rumours of pipework and masonry plug movements when previous attempts had been made to remove the valve.

Blackhall Engineering Investigation

Blackhall Engineering was employed in April 2005 by United Utilities / Welsh Water Authority to inspect and report on the state of the valves. On receiving the enquiry, before even seeing the valves we were able to produce from archive records the manufacturing drawings and an original photograph of one of the original valves as produced in 1923.

Fig 6: Llwyn-On valve in production.

General assessment of old valves.

The primary task of any investigation is to establish whether the main body castings are capable of being reused, whether the structure is sound.

Deterioration in valve castings can take four forms:

- Mechanical wear, due to movement of internal components, usually repairable.
- Erosion, due to mechanical action by grit or debris.
- Erosion, due to low pressure, high velocity cavitation.
- Corrosion. Can sometimes be combined with erosion to form deep pits and accelerated metal loss.
- Graphitization, where the graphite in the iron castings separates out leading to degradation of the metal properties.

Material Analysis.

One of the most important aspects of assessing the valves was carried out as soon as possible. This involved performing a metallurgical assessment of the base material of the main body castings. Other NDE techniques can also be used, such as Ultrasonic, Radiographic inspections etc.

A sample of material was cut from the discharge section casting of the west valve and sent to a laboratory for analysis. The sample was taken from one of the reinforcement ribs between the bolt holes on the flange.

The findings of the laboratory were as follows:

- The casting was of a low strength grey iron with a relatively high carbon equivalent value of 4.5% similar to BS1452 grade 150.
- The casting surface shows some graphitic and general corrosion, down to a maximum depth of 5mm. This is not deemed to be excessive on such a heavy section casting, which would have a corrosion thickness allowance of 16mm added at the design stage.

<u>Internal Examination</u>

The key to the thorough investigation was to prove to be an internal inspection using an endoscopic camera. The findings of this operation were:

- West (Right Hand) Valve.
    - Internal Chamber: The operating mechanisms all appear to be in acceptable condition for an 82 year old valve.
    - Plunger guide: These valves were produced without non-ferrous bearing liners, the cast iron bearing surface was found to be badly corroded and severely pitted. The plunger and pilot valve were found to be capable of achieving full travel.
    - Upstream guard valve: appears to be in good condition.
- East (Left Hand) Valve – Previously over torqued.
    - Internal Chamber: Some damage could be seen to the teeth on both vertical and horizontal racks. The pinion wheel connecting the two racks was missing, the mounting bosses for the pinion shaft on the gear bracket were seen to be sheared off.

<u>Functional tests.</u>

- West Valve.
    - Although the valve achieved full movement the action was found to be far from smooth.
    - There was found to be excessive seat leakage between the 9 and 3 o'clock positions.
    - The plunger was independently lifted in order to achieve better alignment with the valve seat. With the plunger in this position the valve was pressurised. The leakage rate was reduced by 75%.
- East Valve. No functional test could be carried out on this valve due to it being seized, with no means of operating the plunger.

Assessment of Site Access

Measurements were taken to establish the feasibility of removing the valves from the bottom of the valve tower. The only possible option was to remove the valves down the outlet tunnel. A full size mock up was made of the valve from polystyrene flanges mounted on a length of plastic pipe. This was walked through the tunnel to determine whether it would pass all obstructions. It was estimated that the valves could be removed with small localised areas of interference with the tunnel brickwork and outlet pipe intermediate flanges.

Comments on Investigation Findings

West Valve:

- The seat leakage is due to the loss of support and guidance of the plunger afforded by the poor bearing surface condition. The operating mechanism cannot generate sufficient thrust to force the plunger up the seat face to seal.
- There is severe corrosion to valve ancillaries. A cavity has formed behind the discharge section seat ring; this must be repaired before further corrosion exacerbates the situation.

East Valve:

- It was concluded that this valve had stopped functioning and could not be repaired in situ due to the following sequence of events:
  - Infrequent operation, coupled with a poor design of bearing surfaces have caused the valve to seize.
  - The mechanism has been subjected to a very large torque at the handwheel. This has loaded the gear bracket in a tensile direction to a stress beyond its capability and failed. The bracket may have survived a load in a compressive direction.

Recommendations from Initial Investigation

- Remove both valves and conduct a factory refurbishment. The bearings could be brought up to modern standards. The seat leakage eliminated. The valve could be hydraulically tested and re-certified.

Sitework Operational Challenges

Blackhall Engineering were appointed as main contractor to remove and refurbish the East valve in June 2006, and subsequently to perform the task on the West valve in November 2007

The project brought many challenges and difficulties, the most notable of which were:

- Blackhall took responsibility for the Health and Safety on site. This involved managing subcontractors performing a "mixed bag" of site tasks: Hot work, demolition, hazardous chemicals, lifting heavy objects in confined spaces, and the monitoring of possible structural movements. No reportable accidents or near misses occurred during the removal and replacement process.

- The corroded flange bolting was removed by oxy-acetylene burning. The rib stiffeners between the flange bolt holes made the normal operation of grinding through the corroded bolts impossible. The possibility of heat affecting the castings was investigated and a procedure implemented to minimise this.
- The rotten wooden thrust blocks were removed and replaced with epoxy resin. This would ensure a perfectly even pressure being exerted on the bearing area on the side faces of the gate valves. Shuttering was fabricated around the area between the valves and support beams and two pack resin poured into the voids
- Throughout all valve removal operations, any possible movement of the gate valve and pipework was monitored using a micrometer every 15 minutes. No significant movement was detected.
- A free standing lifting frame was built and load tested within the base of the chamber to lift and move the 8000kg valves
- A temporary rail track and supporting steelwork was constructed within the outlet tunnel. This was free standing and did not exert any load on the existing outlet pipe steelwork. The valve was moved on purpose-made bogies along these rails. The rail system was dismantled and moved forward as the valve was moved along the 80m tunnel.
- Some local areas of brickwork within the tunnel walls had to gouged out by around 50mm to allow the valve to pass intermediate flanges along the length of the main dam outlet pipe, such was the tightness of fit of the valve down the tunnel.

Fig 7: Valve on temporary rail system

- The end of the outlet tunnel was beyond the reach of a mobile crane. Therefore the water level in the stilling pond had to be reduced to allow heavy plant machinery access to carry the removed valve from the end of the outlet tunnel to a position accessible by road transport.

Design Improvements.

The first task on the valves on arriving back in Brighouse was to perform a hydrostatic pressure test on the body shell. This was to prove the investigation's assessment that the main castings remain sound.

The valves could then be stripped, blast cleaned and prime painted. Significant internal surfaces such as the seat ring and flange faces were re-machined to restore their surfaces. The cavity above the seat ring was repaired using metal filled epoxy repair compound. The main advantage of performing a full factory refurbishment was that the valves could be both restored and brought up to a higher standard than they were originally.

The design improvements made were notably:

- The gear bracket that had fractured was remanufactured from a casting in a superior material. Design changes were also made to improve its strength. FEA was not available in 1923!
- The vertical gear rack was remade in stainless steel.
- Bronze liners were cast and fitted to the bearing surface of the inlet section.

Fig 8: Gear bracket as found.

Fig 9: Refurbished gear bracket

Fig 10: Plunger cavity as found.

Fig 11: Refurbished plunger cavity

- The plunger was previously made of cast iron with small strips of bronze providing bearing surfaces. The seized plunger had to be machined out of the inlet body such was the level of corrosion between the two components. A complete new plunger was made from stainless steel casting.
- All iron surfaces were coated in a modern three coat epoxy paint system.

Fig 12: New plunger.

Fig 13: Complete refurbished valve.

- The secondary scour valves were replaced with new valves with longer, more accessible operating spindles.
- All bolting was replaced with stainless steel items.

Significant Project Tasks

- Safe working practices were observed at all times throughout the project.
- The hydrostatic pressure testing of the valve shell at 15bar confirms that the castings remain in a sound and safe condition.

- The hydrostatic seat tests proved the valves were water tight and serviceable.
- The valves were successfully reinstalled and commissioned. A fully open flow test was performed, and the resulting flow rate well exceeded the required draw down rate.

Fig 14: Refurbished valve returning into tunnel.

- The valves were now capable of performing their task safely and reliably for a long period of time, at least another 50 years.

Blackhall Engineering has been able to bring a wealth of unique experience and knowledge to this project, contributing to its successful outcome:

- Experience. The four members of staff directly involved in this project hold a combined 138 years of experience with valves and valve applications.
- Historical Records. We hold all the original drawings and records of every Larner Johnson valve produced in the UK since the 1920's.

These factors enabled Blackhall to accurately analyse the problems associated with faulty valves.

CONCLUSION

Valves that may have previously been written off as being irreparable are not necessarily so. In this case the original manufacturer had deemed these valves beyond repair because they did not have modern inspection equipment to enable them to make a better on site judgement. Without these methods of assessing the valves in situ it was the safer option for both the client and the manufacturer to advise that they be replaced with new. Clients can only make decisions based on the information presented to them. However, with the careful and accurate assessment of the valves whilst in-situ, it becomes possible for the client and contractor to commit to the costs of temporary removal, confident in the knowledge that they are worth the retrieval costs.

Modern manufacturing techniques and materials can positively alter the economic feasibility of repairing old valves, typically half the cost of a new valve. As long as the valve shell is still capable of withstanding the test pressures, it is quite possible to repair and even improve on the original valve specification. One other perceived barrier to valve refurbishment is the cost of site labour. In this case the hours would have had to be spent on removing the valves whether they were going to be repaired or replaced. It is quite tempting for a client to keep costs down by replacing an old valve, designed for the duty, with a new but different type of valve not suited to the application. Any such decision may, however, be a severe compromise to performance and reliability.

In this particular case the Larner Johnson valves were economically repairable, they could be removed safely and they are absolutely the right valve for the job. The Larner Johnson valve's ability to cope with high flow velocities at high pressure drops, its inherent robustness and controllability is perfect for this application.

The Repair of Llyn Morwinion Dam

SIMON R HICKMAN (Atkins Ltd)
ANDY K HUGHES (Atkins Ltd)
L STUART DAVIES (Dwr Cymru)

SYNOPSIS. In 2000, observed settlement of the crest of Llyn Morwinion over its outlet conduit, and sediment within this conduit associated with seepage into it, led to concern that the dam was suffering from internal erosion of the core. Initially, a remedial procedure of constructing a concrete plug in the conduit under the core was proposed but, following another inspection, it was recommended that the cause of the settlement be further investigated, the dam made good and seepage minimized. This paper describes the further investigation and recommended remedial works.

INTRODUCTION

Llyn Morwinion reservoir is retained behind a 4m high embankment dam constructed in 1879, although the water is only raised approximately 2.4 m above the lowest natural ground level, the stream level at the toe of the embankment.

The report of a Section 10 inspection of the reservoir on 18 March 2004 noted "significant leakage of sediment-bearing water into the culvert in the vicinity of the puddle clay core". This was first noted in August 2000, as was a depression in the crest within the core zone.

A recommendation was made to fill the conduit with concrete but, following a request for Dr Hughes to carry out a further inspection, it was concluded that an alternative solution might be identified that would not restrict future entry to the conduit and, thus, access for future inspection and repair to the existing outlet pipes of cast iron and uPVC materials under the direction of a Qualified Engineer.

It was recommended that further investigation be carried out and measures taken to repair the embankment where depression had occurred, and identify where leaks into the conduit are occurring and seal these under the direction of a Qualified Civil Engineer.

Ensuring reservoir safety into the future 2008, Thomas Telford, London

The results of this investigation resulted in the identification of the hard horizon as shown on Figure 1.

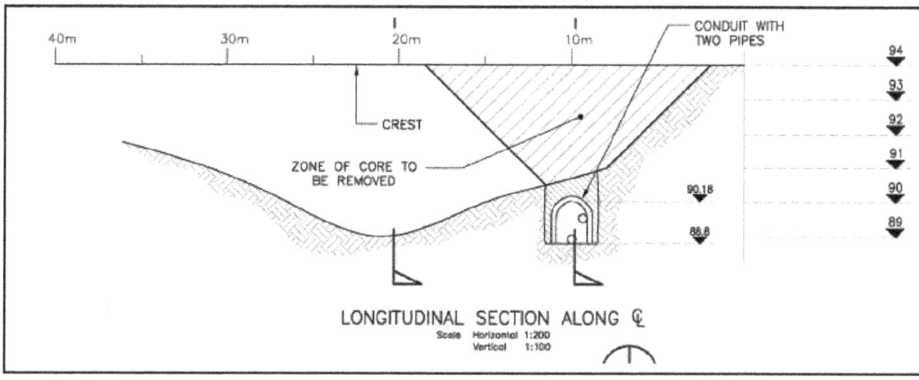

Figure 1 – Longitudinal section identifying hard horizon & conduit

This paper describes the investigation of this leakage and its proposed remediation to ensure that the embankment is made safe.

EMBANKMENT DETAILS & HISTORY OF WORKS

The earliest information of the dam appears to be a drawing prepared in 1977 by Ward Ashcroft and Parkman in 1977 entitled, "General arrangement and section through the dam", but there are no records of the geology of the dam site or the reservoir basin or indeed any other drawings identifying features at the dam.

This drawing identifies an embankment with a 3m wide crest with central 1.8m puddle clay core contained within vertical dry-stone walls. These are supported by a "hearting" at 1 in 3.25 upstream and 1 in 2 downstream slopes surfaced by dry stone pitching. The outlet capacity was increased when the original 300m diameter cast iron outlet pipe from an upstream tower was duplicated in 1983 by a 300mm uPVC pipe passing through the original approximately 1.8m D shaped conduit. The original outlet has since been altered to be a dedicated draw-down outlet.

A draw-down facility is provided by both these pipes although the upstream inlet pipework has been moved upstream within the reservoir basin, increasing the available draw-down by over 400,000 m^3.

The reservoir has a capacity of some 510,000m^3 and a surface area of 113,000m^2, and is a very important source of water for treatment and onward transmission to the Ffestiniog area.

FAILURE FEATURES

The observation in 2000 of sediment in the conduit and settlement of the crest lead to the concern that Llyn Morwinion might be in danger of failure by internal erosion.

Monitoring of seepage the quantity and quality of water into the conduit was initiated, as was deposition into the conduit.

The flow into the conduit has been measured since 2001 and the Figure 2 below demonstrates that the flow rate mirrors the changes in reservoir water level with a rate of approximately 0.32 l/s at TWL. This rate has not increased although there are variable higher rates recorded, up to 0.65 l/s in January of 2004 and 2005 which might indicate that surface rainfall into the downstream shoulder influences the flows.

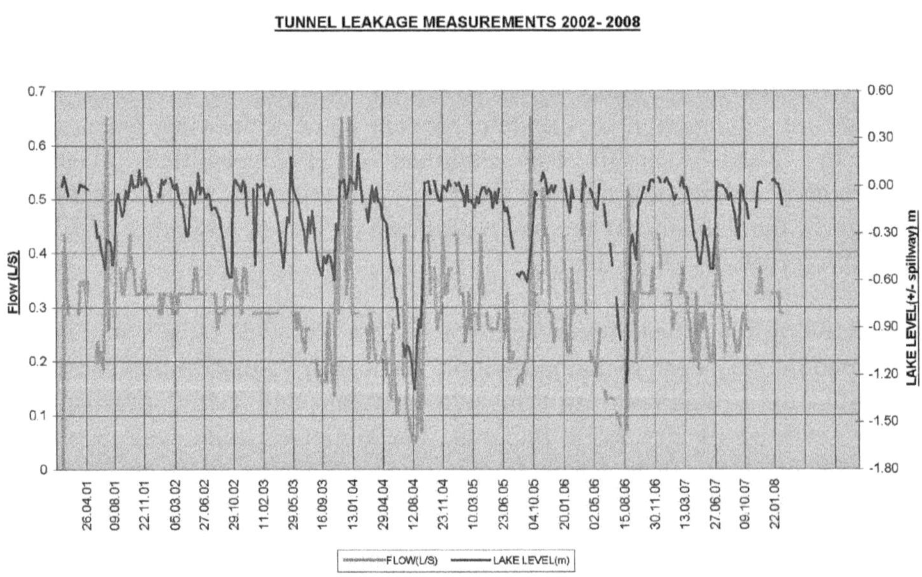

Figure 2 – Leakage into the Outlet Conduit from 2001

A preliminary investigation was carried out with boreholes drilled and piezometers placed in locations indicated in Figure 3.

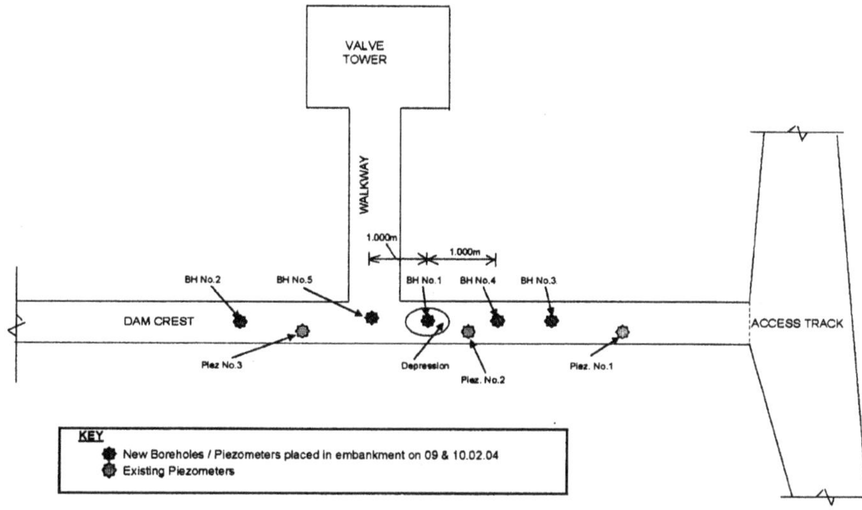

Figure 3 – Plan of Llyn Morwinion - Location of boreholes and piezometers placed in 2001

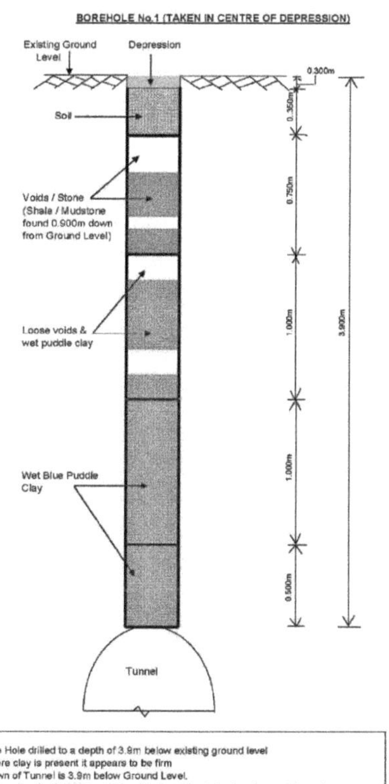

Figure 4 – Schematic log of borehole BH1 in the depression on the crest

A figurative log of the borehole drilled into the depression is given in Figure 4. Worryingly, this identified voids up to a metre below crest level and wet blue puddle clay above the crown to the conduit. The piezometers again mirrored the water level in the reservoir with a rapid response to changes in water level. At the same time survey pegs were installed in the locations shown in Figure 5.

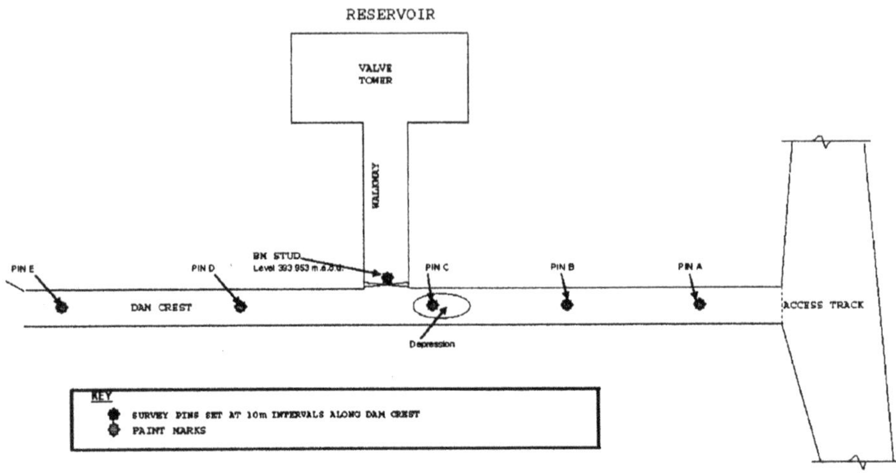

Figure 5 – Location of survey pins placed in the crest in 2001

The levels of these pins have been taken since 2001 and their rise or fall has been charted as shown in Figure 6. Apart from the pin in the depression, Pin C, which shows downward movement, the crest is shown to be not moving. Pin C however shows a continuous year on year fall of some 10mm/year that suggests a loss of material from this zone in the core, into the conduit.

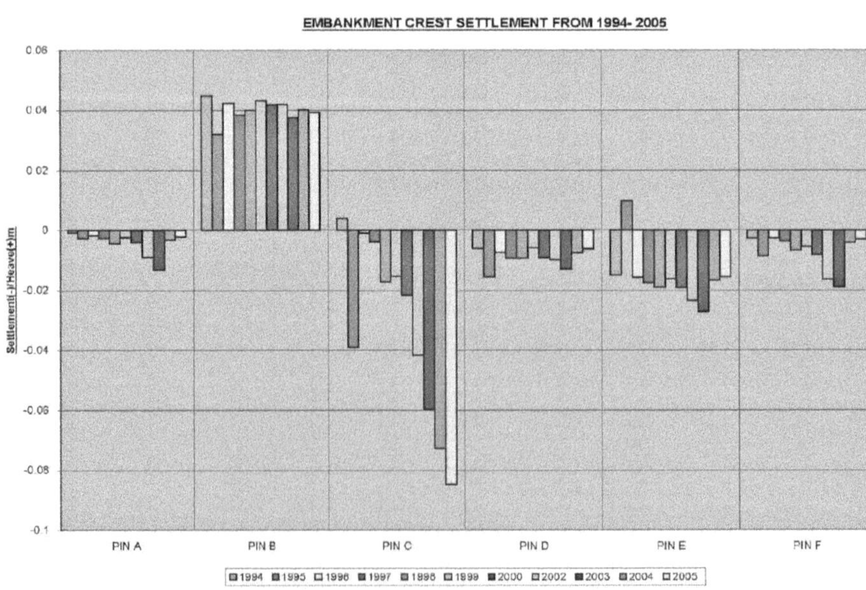

Figure 6 – Crest Settlement 1994 to 2005

Having identified a potential failure mode of internal erosion the Inspecting Engineer made a recommendation to fill the conduit with concrete. The Undertaker was concerned that this recommendation might not be the best solution to the problem and requested advice from Atkins.

Following an inspection, it was recommended that the conduit should not be filled with concrete, that further ground investigation be carried out to identify the route of the leak into the conduit, that this leak be sealed and the embankment be repaired, leaving the conduit open so that the pipes and valves could be maintained. It was considered important to identify the source of the leak and minimise it, rather than probably diverting it elsewhere.

GROUND INVESTIGATION

The results from the earlier ground investigations highlighted the voids and "changed" materials above the outlet conduit shown in Figure 4, but further information was required.

A Ground Investigation (GI) contract was designed to provide information on

- the condition of the core above the conduit to identify whether the sediment carried into the conduit originates from the core
- the depth of the embankment to the foundation
- the material and permeability of the foundation

as well as confirming the properties of the existing core material.

A GI contract under the ICE General Conditions of Contract Measurement Version 7th edition and Specification for ground investigation by Site Investigations Steering Group, 1993 was let with Structural Soils Limited in March 2007. Draft information was provided in April 2007 and a final data report in November 2007.

The location of the proposed boreholes and exploratory pit are shown on Figure 7.

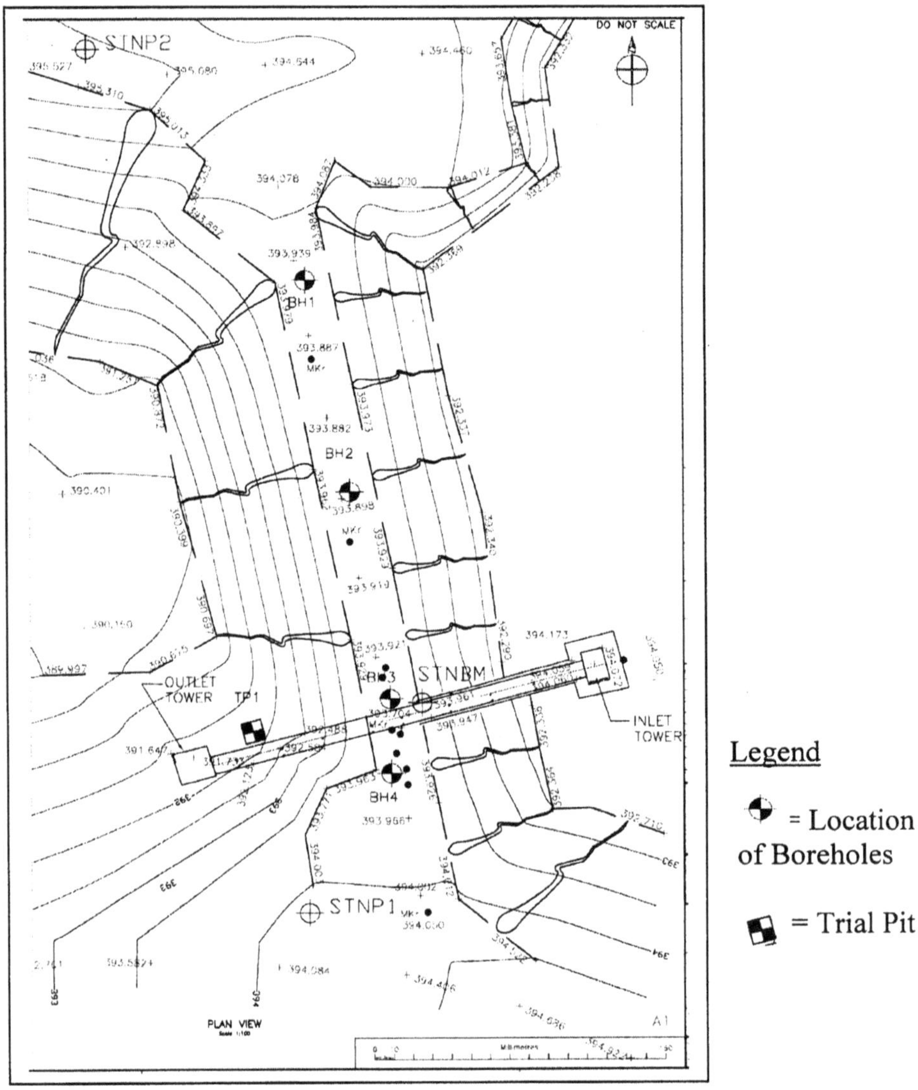

Figure 7 – Plan of Embankment identifying the Location of Boreholes and Trial Pit of the Second Phase Inspection

RESULTS OF GROUND INVESTIGATION

Preliminary logs were provided from which the laboratory testing was defined. The GI found the following:

- the puddle core " .... becomes locally fissured from 2.0m depth" with refusal at 4.7m
- the foundation material is strong narrowly foliated fine to coarse schist with slaty cleavage from refusal to over 11m.
- the trial pit identified a large flat rock covering the conduit with in situ foundation material higher than anticipated.

A log of the trial pit excavated on the downstream section of the conduit below the embankment is given in Figure 8. Whilst the internal condition of the conduit is in good condition, the information from the trial pit introduced uncertainty into the composition and condition of material above the soffit of the conduit as well as that on the sides of the conduit appears to be constructed into an excavated trench in the rock foundation. (The rock profile is shown in Figure 1).

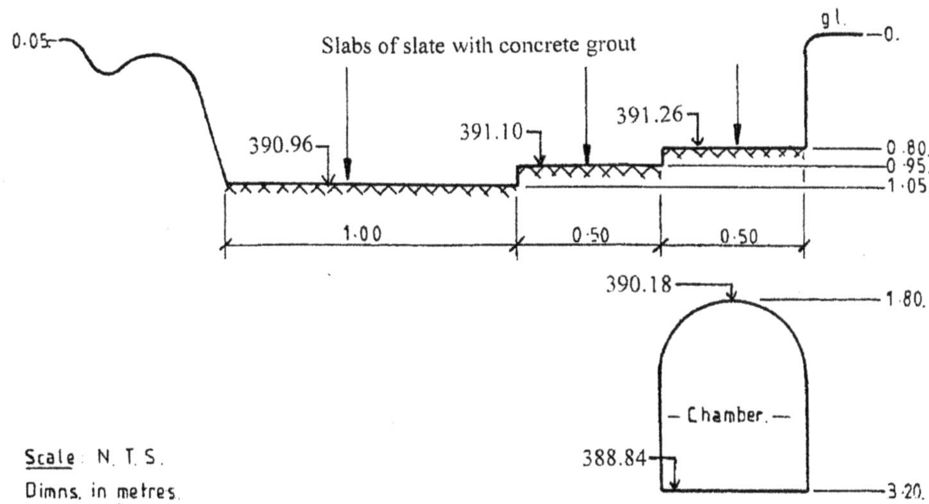

Figure 8 – Log of Trial Pit Downstream of the Embankment

The properties of the clay core from the results of laboratory testing are given in Tables 1 to 3.

Table 1 – Typical Moisture Content and Plasticity Index

Borehole No.	MC %	LL %	PL %	PI %	<425μm %
BH 2	23	40	22	18	99
BH 5	14 to 21	44	22	22	98

Table 2 – BH2: Soil Properties

Dia.	Passing
0.02	87
0.006	63
0.002	31

Fraction	Percentage
Gravel	3
Sand	2
Silt	64
Clay	31

The strength of the clay core is shown to be satisfactory with the following results from triaxial testing.

Table 3 – BH2: Triaxial Compression Tests

Effective Cohesion (kPa)	5
Angle of Shear Resistance (degs)	27.5

The foundation material is described as a grey schist with a Rock UCS = 68MPa and a Point Load Index ($I_{S(50)}$) of 4 samples ranging from 0.2 to 5.86 $(MN/m^2)S_{(50)}$

REMEDIAL WORKS

The findings of the ground investigation confirmed the loss of material from the puddle clay core and identified the foundation of the embankment to be stable.

However, the condition of the soffit to the conduit and the interface between the conduit and the excavated trench in which the conduit was constructed remain unknown. So a function of the remedial works became a procedure for providing this information and the objectives of remedial works were therefore defined to be:

- the replacement of fissured clay material in the puddle core and making good any settlement of the core
- identification of the condition of the soffit of the outlet conduit under the puddle core
- consolidation and reduction of permeability of
  - the soffit of the conduit below the core, and
  - the zone between the conduit and the excavated trench, extending into the in-situ foundation if fractured .

The TWL is 392.3m AOD with the crest level varying from 393.93 m AOD to 393.98 m AOD, i.e. a freeboard of 1.6m. Elevation of the hard horizon immediately left of conduit (looking downstream) was found to be 390.87 m AOD and right of conduit as 390.53 m AOD. The invert of the conduit is 388.8 AOD, which demonstrates that the conduit is constructed into a trench in the foundation rock. There are not records identifying how the conduit was sealed in this trench.

The depth of core to the conduit soffit was identified to be approximately 3.3m and, with the TWL at 392.3 m AOD, the hydraulic head is 1.8m.

The evidence of sediment in the conduit and settlement of the core above the conduit indicates flow into the conduit is carrying material from the core, i.e. through the conduit soffit. However, inflow is from the side of the conduit, which suggests the flow path is through the soffit into the one of construction between the conduit and the sides of the trench in foundation rock.

The puddle clay core requires remediation and its removal would make available the soffit to the conduit for inspection. The soffit would be inspected, and where possible, the zone above the sides of the conduit within the trench. It will be essential to seal this zone to restrict migration of soil particle through the soffit. Where necessary, dental concrete will be placed to provide an impervious foundation to the core when replaced. A simple grouting programme was designed to consolidate openings and seal larger interstices in the conduit soffit and sides.

As with most grouting, a flexible design is necessary with the quantity of pumped grout for a particular hole and zone limited to an acceptable maximum. The proximate conduit, because it is masonry, provides a readily available observation face but it restricts the grout pressure.

However, as eroded material had been transported via the zone to be grouted, it was considered likely that the grout would flow into the route through which the sediment has passed at a low pressure. The design includes drilling boreholes from the soffit of the conduit, shown as “a” to “d” in Figure 9 with injection of cementitious grout at low pressures us as a simple grouting programme.

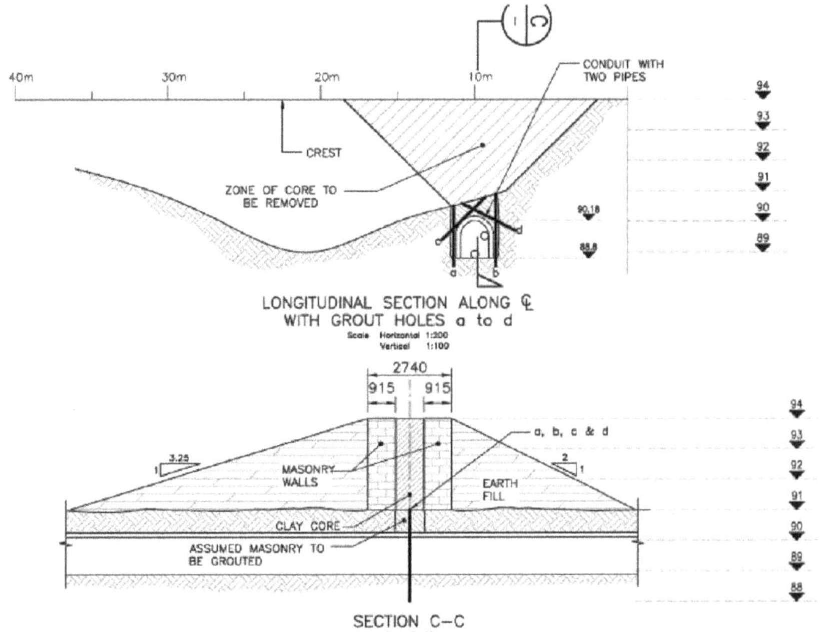

Figure 9 – Longitudinal Section and Embankment Cross-section identifying approximate location and alignments of proposed grout boreholes “a” to “d”

The sequence of grouting and pressures will be determined from water testing. This is expected to require any necessary dental work to the soffit of the conduit to be completed prior to water tests and grouting. An alternative procedure would be to drill the grout holes and backfill the core around placed steel pipes placed into these grout holes. The benefit of the latter would be to provide a head against which to grout but close inspection would be lost.

The remedial works to the core include placing a geofabric between the core and the vertical masonry wall to reduce any possibility of migration of clay particles through the wall. The geofabric has been specified to retain >10μm, the nominal size of clay flocs behind an equivalent D_{10} of 105 μm. Clay removed from the core will be inspected and, where acceptable, re-compacted in the core. Deficiencies in quantities will be made up with imported clay of similar properties.

At the time of writing this paper in spring 2008, negotiations are ongoing with prospective contractors to carry out the work that is expected to be carried out during summer of 2008.

ACKNOWLEDGMENT

The authors acknowledge the supporting data provided by Dwr Cymru and their permission to publish this paper.

REFERENCES

Structural Soils Ltd (2007). *Factual Report on Ground Investigations of Llyn Morwynion.*

The Discontinuance of Dams – it's not easy and it can be expensive!

PROF AK HUGHES, Director of Dams & Water Resources, Atkins Ltd,
MRS ANNE EVANS, Severn Trent Water
MR ROB GALLOWAY, Severn Trent Water

SYNOPSIS. This paper seeks to examine the issues associated with proposals to discontinue dams. The paper will examine the significant number of issues that need to be considered and show that the process of discontinuance can be long, often protracted and expensive. The paper, via case studies of a number of dams which have been discontinued, will demonstrate the issues that have been considered at the sites which are very different. It will illustrate that the sites can be left in different states depending on the location and that the process can often be found to be very acceptable to those who judge.

INTRODUCTION

The discontinuance of dams is an issue that the dam engineering as a profession is having to at least consider as a means of addressing problems at existing dams. The 'discontinuance' of a dam, under Section 13 of the Reservoirs Act 1975 requires an owner to reduce a reservoir's capacity to a volume that is less than 25,000 m^3. In some cases, owners choose to reduce the capacity to say 24,500 m^3 to avoid being subject to the requirements of the Reservoirs Act 1975 whilst some owners will choose to remove or breach the dam completely with no water then being retained. If we move to legislation based on risk and consequence of failure then we could see a move towards more sites discontinued where no water at all is retained.

Often, as part of the process of looking at remedial works at dams to upgrade them to meet current standards of dam safety, it is common to assess risks if 'nothing is done' – the 'do nothing' scenario – but this is normally not permitted because it will be a statutory requirement but also to remove or discontinue the dam and take it out of service and remove it from the ambit of the Act.

Ensuring reservoir safety into the future 2008, Thomas Telford, London

The process of discontinuance would involve using a Qualified Civil Engineer (QCE) from the appropriate panel (AR, NIR, SR) to oversee a project to carry out the works to reduce the capacity/removal of the structure through a process of design, and construction of works. The process would be completed with a Certificate of Discontinuance. On receipt of a Certificate under Section 13 of the Act the Enforcement Authority will remove the reservoir from its register and there is no need for a Supervising Engineer or the services of an Inspecting Engineer.

Historically there has been adverse public option when new dams have been proposed but it is also true that there can be adverse reaction when it is proposed to discontinue a dam. Dams are built for a number of reasons – water supply, flood alleviation, industrial use, amenity etc. In some cases the original use for the dam has perhaps changed/gone which makes the use of the dam redundant. Equally the rationalisation of systems and reduction in use of small resources leads to making a number of sources redundant. Often the geographical location, the degree of public access, land value and possible future land use, how the site is to be left and public opinion etc will dictate the way in which a dam is discontinued.

The issues highlighted above will dictate which 'third' parties need to be included in the consultation process. Some issues which need to be considered include the following:

- If the reservoir is an impounding reservoir and provides any alteration to the protection afforded to the area downstream, then the Environment Agency will often wish to be involved and require analysis to show the effect of the removal of the reservoir on the flood risks downstream. In some cases the EA might even object to the removal if the affects of removal increase the possibility of flood risk. They may even ask for other flood mitigation infrastructure.

- Silt in the reservoir – if there is silt in the reservoir, and recognising that the amounts are likely to be relatively small, this silt will have to be dealt with. It will not be acceptable to discard silt to the downstream area, either during the discontinuance works or once the works have been done. Plans will also be have to be made to deal with silt that is left within the basin for a number of reasons:-

    - to ensure it is not eroded and transferred to the stream downstream
    - that it remains stable
    - that perhaps re-vegetates

    - that is safe in that it allows public access for fishermen etc and it is not polluted

- Public access – thought has to be given to future public access to the site if this is to be allowed or even encouraged. The access will have to be safe, not put people at risk from deep water, soft sediments, steep slopes etc. Consideration will have to be given to public use of any residential body of water and the access they require.

- Materials used – in discontinuing a dam it is usual either to remove the dam completely or to partially breach the dam. The majority of dams in the UK are earthfill embankments and this will involve excavation of earthfill, removal of upstream protection facilities which might be masonry, or concrete etc. In some cases the material is buried or spread in the basin, in other cases the fill is sold, particularly where it is dressed stone, and in rare occasions moved off site where landfill costs will be incurred.

  The final form of the basin may involve a residual water body formed. It is often necessary to reform the original stream in the bed of the reservoir and often opportunity is taken to meander the stream and induce provision for fish, marginal plants etc. Whatever the form of the basin it is common for some sort of erosion protection within the stream bed and therefore in the breach or point at which it crosses any cutoff. The 'best' and perhaps most acceptable schemes are likely to use 'natural' or local materials rather than systems such as wire gabions etc.

Communication

In any civil engineering project communication must be a key element. It is essential once the draft proposal is fairly well defined that the public and land use organisations are involved. This is not only to get those people to 'buy' into the proposals but also to get agreement to the proposals perhaps more importantly have an input to the scheme. They will often have a significant amount of knowledge and experience relevant to the project or indeed may well have suggestions for improvements and changes. Often the dam engineers do not have those skills so the communication process is essential.

At the time of the works there is often much to be learned. It might be that the discontinuance process involves demolition of the structure, which if done in the planned way, can yield useful information on how our old dams were built, the condition of pipes and valves after many years of use, core materials – sampling and establishing core materials etc.

There is an opportunity to inform the public about the site – perhaps about the former use particularly if the dam formed part of our industrial heritage, about the construction of the dams and the features seen on site. The opportunity should be given to industrial archaeologists to study the site prior to and during the breaching process. Communication forms might include pamphlets, story boards, information books, information packs depending on the site – its size and extent and the degree of public access.

CASE HISTORIES

The following case histories describe five projects involving the discontinuance of reservoirs which are very different in nature but which have achieved the objectives set by the client.

Westworth Reservoir

Location	-	Guisborough
Owner	-	Northumbrian Water
Dam	-	Earthfill, 11.3 metres high, 118 metres long
		Reservoir capacity 69,200 cubic metres

Westworth Reservoir was owned by Northumbrian Water and was situated some 3 kilometres south east of Guisborough. It was used for water supply but in 2001 the company decided to look into the possibility of discontinuance.

Westworth dam construction was completed by 1875 to supply water for Gisborough Hall and the town of Guisborough in the Cleveland Hills, close to Teeside. The water supply system was replaced in the early 1970's but the reservoir retained as it was thought to be a useful forestry fire fighting water supply. However by the 1990's vandalism to the dam's structure was becoming a problem and fires were being lit to dry clothes after swimming in the reservoir; to cook etc. thus creating a fire risk.

Westworth dam was built following a Water Order in 1871, but at a revised location chosen later by the consulting engineer Sheriton Holmes of Newcastle and constructed during 1874 to 1875. Westworth's reliable yield was less than 2Mld and in dry weather was not able to support Guisborough's water demand. In 1878 proposals were made to increase the reservoir inflows with catchwaters, subsequently carried out with puddle clay lined channels.

The recorded details gave Westworth's capacity at 69.2 thousand m^3, a surface area of 11,800 m^2 and maximum height of 11.31 m with 105m crest length. The overflow weir was 6.1m long giving 2.45m freeboard to the top of the embankment. There were three 7" draw-offs connecting into the 12" valve shaft pipework and linked into the 12" scour draw-off horizontally to two downstream 12" valves allowing flow to treatment or to the river.

The dam had been inspected routinely since the 1930 Act. The last inspection recommended that the reservoir be taken out of the 1975 Act's juristiction or that remedial works were carried out to allow the reservoir to remain in use. The land owner, Lord Gisborough, did not want any water retaining and accordingly plans were made for the dam to be breached and the area landscaped. North Yorkshire National Parks became keenly interested and wanted the water area retaining as they said it was a migratory bird resting place, although strangely the Supervising Engineer could not remember seeing any birds resting there since his first visits in 1978.

The downstream face had a slope of 1 in 2.2 with a berm about 3 m wide at one third height. The upstream face was covered by stone pitching and had a slope of 1 in 3. The crest had a width was 4 m wide and was supported by a substantial vertical stone wall between 1.0 and 1.5 m high.

The options studied into the future of the reservoir included:-

- repair and retention
- discontinuance by breaching the full height of the embankment
- discontinuance by partial breaching the full height of the embankment
- discontinuance by partial filling of the reservoir
- abandonment by the construction of an embankment diversion pipe or culvert.

There was an overflow at the left hand end of the dam some 6.1 m wide with a stone channel running away from the river. A masonry valve shaft was located at the toe of the upstream slope at the mid part of the reservoir.

The condition of the dam was adequate but major works had been identified including:-

- complete refurbishment of the valve tower with the inclusion of an access bridge
- lining of the scour pipe
- repairs to the pitching and inclusion of rip-rap protection
- repair the spillway

An ecological and environmental study was undertaken. The study found that the aquatic habitats of the reservoir were somewhat impoverished due to acidifications and it supported an extremely limited fauna with no aquatic macrophytes. The key habitat around the reservoir was upland heath with areas of wet heath and heather moorland. Upland heath is a key habitat in the UK Biodiversity Action Plan whilst wet heath is a priority habitat in the EU Habitats Directive. The area is also designated as 'criteria moorland' under the 1981 Wildlife and Countryside Act as part of the National Park Plan for the North Yorkshire Moors.

Objections to Discontinuance

As it happened no one from North Yorkshire National Park (NYNP) had visited the reservoir and they had made an uninformed comment – later they admitted the water did not support either aquatic plants or fish and so was no attraction to birds.

Then the local County Archaeologist wanted the whole site preserving (again he had never visited the site) and had visions of a gravity buttress dam with granite blocks. After being shown the site he concluded that it was not what he thought and was happy with a descriptive panel showing it 'as it was' and leaving the valley abutments to indicate where the dam was. Clearly the breaching of the dam and restoration of the site would result in the loss of open water habitats and marginal habitats but because the reservoir had relatively limited nature conservation value it was recognised as not being significant.

The recommendations of the various parties were:-

- minimise construction impacts on adjacent heathland by prescribing working area
- create a land form which retains wet areas alongside streamside
- establish heather and grassland on the site by transfer of cuttings
- control any bracken invasion of the new landform.

Demolition eventually commenced in summer 2002 and progressed well once the contractor learnt to deal with clays and not top soil. The dam was breached by autumn 2002 and the stub ends of the dam trimmed with a toothless excavator bucket. This revealed the well-formed puddle clay core which was still well defined, very pliable. The core width increased by about the classic 1" horizontally for 1' depth on each face. However of much more interest was the defined shape of the original dam and the later clay addition as the dam was raised. The first dam's upstream toe was also discovered with the original stone valve tower base. This tower stonework had been re-used to form the later tower and only the original base was left within the original embankment.

The valve tower was partially dismantled to make it safe and the pipework removed. The pipe material was in excellent condition although being vertically cast illustrated clearly the incorrect alignment of the inner and outer moulds. The valves were to James Nasmyth patent (developed in 1839 at the request of the East London Water Company) and remained in operational condition.

Reinstatement of the reservoir area was confined to spreading and grading the dam material. No 'artificial' seeding was carried out to encourage native species to colonise the area. This led to the spread material being exposed for a long time – and still not fully green – so the material became incised as stream paths developed with the eroded fine material deposited in the original stream beds that became quite fertile. Relatively lush growth has established in the streams retaining in flowing material and now water moves in a stream-delta formation towards the new outlet weir.

The cost of the works was of the order of £100k.

Fig 1: Disused Westworth Reservoir

Lightwood Reservoir

Location	-	Buxton, Derbyshire
Owner	-	Severn Trent Water
Dam	-	Earthfill, 13.4 m high, 105 m long Reservoir capacity: 73,363 m^3 (large reservoir); Less than 25,000 m^3 (smaller reservoir).

Haswell Consulting Engineers were commissioned by the Undertaker, Severn Trent Water (STW), in March 2004 to investigate the feasibility of decommissioning Lightwood Reservoir (actually 2 reservoirs) and removing it from the ambit of the Reservoirs Act 1975.

Lightwood Reservoir was situated on the northern outskirts of Buxton, Derbyshire, just outside the Peak District National Park. The reservoirs consisted of a cascade of an upper and lower reservoir, divided by an earth embankment, with connecting weir. Water passed through the upper reservoir flowed over the weir into the lower reservoir. Spillage from the lower reservoir took place over the main outlet weir which discharged into the bywash channel. The lower reservoir was impounded by an earth embankment which was brick lined on the upstream face and which extends above top water level to form a wave wall. The original purpose of the dam was to impound water, for water supply, for the town of Buxton.

The Reservoirs were fed by Hogshaw Brook via a penstock into the top (small) reservoir. A bywash channel existed to the Western side of the reservoir to take excess flow when the reservoir was full. The construction of the reservoirs was concrete lined with brick facing on the lower basin & stone facing on the older upper reservoir. (Fig. 2)

Fig. 2: Upper reservoir looking in a southerly direction

Problems

The issues associated with the discontinuance of Lightwood Reservoir are detailed below.

Social/Environmental

The reservoir was an attractive feature in the local landscape, and was used for legitimate purposes such as dog exercising and walking. However, the water bodies also attracted less desirable activities, such as outdoor parties, camping, and swimming with associated problems such as alcohol and drug abuse.

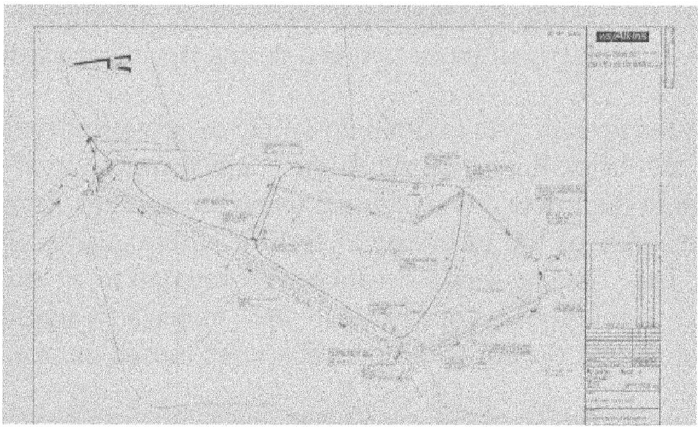

Fig. 3: Lightwood Reservoir Plan

The local residents in the roads leading up to the site were worried about the loss of habitat from the complete disappearance of the water bodies, but also were aware that if the reservoirs were removed, the anti-social behaviour would be greatly reduced, if not cease altogether. They therefore supported the discontinuance, but requested that some water remained for the birds and other form of wildlife which inhabited the reservoirs.

Fig. 4: Lower Reservoir Construction

Design considerations

The Landscape Architect to the consultants, in consultation with the ecologist, agreed that the normal approach to reservoir discontinuance (ie: to just form a V-notch in the dam wall), would not be appropriate in this case, and that the entire base and sides of the reservoirs should be demolished and the whole site returned to a natural valley.

Due to the presence of a significant number of water-loving bird species and before demolition of the main reservoir, a small pool was formed at the top of the site within the curtilage of the small reservoir to allow birds to return for breeding (some actually continued to breed during the major earthmove!)

The scope of the project was to drain down the existing reservoirs, break out the concrete and brick liners, demolish the water tower and valve chambers, cut a V-notch in the lower dam wall, and using the arisings from the dams to re-profile the sides of the reservoir to form a natural valley shape, with a stream at the base feeding 4 pools, which were created to encourage wildlife and aquatic vegetation. (Fig.5) All the materials arising from the earthworks including the concrete and brick were buried on site.

Fig. 5: View from position of top reservoir looking south

Scheme recognition

This discontinuance has won 2 awards since its completion. The first was a platinum award at Severn Trent Waters' Corporate Responsibility Awards in 2006, and the latest was a Commendation at the Landscape Institute's Awards in 2007. The judges for the Landscape Institute Awards, which is an International Awards Programme, commented that *'It artfully removed a negative activity by re-establishing the power of the landscape. Low budget, high impact. Demonstrates the subliminal art of landscape.'*

Barbrook & Ramsley Reservoirs

Location	-	Nr Chesterfield, Derbyshire
Owner	-	Severn Trent Water
Dams	-	Earthfill, 10m high at Barbrook/ 8.65m high at Ramsley, Reservoir capacities 310,000 m^3 (Barbrook) 76,800 m^3 (Ramsley)

Barbrook Reservoir is situated on Big Moor, approximately 12 km northwest of Chesterfield,and had a catchment area of 3.4 km². (Fig. 6) The reservoir, formed by two earth embankments, had the main dam running NE/SW across Bar Brook, and the side dam at right angles to it, running SE/NW. Both dams had a central puddle clay core. A masonry step-overflow weir was situated at the north east end of the main dam, with the spillway channel running down the northeast mitre where it joined the tunnel tailbay channel. (Fig. 7).

Draw-off arrangements were incorporated into a cast iron segmental valve shaft built into the upstream face of the main embankment (Fig. 5). Three levels of draw-off were connected to a downpipe and a common draw-off main laid within a cast iron, segmental tunnel, which has now been capped off. The main had been severed at the downstream end of the tunnel where water discharged into the tailbay and brook, which ran into the downstream and still existing Little Barbrook Reservoir. The iron and steel components of the tunnel shaft had become so severely corroded that safe operation of the draw-off valves was no longer possible. At the foot of the dam was the site of a water treatment works which was demolished many years ago. The reservoir had not been used for water supply since the mid seventies. The reservoir was also located within the Peak National Park and had been designated a SSSI.

Description of Works

The work involved the breach of the main embankment by excavating a 'V' shaped full height void and demolition of existing structures. At the deepest point, a 4m wide channel was formed to take the stream flow, with the sides of the breach graded to slopes of 1:3. The work comprised of the execution of the following:

- Protection of existing services
- Demolition of the Draw Off Tower and Walkway
- Excavation in the reservoir basin
- Excavation in the reservoir embankments
- Re-profiling of reservoir embankment
- Landscaping works

Project Aims

- Discontinuance
- Habitat Creation
- Flora and fauna
- Sustainable Result

- Minimum Maintenance
- Minimal impact of construction

Project methodology
Silt traps installed downstream.
Reservoir drained over a period of 4 weeks.
Water monitored for sediment content.

Fig. 6: Valve tower and Bridge

Fig 7: Masonry step overflow weir

In order to maintain the flow of the Bar Brook during the excavation work a 'V' notch was cut to one side of the draw off tunnel and the stream allowed to flow through the tunnel. This was sealed upon completion.

The course of the brook was allowed to form naturally through the reservoir basin (Fig 8). From historical records the brook appeared to return to a similar course to that which existed before the dam was built.

Fig. 8 Course of stream allowed to follow natural course

Fig. 9: Stone weir discovered buried

There was a flexible approach to the work on site. The stone weir shown below was discovered buried in the silt after work started. It was decided to utilize it as part of the reinstated water course and to form a pool near the head of the reservoir.

Fig. 10: Pools created in reservoir basin

As part of the habitat creation work clay excavated from the core of the dam was used to create pools in the reservoir basin. (Fig. 10). One of the aims of the project was to retain and reuse materials on site as far as possible. In the event only scrap metal from the draw off tower and some timber had to be removed from site.

Ramsley Reservoir (1880 to 2003) is situated on the edge of Ramsley Moor, approximately 10 kilometres northwest of Chesterfield, Derbyshire, Ramsley had a small, direct catchment area of 52.6 hectares. Additional inflow came from water imported via a 300 mm (12") diameter earthenware pipeline from Little Barbrook Reservoir, which has now been capped off, some 2.5 kilometres away. The earth embankment dam had a puddle clay core with the upstream face of the dam having stone pitched armouring. The embankment had settled unevenly during its life and there were some visible signs of slight but continuous seepage. Changes to the overflow and spillway were made in 1990 and a sleeved outlet pipe was installed in 1985. Despite these improvements, the reservoir, in its present form, could not meet today's standards and it was uncertain whether it would withstand an extreme flood.

The reservoir was originally used as a source of supply to Smeekley Treatment Works, Crowhole Reservoir and Linacre Upper Reservoir. Smeekley has since been abandoned and Crowhole discontinued. The reservoir was also located within the Peak National Park and had been designated a SSSI.

A single contract was let for the discontinuance of both Barbrook and Ramsley reservoirs. The project aims at Ramsley were similar to those at Barbrook but on a smaller scale. (Fig. 11). It also did not have the problem of having a sizeable water course running through the basin.

Fig. 11: Pools created in reservoir basin

Stanley Moor Reservoir

Location	-	Buxton, Derbyshire
Owner	-	Severn Trent Water
Dam	-	Earth embankment with clay apron on Upstream face, 13.7m high, 630m long Reservoir capacity 244,100 cubic metres.

Stanley Moor Reservoir was a small, rectangular, impounding structure, formed on three sides by earth embankments and founded partly on limestone and partly on millstone grits. The whole of the inside was lined with puddle clay with a protective layer of masonry pitching. The reservoir had not been used as a water supply source for over twenty years. The principal feeds were two nearby streams, to the west and to the north; the flow being intercepted and directed through aqueducts to the intake chute, by the operation of "leaping weir" sluices.

On first filling, appreciable leakage occurred due to defects in the clay blanket. Strenuous but unsuccessful attempts were made to overcome the inherent problem caused by the underlying material being honeycombed by solution holes and caverns. Eminent dam engineers of the period, including Sir Alexander Binnie & William Binnie, were all involved by giving their advice or by making statements, as expert witnesses, in the action taken by Buxton Council against the Engineers. A statement from one of the experts that *"I have never seen better work than that showed in the Stanley Moor Reservoir; and I have never seen a site so plainly and almost hopelessly bad"* seems to summarise all the opinions voiced at the time.

Finally, the working top water level was reduced by 10 feet and, in 1946, the original overflow weir and spillway was superseded by the construction of a low level weir channel and 36 inch overflow culvert, passing through the embankment. The capacity was thus reduced from 450,000m³ to 275,000m³. The works comprised of:

- Breach of the embankment by excavating 26,000m^3 of material.
- Demolition of Draw Off Tower, Walkway and reinforced concrete Draw Off Culvert.
- Demolition of Overflow Structure
- Landscaping

Contract value £178,000

The embankment was breached by cutting a 'V' notch in the embankment at the position of the draw off tower and culvert and forming a channel through the notch to allow all incoming waters to exit the reservoir basin into the spill channel at the toe of the dam (Fig. 12). Before work could start an inlet diversion ditch had to be dug to divert the inlet water from the inlet weir to the spill channel (Fig. 13).

All the excavated material and masonry lining resulting from the formation of the notch was retained on site to line the bottom of the reservoir, pools and scrapes for habitat creation and embankments within the reservoir basin. The masonry lining removed from the area of the 'V' notch was used to line the new channel and the surplus buried under the mounding formed from the excavated material. It was not considered economic to sell the surplus stone.

Fig. 12: Spill channel

Fig. 13: Inlet diversion ditch

('V' notch location)

Fig. 14: View of the reservoir from the west showing the 'V' notch, pools and embankments formed in the reservoir basin.

SUMMARY AND CONCLUSION

This paper illustrates via case histories that the discontinuance of dams is often not a cheap or easy option.

It is clear that when a dam is discontinued it can be carried out in a variety of ways and the way in which this is done will often de dictated by the final result and land use that is required. It will always be necessary to communicate and consult with the public and other interested parties to achieve the 'most acceptable' result.

The Staged Construction of Imang Dam

M.J. HILL, MWH Brunei Darussalam
I.C. CARTER, MWH, High Wycombe, UK
I. DAVISON, Mott Macdonald, Altrincham, UK
R.C. BRIDLE, Dam Safety Ltd, Amersham, UK

SYNOPSIS. Imang Dam and reservoir was constructed between 1995 and 1997 for irrigation purposes and is located on the Sg[1] Imang in the Brunei-Muara District of Brunei Darussalam. The dam has a maximum height of about 10m, length of 420m and contains a reservoir of $10Mm^3$, which is supported by a $14km^2$ secondary rainforest catchment. The scheme was originally designed to serve local irrigation requirements of about 700 ha of double cropped rice.

A major feature of the design of the embankment dam was the soft clay foundation and wide stability berms. The dam was originally designed to be constructed in one stage following treatment of the foundation soils to accelerate consolidation settlement and increase shear strength. However, the foundation materials did not gain sufficient strength during construction to allow the embankment to be constructed to its full height and interim measures were implemented to allow first filling and operation of the reservoir. Following a period of foundation consolidation settlement and strength gain, the embankment crest was raised to its full height in 2006. This paper describes the design and staged construction of Imang Dam.

BACKGROUND

Imang Dam was first proposed in 1975 to support irrigation of the Mulaut area for mechanised cultivation of rice padi. The first series of field investigation and feasibility studies were carried out by MWH by 1982 and these early studies highlighted the difficulties in constructing a dam embankment of more than about 6m in height due to the deep estuarine and swamp deposits covering the dam site. Further investigation and studies were concluded in 1990 and the construction of the dam and reservoir works commenced in 1995.

---

[1] Sungai (Sg) – river or stream

MAIN FEATURES OF THE SCHEME

Imang Dam is formed by a 10m high earthfill embankment with wide stability berms. The Full Supply Level (FSL) is 11.5m BSD (Brunei Standard Datum) and the nominal dam crest level is 13.9m BSD. The foundation comprised soft to very soft alluvium, swamp deposits and soft estuarine clays to depths of almost 28m and it was neither technically nor economically feasible to remove and replace these soft materials. A review of embankment dams on soft ground was carried out and a wide bermed embankment design was developed that incorporated much of the experience gained by MWH in Brunei during the 1980s at Benutan Dam.

The dam is approximately 420m long and incorporates a 35m wide ogee shaped overflow / spillway and outlet structure at the right abutment ridge. A further auxiliary spillway with a 55m wide overflow weir is located further along the right abutment ridge to cater for flood events in excess of 200-year return period. The combined overflow system is designed to pass the PMF discharge of 383 cumecs and the arrangement was tested by a physical hydraulic model.

The outlet structure was used for diversion of the Sg Imang during construction of the embankment and later fitted with a series of four penstock gates to control irrigation water into a canal distribution system, which was to convey water to the irrigation areas. However, the scope of the canal system was curtailed and a recent scheme has installed a dedicated pipeline feeding the nearby plantations that have been developed. The general arrangement of the dam and ancillary works is shown on Figure 1.

DAM SITE GEOMORPHOLOGY AND GEOLOGY

The dam site is situated in the Bukit Siudam range of hills and lies on the Miri Formation rocks comprising laminated siltstones, silty mudstone and fine-grained sandstones of marine origin. The beds lie on the western flank of the Jerudong anticline and dip at up to 40° adjacent to the Mulaut plain. The topography is controlled by the structure of the underlying Miri Formation rocks. The beds strike uniformly at N18° E forming a series of parallel strike ridges and cuestas (dip slopes). Faulting is shown on the published geological map of the area but there was no evidence for any major fault in the immediate vicinity of the dam site or reservoir basin.

The valley floors consist of alluvium occupying deeply incised stream channels that were eroded during the Pleistocene glacial maximum and subsequently filled with sediment as the sea level rose to the post-glacial maximum of about 1.8m BSD. Estuarine alluvium was deposited at the peak of the marine incursions over extensive areas of the Brunei valley including the Mulaut plain.

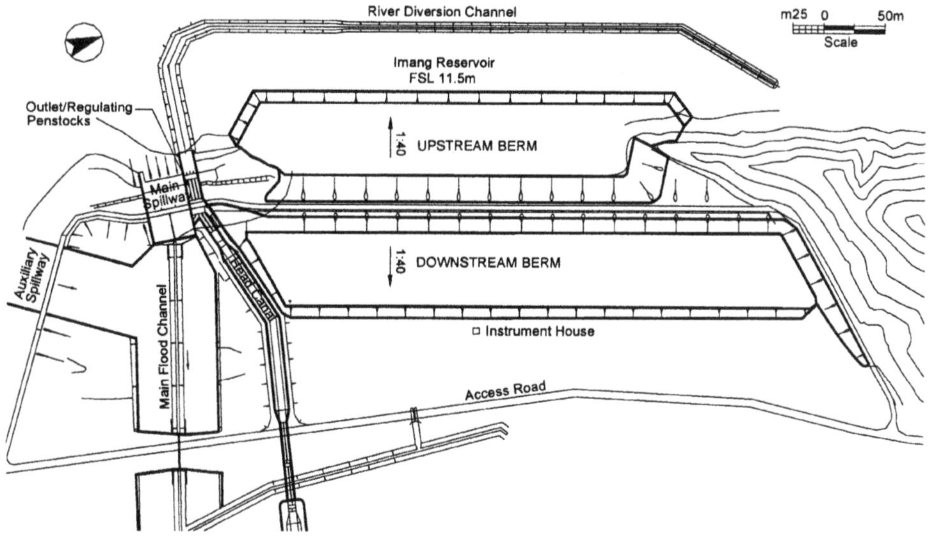

Figure 1 General Arrangement of the Dam and Ancillary Works

The abutments to the dam are provided by narrow ridges of sedimentary rock, which follow the strike of the strata, SSW to NNE. The valley floor between the ends of these ridges is formed by a broad flat alluvial floodplain 290m wide interrupted by a 30m wide outcrop of fine grained sandstone. Two river channels pass one to either side of this outcrop and meet about 80m downstream of the dam.

The maximum depth of alluvium encountered in the boreholes beneath the left and right hand channels was about 21m and 28m respectively. The superficial soils infilling the valley floor were classified into three distinct strata in descending order of depth were:

Riverine Alluvium: Generally loose, clayey sandy silt and very soft, silty sandy clay, of low to intermediate plasticity. Average thickness of 3.4m (2.4m to 4.3m thickness).

Swamp Deposits: Generally very silty, soft to firm, brown grey to dark grey, clay of high plasticity with abundant fragments of decaying wood and lenses of peat. Average thickness of 5.5m (2.4m to 8.8m thickness).

Estuarine Sediments: Generally very silty, sometimes very sandy, soft to firm, brown grey to dark grey, clay of variable plasticity (high at the top of the layer and reducing to low at depths of 16m and more below ground). Average thickness 8.5m (up to 14.9m thick).

The valley sides are generally covered by a mantle of colluvium consisting of sandy clay and silty clays to depths of about 1m. Bedrock consists of Miri Formation of laminated brownish grey siltstones, grey silty mudstones and fine-grained sandstones. The beds dip westward (upstream) at angles between 30° and 40°. The beds strike uniformly at N 18° E. The simplified geological section through the dam is shown on Figure 2.

Site investigations were carried out during the feasibility stage in 1980 and at the design stage in 1989/90. Both investigations included boreholes, trial pits and laboratory testing. The design stage investigation concentrated on establishing reliable in situ shear strength data for foundation shear strength estimation and included in situ vane and self-boring pressuremeter tests.

EMBANKMENT DAM DESIGN

The design concept for the dam was dictated by the strength and thickness of the alluvial soils, which occupied the valley floor. The alluvium was soft, weak and highly compressible and the preliminary design considered two options for construction of the dam embankment. These were:

Option 1: excavation of the alluvium down to bedrock and replacement and compaction of suitable materials.

This option would have entailed excavation of the alluvium to bedrock and replacement and compaction with suitable materials below the groundwater table. The depth of the weak alluvial soils was prohibitively deep (up to 28m) and the volume of foundation material would have been greater than the embankment fill volume by ten-fold. Also, the success of a dewatering scheme required to facilitate such a design was also questionable. Therefore, this option was ruled out on cost and technical grounds.

Option 2: Construction of an embankment on the alluvial foundation using wide stability berms and foundation treatment to accelerate consolidation of the weak soils.

This option entailed stripping the upper surface of the alluvium and construction of a wide bermed embankment directly on the existing foundation materials. This design concept had been successfully used at Benutan Dam and incorporated 60m wide upstream and downstream stability berms. The alluvial soils would be improved by installing pre-formed 'wick' drains under the central section of the dam to improve drainage and accelerate consolidation settlement, thereby improving the strength of the weak foundation. The preliminary design developed these concepts and was later validated using finite element techniques.

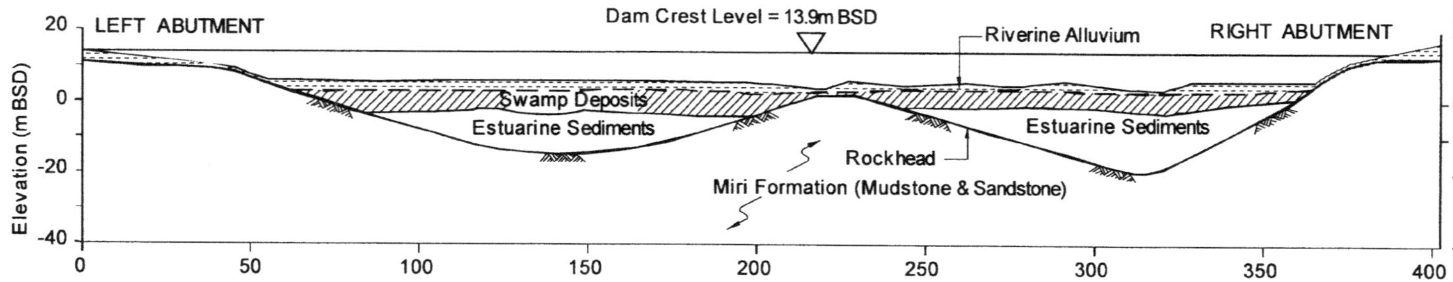

Figure 2 Simplified Geological Section along Dam

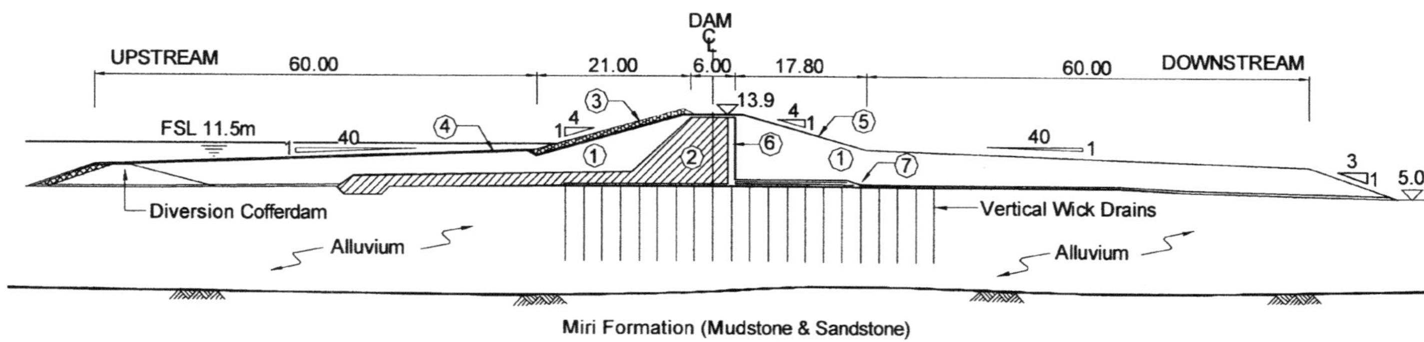

Figure 3 Cross section of Dam

Legend: (1) Shoulder Fill (2) Clay Core / Clay Blanket (3) Rip Rap Slope Protection (4) Beaching Slope Protection (5) Topsoil and Turf (6) Wall Drain (Fine Filter) (7) Drainage Blanket & Finger Drains

Preliminary Design

The dam section was initially designed by limit equilibrium stability analysis using undrained parameters for the foundation soils. Measurement of undrained shear strength of soft and very soft clays is difficult in practice and a number of in situ and laboratory tests were conducted as part of the ground investigations to estimate the relationship between undrained shear strength and depth through the alluvium (i.e. in situ vane shear, self boring pressuremeter, undrained triaxial tests). These were also compared with empirical relationships such as that proposed by Mesri (1975) for normally consolidated clays, $C_u = 0.22\sigma_{p'}$, where $\sigma_{p'}$ is the preconsolidation pressure.

The undrained shear strength profiles derived are shown on Figure 4. The design profile (1) was adopted for initial design purposes and was reviewed and further modified (2) as part of the Finite Element analysis (described below) to incorporate additional ground investigation data and a critical review of the test procedures and results. The shear strength of the alluvium was further estimated during construction (3) using the Mesri relationship and observations of pore water pressure from the foundation piezometers.

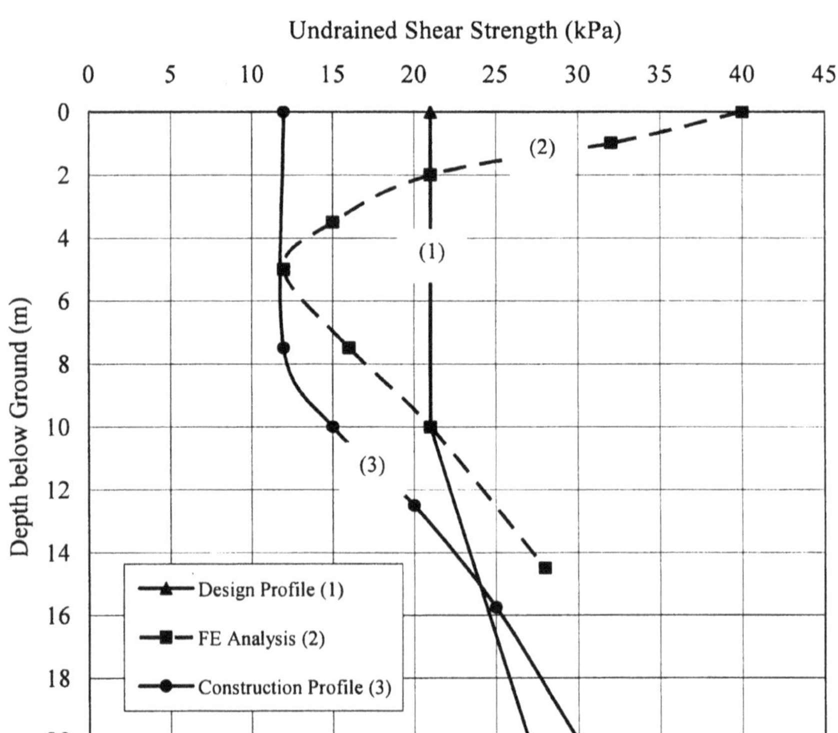

Figure 4 Undrained Shear Strength Profiles

The embankment fill material effective strength parameters of c'=0, ϕ'=30° and bulk density of 20 kN/m^3 were adopted in the initial design and a conservative pore pressure ratio (pore pressure/weight of fill above), r_u of 0.4 was assumed. The results of the stability analysis indicated that stability during construction would be the critical condition with the lowest factor of safety calculated for the upstream shoulder.

Factors of safety would improve following reservoir impounding in the long terms as the foundation gained strength through dissipation of porewater pressures and consolidation settlement. A range of dam crest heights, berm widths and berm height configurations were investigated, with the final arrangement shown on Figure 3. Both upstream and downstream berms were 60m wide with the diversion cofferdam formed in the upstream berm.

Finite Element Analysis

The limit equilibrium stability analysis indicated that the dam section would be satisfactory at the end of construction, but did not model the effect of any dissipation in excess pore pressures in the foundation soils or any strength gain during construction.

Finite Element (FE) methods were therefore used during the detailed design stage to review the soil properties, geometry and predicted performance during construction. Geotechnical Consulting Group (GCG) carried out this work using the Imperial College Finite Element Program (ICFEP).

The shear strength of the alluvium had the overriding effect on stability of the embankment and the shear strength profile adopted in the FE analysis was modified following further review of the original test data and additional in situ measurements carried out during the detailed design stage. The revised profile is shown on Figure 4 and has a minimum shear strength of 12 kPa at about 5m depth. This profile was considerably lower than that assumed in the preliminary design (minimum: 21 kPa) but, given the difficulties in obtaining reliable measurement of shear strength in very soft clays and the implications for stability, is was considered to be a safe and reliable assumption.

The analysis showed that vertical displacement under the central section of the dam would translate to horizontal displacement under the upstream and downstream berms with lateral spreading of the foundation / embankment. The central 'pimple' of the embankment was expected to cause local stress concentrations in the foundation below the central and upstream section. Foundation drainage was therefore provided in this location in the form of vertical 'wick' drains in order to improve drainage, accelerate foundation settlement and increase foundation shear strength.

The finite element studies concluded that the embankment could be constructed in a single lift, although there would be little margin against failure at the end of construction. The computed factor of safety at end of construction would be about 1.1 and following impounding of the reservoir would increase to about 1.6 in the long term. Stability during construction would therefore be marginal and an observational approach was adopted based on an extensive instrumentation scheme. The final geometry of the dam embankment and foundation treatment is shown on Figure 3.

EMBANKMENT MATERIALS

The embankment incorporates a central rolled clay core and upstream blanket as the waterproof element with earthfill shoulders and wide berms. The colluvium and highly weathered mudstone rock members were used to form the clay core and blanket, following some upward adjustment in moisture content so as to produce a dense fill with 98% relative compaction and a shear strength of between 70 and 100 kN/m^2. This required placement moisture content in the range 21-25%, which was generally a few percent above the natural moisture content in the borrow areas.

The core material was watered, leveled in 250mm thick layers and compacted using pad foot compactors and produced a satisfactory dense fill with less than 4% air voids.

The shoulders were formed from weak rock members of the Miri Formation arising from the required excavation for the diversion / spillway structures and also from a borrow area located within the reservoir basin. The mudstone broke down readily and was compacted using pad foot compactors, while the sandstone members were compacted using smooth vibratory drum rollers.

There were no suitable local sources of graded granular materials or rockfill materials for the internal filters or rip rap slope protection and these were all imported to the site. The 'critical filter' wall drain located immediately downstream of the core was connected to a base drainage blanket, which extended to the toe of the central downstream slope. This was a 'sandwich' construction of coarse drainage material within fine filter that drained under the downstream berm via a series of 'finger' drains of similar construction.

The foundation under the central and upstream section of the dam was treated by installation of a grid of vertical 'wick' drains taken to 10m depth on a 2.5m grid spacing. These consisted of a proprietary corrugated plastic internal member wrapped with geotextile filter fabric. The heads of the wick drains were connected to a sand drainage blanket at ground level to provide a drainage outlet and reduce seepage path lengths.

EMBANKMENT PERFORMANCE DURING CONSTRUCTION

The dam was instrumented along two principal sections with a total of forty-four vibrating wire piezometers in the foundation and twenty-two in the embankment. Settlement and displacement was also monitored by a series of nine combined inclinometer/extensometer tubes set into bedrock.

A trial embankment was constructed as part of the upstream diversion cofferdam and was later incorporated into the upstream berm. This enabled the performance of the 'wick' drains to be observed. These trials showed that areas adjacent to the 'drained' areas did not drain or consolidate and cracks formed at the toe of the embankment. The final wick drain spacing was adjusted with the deepest closer spaced drains under the highest section of the dam.

Porewater pressures were monitored as the fill levels increased and piezometric pressures responded closely to fill height, as expected. Excess pore pressures within the foundation increased with depth with the 'wick' drains being particularly effective at shallow depth (less than about 5m depth). Calculated pore pressure coefficients, r_u, at depths greater than about 5m depth were variable and ranged between $r_u = 0.55$ and $r_u=0.85$. Typical plots from the piezometer readings are shown on Figure 5.

Total settlement during embankment filling was not excessive and ranged from about 300mm under the downstream berm to 700mm under the dam crest (an indication of limited pore pressure dissipation). Horizontal deformations were up to 170mm in the downstream direction and were greater than predicted by the finite element analysis.

The Contract included provision for a shutdown period to allow the instrumentation to be monitored when no additional fill material was placed to assess the performance of the dam and its foundation. The fill level reached about 12m BSD, 7m above foundation level, when the shutdown period was invoked. The piezometers located within the drained section showed quicker dissipation of porewater pressure (approximately 7mm to 13mm per day), than those located in the undrained sections (less than 5mm per day).

The conclusion drawn from the instrumentation data was that the loads imposed on the foundation clays by the dam was mobilizing a greater proportion of the available strength than had been expected. This warranted a cautious approach to completion of the embankment.

Following discussions with the Client, it was agreed that the embankment should be filled to 13.25m level (approximately 2m short of the design height at the end of construction) with the remainder of the dam crest completed at a later stage, when the foundation pore pressures had dissipated sufficiently to increase strength of the foundation.

The remaining fill was placed in a strictly controlled manner with limitations placed on the contractor to place fill to a specific zoning sequence. Monitoring was continued after placement of fill in each zone and results provided to the designer before proceeding with subsequent layers.

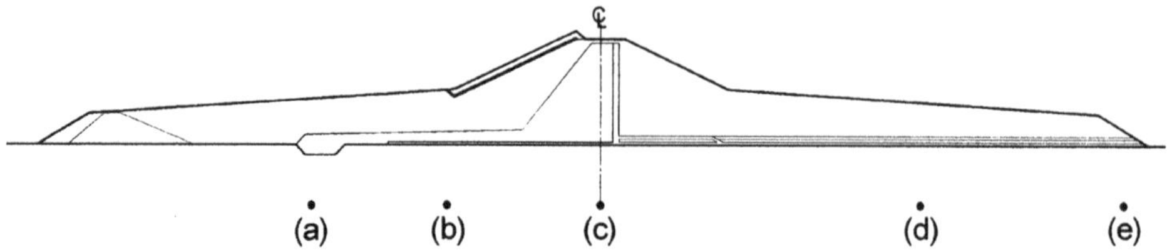

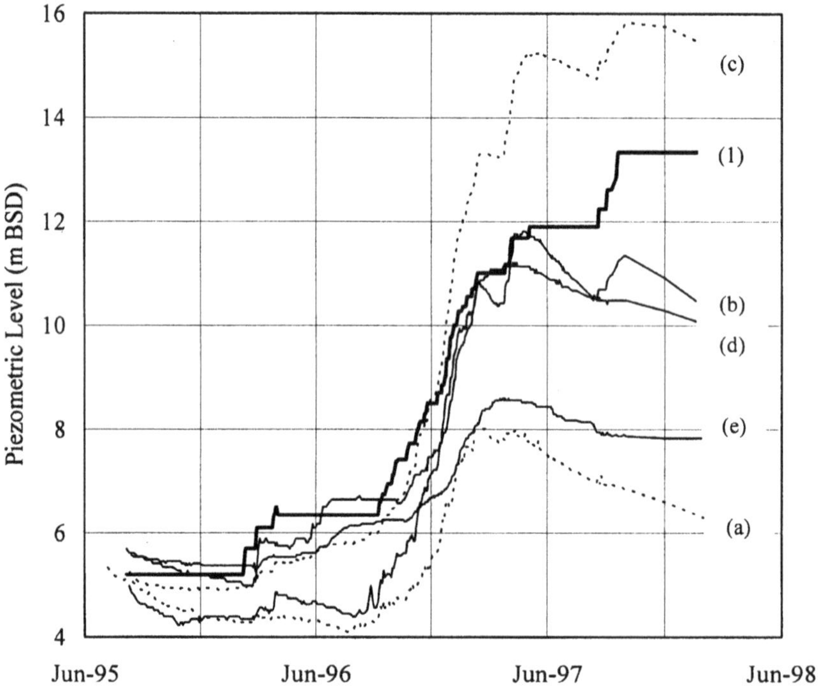

Legend:-

(1) - Fill Level at Dam Axis

(a) to (e) – Foundation Piezometer Locations

Figure 5 Foundation Responses to Embankment Loading

INTERIM DAM CREST ARRANGEMENT

A review of the flood hydrology was carried out with the Full Supply Level reduced by 220mm to 11.28m BSD by omission of the pre-cast overflow crest blocks (giving a gross freeboard of 1.97m). The flood capacity for the dam with the dam crest at 13.25m BSD was estimated to be about 0.5 PMF, which was deemed to be acceptable as a short term measure.

In order to limit the load applied to the foundation a revised crest detail was developed, as shown on Figure 6, which provided for raising the crest to its design level in the future. Core material was stockpiled for later use and the works were finally completed and reservoir impounded in 1998.

FINAL DAM CREST ARRANGEMENT

Monitoring of the instrumentation was not continued by the owner after commissioning of the reservoir to its interim level in 1998, although data from the piezometers and settlement instruments was obtained in 2004 prior to procurement of the works to complete the dam crest.

This information indicated that settlement had been substantial (in the order of 1.0m) and foundation pore pressures had reduced by between 4m and 7m. Therefore, it was clear that pore pressures in the foundation had dissipated and significant consolidation settlement had occurred and was now safe to raise the embankment to its original design height of 13.9m (min). A settlement allowance of up to 500mm was provided where the foundation soils were thickest to allow for future settlement and the embankment was constructed to an inverted 'w' profile to match the foundation thickness with a maximum construction level of 14.4m BSD. The details of the crest raising are shown on Figure 7.

The existing dam crest level was stripped and leveled and the core zone was exhumed to assess its condition. A key trench was excavated and the core material was found to be in good condition. No sign of desiccation of the existing core material was evident. The core and wall drain were raised to the same specification as the original works, with the upstream slope protection re-graded and extended to the new dam crest level. The reservoir level was drawn down to approximately 7.5m BSD during the dam crest raising works. The completed embankment is shown on Figure 8.

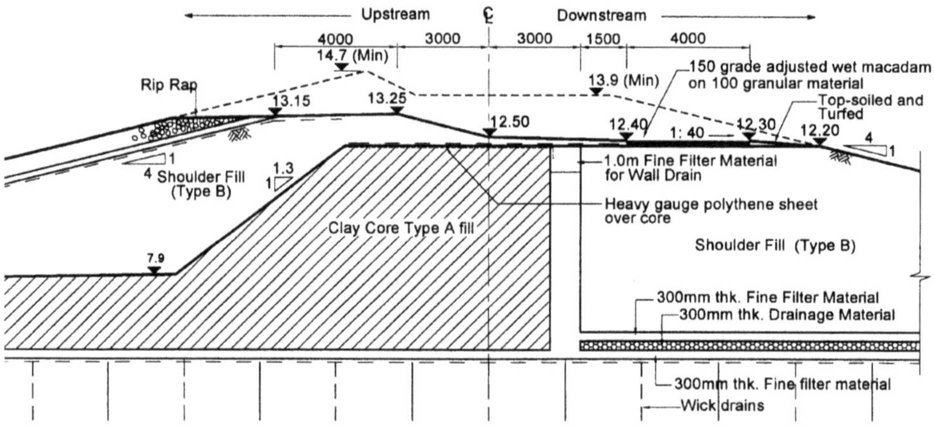

Figure 6 Interim Dam Crest Detail

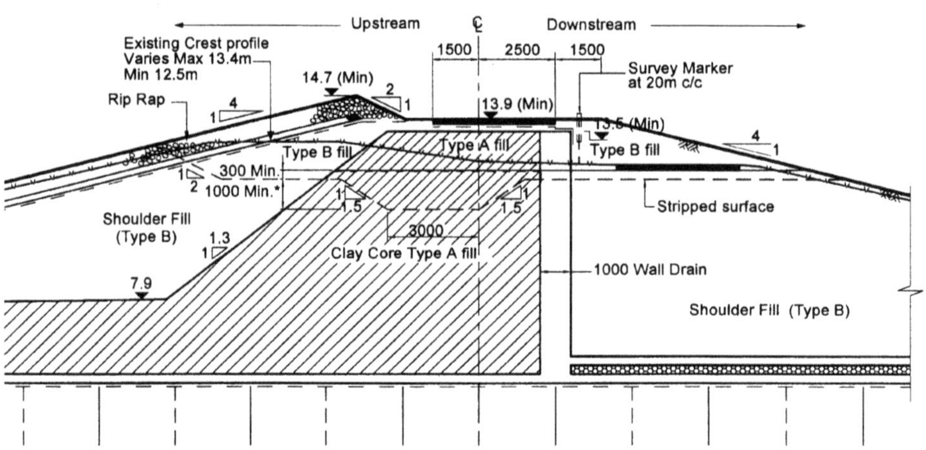

Figure 7 Final Dam Crest Detail

Figure 8 View of Completed Embankment in 2006.

CONCLUSIONS

Imang Dam is formed on a weak and highly compressible clay foundation and special design features and close construction monitoring were employed to ensure that the embankment did not fail during construction. In particular, the following points are noteworthy:

i) The embankment incorporated wide berms to counteract rotational failure through the soft alluvial foundation. Drainage conditions within the central part of the upper alluvium were improved by installing a series of vertical 'wick' drains. These were useful in dissipation of excess pore pressures beneath the central part of the dam.

ii) Measurement of shear strength of weak clays for use in stability analysis is difficult in practice - especially in hot tropical jungle environments. A number of in situ tests, laboratory tests and empirical estimates were carried out at Imang Dam to derive a shear strength profile for the foundation materials. The final profile adopted for design was biased towards the in situ vane test results and the Mesri formula.

iii) The instrumentation data obtained during construction indicated that the foundation pore pressures were high and showed little or no dissipation, other than in the drained sections. It was concluded that the load imposed by the embankment on the clay foundation was mobilizing a greater proportion of the available strength than had been expected and completion of the embankment in a single stage would not be advisable. Following a shutdown period to review the performance of the foundation with no additional loading, the embankment crest was finished some 2m short of its design height and the spillway crest blocks omitted to maximize flood freeboard.

iv) Although the embankment was not completed in a single stage as originally planned, the interim crest details and omission of the spillway crest blocks provided acceptable flood protection in the short term (approximately 0.5 PMF) and enabled the reservoir to be safely impounded to near its design level.

v) Following a period of operation of the reservoir, the foundation pore pressures dissipated considerably resulting in large settlements of up to 1.0m, where the foundation thickness was greatest. The dam crest was raised in 2006 by 2 - 2.5m to its design level and the spillway crest blocks installed bringing the reservoir flood protection to the PMF level.

ACKNOWLEDGEMENTS
The authors wish to extend there thanks to the Department of Water Services, Public Works Department, Brunei Darussalam for permission to publish this paper. Thanks are also due to Professors Vaughan and Hight of Imperial College, London and to their colleagues at GCG.

REFERENCES

Cole, R.G, Carter, I.C & Schofield, R.J. (1994). *Staged construction at Benutan Dam assisted by vacuum eductor wells.* 18th ICOLD Congress.

Watson Hawksley Asia. (1990) *Imang Dam and Reservoir Preliminary Engineering Design Report.* Unpublished.

Geotechnical Consulting Group. (1993). *Imang Dam Finite Element Analysis.* Unpublished.

Mesri, G., (1975) *Discussion of New Design Procedure for Stability of Soft Clays.* ASCE Journal Geotech Engineering Division, 101 (GTA).

Leroueil, S., Magnan, JP. & Tavenas, F. (1990) *Embankments on Soft Clays.*

Inoperable reservoir valves - why wait for the inspection?

P. ENSTON, Hydra-Ject

SYNOPSIS. Until recently Undertakers have had limited options for dealing with these problematic valves and usually they would only 'appear on the radar' and therefore be identified as requiring remedial action during statutory visits by a Panel Engineer.

This is understandable and until recently was unavoidable as conventional solutions are costly and disruptive perhaps requiring the use of divers or the need for a complete reservoir drawdown. Environment Agency requirements mean that a considerable amount of time and expense is involved in planning such an exercise in order to facilitate for example, a reservoir drawdown in order to enable a valve exchange.

In this day and age where risk management and climate change forces people to consider the consequences of the 'unthinkable' then this approach demands reconsideration.

INTRODUCTION

Rhaslas reservoir is a 60 million gallon reservoir in Ebbw Vale, South Wales, the scour valve was seized shut. Previous attempts at operating the valve resulted in the phosphor bronze spindle being sheared leaving the gate firmly wedged into the valve seats. Following an inspection under Section 10 of the Reservoirs Act 1975, remedial works to the valve were recommended in the interests of safety to make the valve operable.

As the reservoir no longer served as an industrial source of water, serious consideration was being given to decommission it and that would have resulted in a substantial and costly civil engineering project.

H-GRIP: A NEW SOLUTION

At this point the Undertaker became aware of a novel solution for seized valves known as H-Grip. Before this equipment could be attached to the valve, it was necessary to carry out remedial works to replace the broken spindle. This was achieved by firstly redressing the gate and replacing the carrier blocks for the spindle nut.

Figure 1: Rhaslas reservoir embankment – scour valve chamber

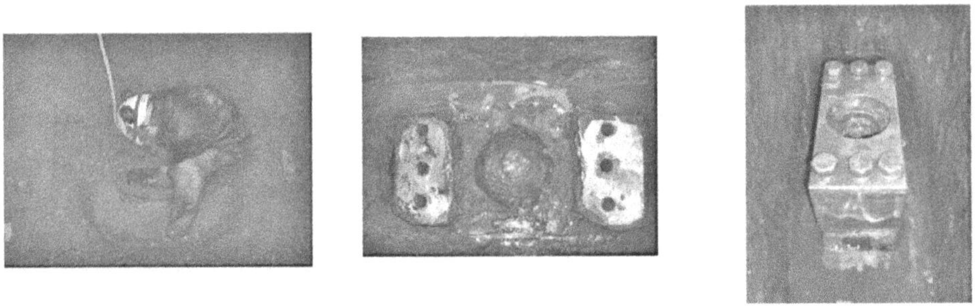

Figure 2: Valve gate prepared for attachment of new spindle nut carrier blocks

Figure 3: Typical H-Grip set up to release a seized valve

Following this, the bonnet and new spindle were reassembled and then H-Grip was attached. Within a matter of minutes the valve gate was made free and could be cycled with great ease. The risk associated with having a large reservoir that could not be drawn down through the scour valve had been eliminated – in an instant.

CONCLUSION

In recent years scores of such seized reservoir valves in the U.K. have been successfully made operational using this technique – to date there have been no instances of failure to make these valves work – as good as new. Significantly, they have all been carried out without shutdowns, drawdowns, lengthy planning exercises or any disruption whatsoever.

However, almost all of these risks were initially identified by the Inspecting Engineer's Report which drew attention to the remedial actions required.

Perhaps in this day and age where extreme weather events of the summer of 2007 brought reservoir safety into sharp focus, it is now time to rethink the traditional approach for dealing with seized reservoir valves and to recognize the important role they play in the safe management of reservoirs.

Put simply: inoperable reservoir valves – why wait for the inspection?

Queen Mary and King George V Emergency Draw Down Schemes

BRYN PHILPOTT, Thames Water Utilities Ltd
YINKA OYEYEMI, Jacobs UK Ltd
JOHN SAWYER, Jacobs UK Ltd

SYNOPSIS.
Thames Water owns a large number of non-impounding reservoirs. These have unusually long embankments with narrow clay cores and gravel shoulders, which could be susceptible to piping failure and erosion. They are typically located in urban areas. Thames Water has therefore reviewed its policy regarding emergency drawdown capacity and started an improvement programme on the most critical reservoirs.

The Queen Mary and King George V reservoirs are two of these critical non-impounding reservoirs with continuous embankment lengths of 6.3km and 6.5km and storage volumes of 30 Mm^3 and 12 Mm^3 respectively. Following the statutory inspection of these reservoirs, a safety recommendation was made which required a significant increase to the existing emergency draw down capacity of each reservoir. This required Queen Mary reservoir to achieve 0.75 metre emergency draw down from top water level within 24hrs and King George V reservoir to achieve 1m draw down in 24hrs. Studies instigated as a result looked at various options of achieving the draw down with the final solution involving the design and construction of twin siphon pipes fitted with submerged discharge valves.

It is believed that a number of UK reservoirs may require additional emergency draw down capacity as measures in the interest of safety, as defined within the terms of the Reservoirs Act 1975, over the next few years. This paper looks at the options considered before arriving at the final solution. It also discusses challenges in the design and construction of the schemes.

EMERGENCY DRAW DOWN REQUIREMENTS IN THAMES WATER

The requirement to provide facilities to enable large storage reservoirs to be drained quickly in the event of an emergency is well understood. However the determination of the minimum requirements in this respect is less clear.

This paper sets out Thames Water's rationale behind its target policy to provide 1 m per day emergency draw down capacity to all of its storage reservoirs and provides details of projects to increase the capacity at Queen Mary and King George V Reservoirs.

Thames Water's Storage Reservoirs

Thames Water has a stock of 23 storage reservoirs within the terms of The Reservoirs Act 1975. The majority are Non Impounding and largely of similar construction detail. Table 1 shows the key features of the total stock as follows: -

Table 1: Key features

Feature	*Range*
Age Range	35 to 175 years old
Total capacity (approx.)	220 Mm^3
Embankment shoulders	River Terrace Gravels
Watertight element	Thin Clay Core Wall
Embankment Heights	5 m to 25 m
Total Embankment Length	77,500 m

Risk Assessment

The principal risk associated with Thames reservoirs revolves around their construction. The thin clay cores (without designed filters) leave them potentially susceptible to hydraulic fracture and internal erosion in certain circumstances. The embankment shoulders, being of erodable gravel, are likely to have a shorter than average time to failure once a leak develops.

Hazard Assessment

The majority of the reservoirs are located in and around London and are mainly category A reservoirs. These are all located near to heavily populated areas with significant infrastructure.

Other Issues

Whilst Thames Water undertakes routine reservoir surveillance to all of its reservoirs, the nature of the geology, embankment construction and length, means that early stages of leakage can be difficult to detect visually. Instrumentation on a 6 km embankment is not likely to be representative of the total situation. Therefore the time available to react to an incident when detected is likely to be reduced when coupled with the time needed to mobilize resources to commence an emergency drain down.

The significant length of the reservoir embankments is also likely, on a purely statistical basis, to increase the probability of failure of Thames Water's reservoirs when compared with other undertakers in the UK.

Consideration and Constraints when determining EDD rate

- The large surface areas of the Thames Water reservoirs results in high volumes to be disposed of. There clearly needs to be a sizable receiving body of water to take the required discharge rate. Unlike the majority of UK reservoirs, Thames Waters reservoirs are non-impounding and are not always very close to the River Thames or Lee.

- Whilst all the reservoirs have existing inlet and outlet pipe work or tunnels connected to water treatment facilities, they do not always have scour pipe work, which runs directly to waste. Careful assessment needs to be made as to how much of the stated treatment capacity of the works can be relied on at all times. A large algal bloom, for example, can significantly reduce works capacity.

- Some existing outlet routes, if opened fully in an uncontrolled manner, may result in residual flooding downstream due to the inability of the receiving water to take the high flows. Limited residual flooding may be preferable to the consequences of a major dam breach but the undertaker must clearly understand the risks and potential consequences when looking at risk mitigation.

- The selected draw down rate needs to take into consideration the permeability of the embankment shoulder material to ensure that rapid draw down does not adversely impact on the stability of the internal embankment by generating high pore water pressures within the internal slopes. Thames Water has assessed this with respect to the embankment shoulder materials at its reservoirs and do not envisage an issue at the rates required.

Existing draw down capacities at Thames Water's storage reservoirs

There currently appears to be little by way of industry guidance and in the past figures of the order of 300 mm per day have been accepted. In recent years, larger figures, between 0.75 m and 1 m per day, have been required by Inspecting Engineers.

Thames Water has modelled the potential draw down capacity utilising all existing pipework connections. Table 2 shows a summary of the modelled draw down capacity ranges as of 2007 and differentiates between existing supply routes and dedicated scour facilities.

Table 2: Summary of draw down capacity

Emergency Draw Down Capacity Range	*No of Reservoirs Achieving Stated Capacity*	
	Treatment & Waste	Waste Only
> 1 m / day	16	7
0.8 – 1 m / day	4	0
0.5 – 0.79 m / day	1	6
0.3 – 0.49 m / day	1	7
< 0.3 m / day	1	3

It can be seen that the exclusion of treatment capacity significantly impacts on the ability to draw down the reservoirs quickly. Two of the larger reservoirs were assessed as having low emergency capacities after allowing for treatment capacity. It is Thames Waters policy to provide a target of 1 m per day at all its reservoirs and reduce its reliance on the availability of treatment capacity. This paper describes the design and construction of new fully dedicated facilities to improve the emergency draw down to waste at Queen Mary and King George V Reservoirs.

Queen Mary and King George V Reservoirs

Queen Mary Reservoir, located at Ashford in Middlesex, was completed in 1925, following a suspension of construction during W.W.1. It was believed to be the largest storage reservoir of its kind ever constructed at that time. King George V was completed in 1913 and is located at Chingford in the Lee Valley. Both are of similar construction. Table 3 gives the principal details of the reservoirs.

Table 3: Principal details

Feature	*Queen Mary*	*King George V*
Max. height of embankment	12 m	10 m
Length of embankment	6324 m	6532 m
Total capacity	30.4 Mm^3	12.4 Mm^3
Surface area at TWL	2.863 km^2	2.168 km^2
Inflow capacity (controlled)	770 Ml/d	375 Ml/d

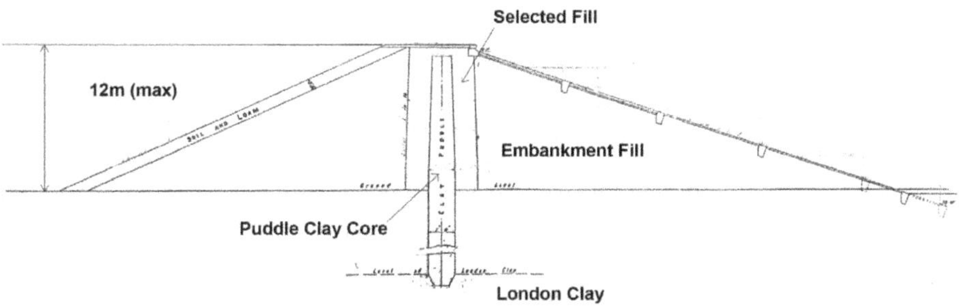

Figure 1: Section through Queen Mary (King George V is similar)

Following periodic inspections in 2005 at Queen Mary and 2006 at King George V reservoirs, the Inspecting Engineers assessed that the existing draw down capacity of 0.25 m and 0.35 m per day respectively was inadequate. The requirement to improve the emergency draw down capacity to 0.75 m and 1 m per day respectively was recommended in the interests of safety, as defined within the Reservoirs Act 1975.

In the case of Queen Mary it was decided to provide facilities to discharge the full 0.75 m per day to waste whilst the existing 0.25 m per day would be available to assist the rate in line with Thames Waters policy objective of 1 m per day draw down.

OPTIONEERING

Queen Mary Reservoir

Thames Water had initially considered internally various options for achieving the required drawdown rate for the Queen Mary reservoir. The preferred option was to install two 1600mm diameter siphons to pass over the reservoir embankment and discharge via a new concrete culvert into a new concrete spillway and into the existing Laleham Aqueduct. The study considered using butterfly valves or cone valves for flow control at the downstream end of the siphon.

Jacobs was employed to review Thames Water's design proposals and to finalise the design. Jacobs confirmed the suitability of using the 1600mm diameter siphons and looked at various options for achieving the proposals within the constraints of the site. The site constraints governed the location and discharge point of the siphons.

Location of Options

The main location options considered for the Queen Mary reservoir emergency drawdown related to the pipeline route and the discharge point of the siphon outlets into the Laleham aqueduct. The choice of location was also governed by a number of site constraints, which include the following:

- Existing service pipes and mains.
- High Voltage overhead cables
- The ability of the Laleham Aqueduct to take the required flow and velocity without scouring
- Working in water within the reservoir, while maintaining the normal operating water level
- Limited cutting into the embankment to protect the integrity of the puddle clay core
- The 'River Ash' (a stream) adjacent to the embankment
- Maintaining access to the treatment works

The substantial number of constraints led to the decision to extend the siphon as an above ground pipeline to the point clear of all services. The concrete channel was reduced to a relatively short structure connecting the stilling well to the Laleham Aqueduct.

King George V Reservoir

The existing outlet works consist of an inner and outer tower, either side of the embankment core, linked by a 1.44m I.D tunnel below the core. The current discharge capacity is around 0.35m per day. Jacobs looked at options for achieving the required 1m/day drawdown rate. The options considered two fundamental approaches to achieving the emergency discharge of the reservoir: providing a new facility or modification of the existing outlet. The options considered using these approaches include the following:

New Facilities:

Option 1 – New siphons over the embankment discharging into the river Lee diversion via a stilling basin.

Option 2 – New tunnel beneath the embankment connecting two new shafts located within the reservoir and beside the river Lee diversion.

Modifications to existing outlet:

Option 3 – New siphon to connect existing outlet towers. The siphon would work in the same way as in option 1.

Option 4 – New siphon over the embankment connecting either the existing inside outlet tower or the bottom of the reservoir to an existing 1500mm diameter outlet pipe.

Option 5 – A new 2000mm diameter tunnel connection through the embankment connecting two new launch and reception shafts, which in turn are connected to the existing outlet towers.

Option 6 – Enlargement of existing tunnel between the outlet towers to a diameter of 2500mm.

As with the Queen Mary reservoir, the preferred option for achieving the drawdown was option 1 which is shown in the schematic below.

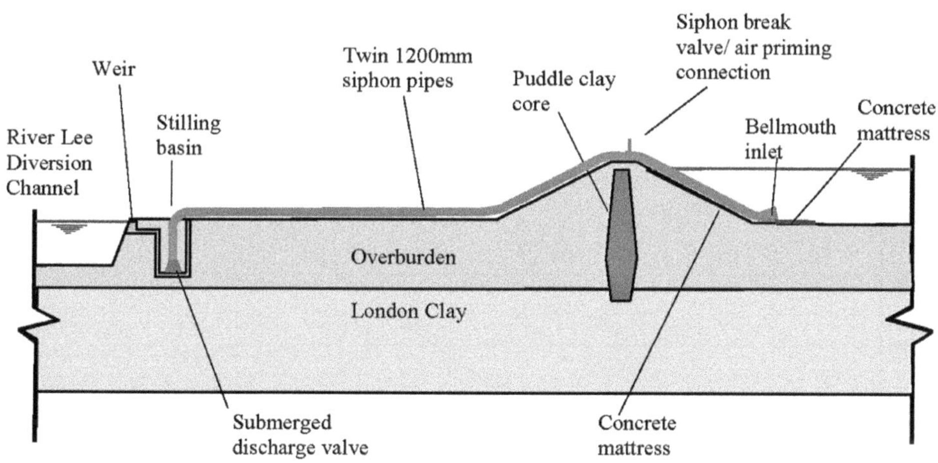

Figure 2: Schematic of the King George V Siphons

The preferred discharge water body was the River Lee Diversion on the east side of the reservoir as the alternative River Lee Navigation had limited hydraulic capacity.

DESIGN

One of the main criteria of the project was that the design had to be robust. This was because of the assessed potential for damage to the surroundings as a result of potential embankment failure and the infrequent use of the siphons. The main design elements, which had to be able to stand the test of time, include delivery of the hydraulic requirements, the siphon pipe material, the water priming system, the vacuum priming system and the flow control device.

Hydraulic Modelling
A mathematical model of the siphons system was developed to deliver the drawdown requirements set by the respective QCE's. This meant delivering a flow of 24.8 $m^3/_s$ and 20.8 $m^3/_s$ for Queen Mary and King George V reservoirs respectively. The velocities in the siphon pipes were similar for both projects, in the order of 6.1m/s. The choice of valve to dissipate the energy generated by the velocity was therefore critical and is discussed later.

A discharge curve was generated showing total outflow through the siphons vs reservoir water level. An empirical relationship between flow velocity and inlet submergence for the onset of vortex formation was used to derive a second discharge curve. This shows the allowable discharge vs reservoir water level at the onset of vortex formation. The combined curve is shown below.

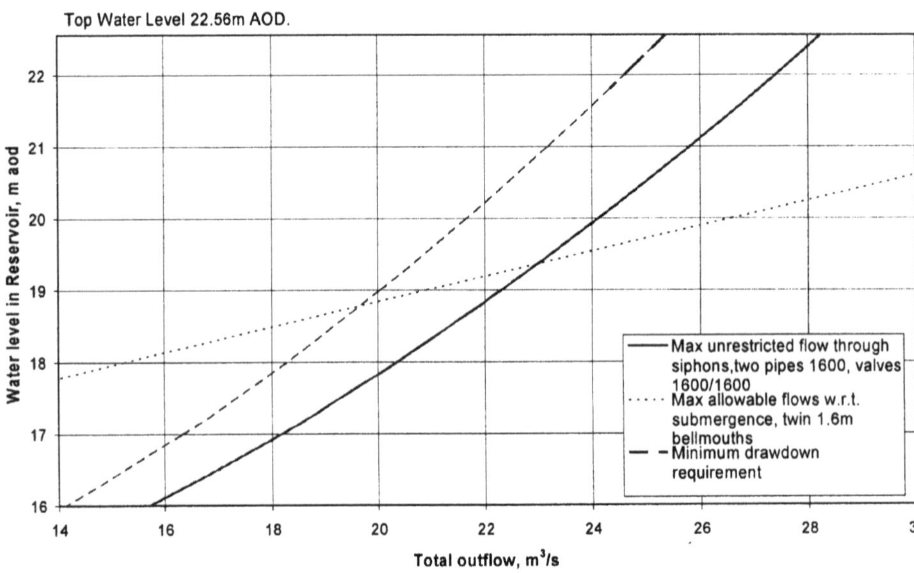

Figure 3: Queen Mary Reservoir siphons hydraulics as designed

The depth of water of the Queen Mary reservoir at the location of the siphons was 9.6 m. From the chart above, the effect of vortices is noticed at a depth of 3.06 m below top water level for a standard 1600mm diameter bellmouth. In order to improve on the depth at which vortices start to form, a non-standard bellmouth was designed which improved the vortex formation depth to 5.8 m. This however proved difficult and expensive to build. As a result, the QCE accepted a standard bellmouth with its limitations. During construction it proved possible to locate the inlet at a greater depth, thus reducing the constraint in operation.

Pipe Material
The preferred choice of pipe material over the embankment was one with minimal risk of joint failure so as not to break the siphon. Therefore a welded steel pipe with epoxy coating was chosen. During contractor pricing for the Queen Mary reservoir, the contractor suggested changing the pipe material to Ductile Iron as he could offer some savings from the change while maintaining the integrity of the design. This was accepted by the client for the non-vacuum section and the pipe material was changed to Ductile Iron downstream of the embankment.

Water and Vacuum Priming
The options considered for priming of the system included water priming and vacuum priming. Water priming while being a robust and reliable system would require an upstream valve on the siphon pipes, located within the reservoir. The valve was considered to be difficult to install, maintain and operate. Vacuum priming on the other hand has the advantage of not requiring an upstream valve but may be slow for the large volume of pipe to be filled and susceptible to non-sealing of the siphon system. Vacuum priming would also require pipes with thick walls to resist the vacuum pressure. The designed system was a combination of water and vacuum priming systems as this combines the advantages of both systems. This process was to involve two stages of priming.

The first stage would involve water priming of the system with the pipes downstream of the crest being filled to the reservoir crest level. Various options for water priming of the siphons were considered. These included using existing reservoir pipework or pumps, portable pump on the embankment crest and a new pumping station independent of the existing works. The latter option was chosen as the water priming can be undertaken as a first stage of a suspected emergency. The new pumping station would use water from the Laleham Aqueduct for water priming the Queen Mary reservoir. For King George V reservoir a secure supply was identified, which avoided the need for a new pumped supply.

The second stage of priming would involve vacuum priming to fill the remainder of the system. This would be carried out by using vacuum pumps on the pipes at the crest of the reservoir. The vacuum priming system would also serve to remove air which collects in the siphon during the operation of the system.

Control Valves
The flow and velocity generated to achieve the required emergency drawdown for the two reservoirs were relatively high. At the point of discharge the local velocity is over 12 m/s. It was therefore essential to be

able to provide a robust flow control device at the discharge point to the Laleham Aqueduct and the River Lee diversion for Queen Mary and King George V reservoirs respectively. The flow control also had to dissipate energy in order to minimise risk of damage to the discharge water bodies.

Figure 4: Submerged discharge valves awaiting installation.

The flow control chosen was a vertical sleeve type Submerged Discharge Valve. These have been used successfully on dam outlets and would be able to provide the required control and energy dissipation. Two options for the material of the submerged discharge valve were considered; fabricated steel and cast iron construction. The client chose the more robust cast iron construction.

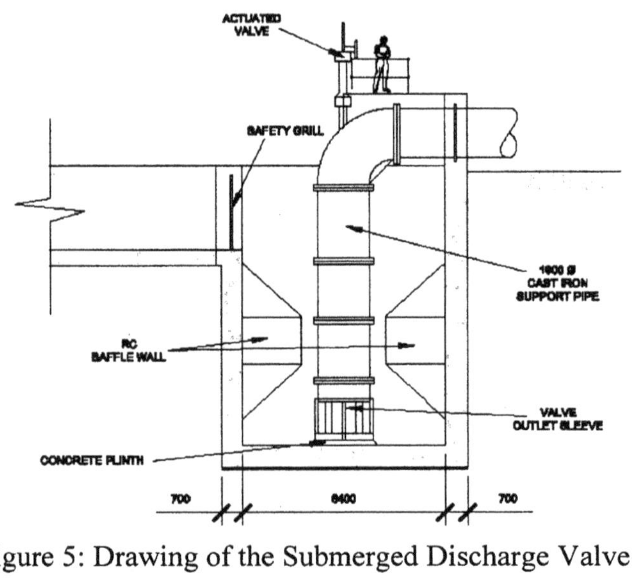

Figure 5: Drawing of the Submerged Discharge Valve

Pipelaying in the reservoir

One of the constraints at both reservoirs was the operational requirement of Thames Water to keep the reservoirs full while the pipelaying was carried out. Hence it was not possible to construct accurately aligned pipe saddles on the inner face of the embankment. The challenges for both Queen Mary and King George V reservoirs were however slightly different.

The inside face of the Queen Mary embankment at the siphons' location is finished with a wave wall and concrete facing all the way down to the bottom of the reservoir. The concrete facing provided a firm surface on which the pipe supports could be laid. The pipe support design involved the use of inflatable grout bags, which could be positioned under the pipes and injected with grout to even the loadings on the embankment. The bags were also used to provide load on the pipes at the vertical and horizontal bend to the bellmouth at the bottom of the reservoir. The load was required to anchor the pipes at the bends.

King George V reservoir, unlike Queen Mary reservoir, is constructed with a brick wave wall at the crest and a concrete lining, which only extends to a distance of about 2m down the slope of the embankment face. Below the concrete and down to the bottom of the embankment, the slope is unprotected and comprises the gravel embankment fill and silt settling down on top of the gravel. This face of the embankment may potentially be vulnerable to erosion as a result of wave action on the water surface when the reservoir water level drops below the concrete facing.

The design solution to protecting the upstream face of the King George V reservoir was to provide interlocking concrete mattresses on top of the exposed face which were tied back using stainless steel cables to the new reinforced concrete support to the pipes at the crest.

A similar grout bag arrangement to the QM reservoir was then used to support and anchor the siphon pipes.

COMMISSIONING AND OPERATION

Commissioning is planned to demonstrate successful operation to the Q.C.E. for final sign off and issue of a section 10(6) certificate. The need to fill and prime the siphons expediently is a requirement along with the need to demonstrate that the required flow rate is passed.

The flow rate will be determined by the use of temporary strap-on flow meters. These will be used to validate the design curves.

It is envisaged that routine testing will be undertaken at the six monthly Supervising Engineers examination visits, demonstrating the operation of one siphon every six months. This will help to ensure that the plant is checked as being operable and that all personnel involved remain familiar with the installation.

CONTRACT AWARD AND CONSTRUCTION

Contracts for the construction of both projects were awarded under the NEC conditions of contract.

The contract for Queen Mary was awarded in April 2007 to Barhale Construction Ltd. The submerged discharge valves were pre-ordered by Thames Water to ensure that the construction programme could be met. Work started on site in May 2007 and there were a number of construction challenges to overcome.

The requirement to keep the reservoir in operation meant that the upstream pipework had to be installed in the wet. This was assembled on a wide section of the embankment crest at the opposite side of the reservoir. A 1000 Tonne crane was needed to lift the assembly into the reservoir from where it was floated across some 2.5 km of water and lowered onto the internal embankment shoulder.

Figure 6: Pipework assembly being floated across Queen Mary Reservoir

The pipework on the downstream face was to be founded on a reinforced concrete slab, partly to spread the loads and partly to act as a channel to pass any leakage without impacting on the embankment should one develop during operation. In order to construct the slab, a 1 metre layer of loam had to be removed and replaced with compacted granular backfill. This was undertaken in sections to minimise the impact on the embankment. Particular care was needed when crossing the surface of the clay core wall.

A significant element of the work is associated with the stilling chamber and weir connecting into the existing intake channel. The chambers were constructed within a deep steel sheet pile cofferdam. The management of ground conditions and high water table along with periods of elevated river levels all contributed to programme delays.

At the time of writing the Queen Mary project is due to be completed during March 2008.

The contract for King George V was awarded to Morrison Construction Ltd in December 2007. This project is due to be complete before the end of December 2008.

A new geomembrane system in construction of fill dams: concepts and 2 case histories

A.M. SCUERO, Carpi Tech S.A., Chiasso, Switzerland
G.L. VASCHETTI, Carpi Tech S.A., Chiasso, Switzerland

SYNOPSIS. A new geomembrane system, totally deformable and capable of following large settlements and differential movements, has been developed to waterproof new fill dams. The geomembrane liner is anchored by PVC elements embedded in the fill as it is being raised. This unconventional design, presented at the 2006 ICOLD congress in Barcelona, has found recent field application in two projects, both in Iran: Kohrang dam and reservoir in 2004, and phases BII and CII of Sar Cheshmeh tailings dam in 2007-2008. The paper discusses the general concepts of the system, and reports on Kouhrang and Sar Cheshmeh projects.

INTRODUCTION

The use of impervious geomembranes to waterproof fill dams is about half a century old, the first installations dating back to 1959. Since 1959, geomembranes have been widely used in fill dams, in new construction as well as in rehabilitation of asphalt concrete facings and of concrete facings (CFRD). As discussed at a previous BDS conference (Scuero & Vaschetti 2004), geomembrane systems can provide substantial advantages in new construction of fill dams as compared to traditional waterproofing systems, because they avoid the problems related to lack of suitable materials, to adequate installation of multiple defence lines of waterstops, to deterioration of waterstops, to connections of the core or of the asphalt concrete to the ancillary concrete structures.

Another advantage of geomembrane systems is related to construction times. With traditional systems, the impact that the installation/construction of the face slabs, or the placement/compaction of the impervious core, can have on the overall construction schedule, and the complexity of the techniques needed to construct the waterproofing system, are elements to take into consideration when evaluating times of completion. In dams with a clay core or with asphalt concrete core, a crucial point is that the construction of the dam body and the construction of the central core being strictly related, the constraints imposed by the weather conditions, or any disruption in the placement of the filter material, will affect the whole rate of construction of the entire dam body. On the contrary, installation of a geomembrane system can be scheduled in conjunction with the general programme of construction, and is practically unaffected by weather.

Concerning future performance, as compared to traditional methods a PVC geomembrane, with >230% elongation capability, will maintain watertightness in presence of settlements and differential movements that on the contrary could in CFRDs cause fissuring / cracking of the face slabs and possible failure of the waterstops, and in asphalt concrete facings cause disruption in the connection of the asphalt facing to the concrete structures. In dams with an impervious core on the other hand, the efficiency of the core heavily relies on the efficiency of construction (too often related to dry weather conditions) and on the skill of the contractor.

NEW DEFORMABLE GEOMEMBRANE SYSTEM

A major issue for the waterproofing element in a fill dam is its capability to accommodate the movements that are typical of this type of dam. This issue has directed the development of a new geomembrane system to be applied in construction of fill dams. The concept of this unconventional design, presented at the 22nd ICOLD congress in Barcelona (Scuero & Vaschetti 2006), is to exploit at best the elongation characteristics of geomembranes. The result has been a geomembrane system that allows construction of a safe dam more easily, in shorter times, and at lower costs.

The rationale of the system

The concept is to provide imperviousness to the dam by a system that is totally deformable: the waterproofing liner and its face anchorage system are made with the same material, a highly flexible and deformable PVC material. The cross section of the dam has the following simple layering:

- Dam body. Zoning is not strictly required, one type of fill material can be used
- Drainage layer, as in fill dams but of reduced thickness
- Base/anchorage layer. This single layer forms the transition between the dam body and the watertight system, incorporates the anchorage system for the waterproofing liner, and can also incorporate the drainage layer
- Waterproofing liner: a PVC geocomposite
- Cover layer (shotcrete or unreinforced concrete) optional.

The waterproofing liner

The waterproofing liner is a composite liner (geocomposite) incorporating the impervious element (the PVC geomembrane) and a multi-function layer (the geotextile, coupled to the geomembrane in the factory). The geotextile increases puncture resistance and dimensional stability, eliminating the potential for wrinkling and sagging due to temperature variations. The high friction at the interface geotextile/base layer facilitates the positioning of the geocomposite and increases stability in respect to sliding.

The geomembrane liner is fabricated in a factory, under ISO 9001 procedures, assuring constant computer-controlled properties.

The anchorage system

The face anchorage system (patented) of the waterproofing liner is integrated in the dam, in the form of flexible PVC geomembrane elements ("wings"), about 0.5 m wide, embedded in the base/anchorage layer and anchored to the dam body by the ballasting action of the base layer itself. Part of the wings is left protruding from the upstream face of the dam, so as to form a continuous 0.5 m wide anchorage strip. Spacing between anchorage bands depends on service conditions; the most severe design load is generally the velocity of the wind.

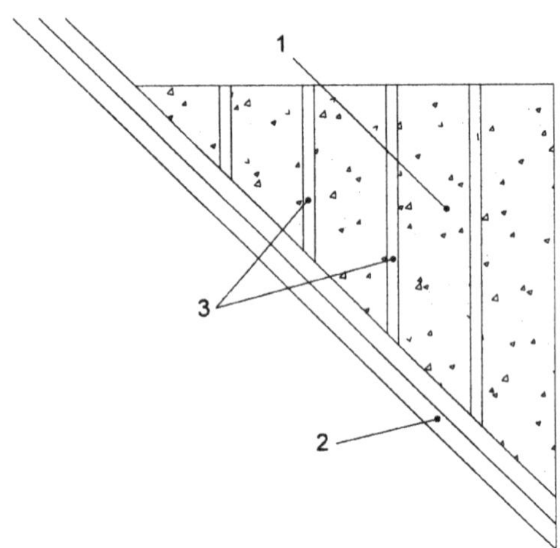

Figure 1. (1: Base/anchorage layer, 2: Plinth, 3: PVC anchor strips)

The base layer for the waterproofing geocomposite can be gravel stabilized with lean concrete placed as the fill is being raised, and then smoothed and compacted with a vibratory plate mounted on the arm of an excavator, as done at Bovilla 91 m high rockfill dam in Albania in 1996. In alternative, the base layer can be made by extruded lean concrete curbs. This method, which is being increasingly used in many fill dams of recent construction, has the advantage of constructing in short times a fairly regular stable slope not erodible by rainwater, that can be also steeper, and has constant characteristics over the entire upstream face.

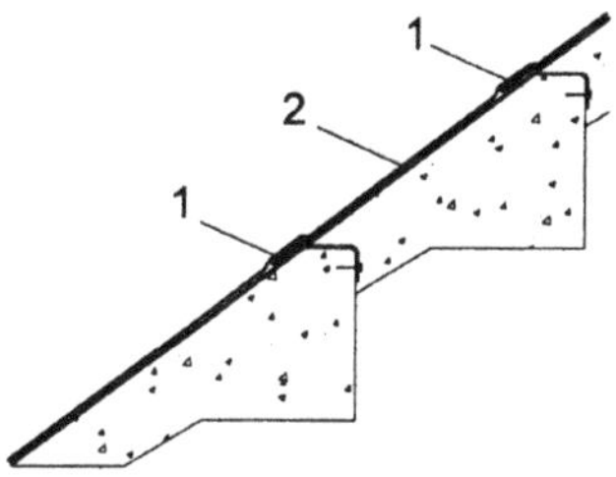

Figure 2. (1: PVC anchor wings, 2: Waterproofing PVC geocomposite)

The PVC geocomposite sheets (or larger panels prefabricated in width adequate to span the space between vertical anchor strips) are deployed over the base layer, and anchored to the PVC wings by heat welding. Adjacent PVC geocomposite sheets/panels are joined by watertight vertical welds, creating a continuous watertight liner over the entire upstream face of the dam.

The perimeter anchorage is made by a watertight seal designed to follow the differential movements between the deformable fill and the rigid concrete appurtenances (plinth, intake tower, spillway, etc.), without failing. The perimeter seal can be of the compression type or of the embedded type. At plinth, the PVC geocomposite is watertight connected to the concrete which is connected to the grout curtain.

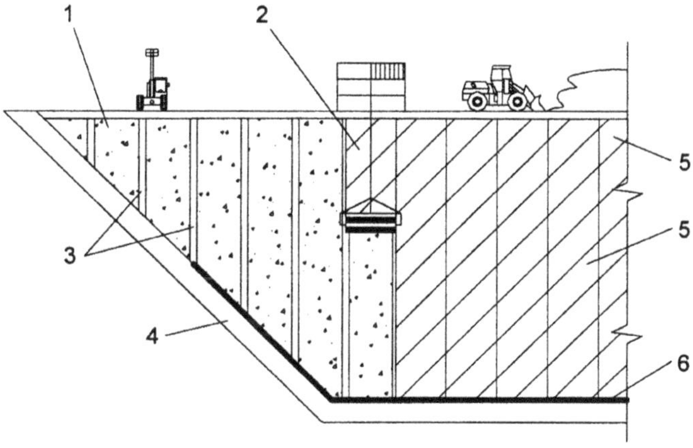

Figure 3. (1 : Base/anchorage layer, 2 : PVC geocomposite panels, 3 : PVC anchor strips, 4 : Concrete plinth, 5 : Geocomposite panels welded to anchor strips, 6: Watertight perimeter seal)

ADVANTAGES OF THE SYSTEM

Being prefabricated in the controlled environment of a factory, under ISO procedures, all components of the waterproofing system (PVC liner, PVC anchor wings, components of the perimeter seal) have pre-established and constant computer-controlled properties. Properties are verified upon arrival of materials at site; weather conditions at installation do not alter the materials, joints and perimeter seals are verified for watertightness with standard methods. Hence the characteristics and the final quality of the waterproofing system are pre-established and unaltered during the entire installation process.

Different from any other existing method, all the components of the waterproofing system are fully deformable and allow elastic recovery. The impervious liner and its anchorage system are capable of following the deformation of the dam and of accommodating all possible differential movements that may occur during service.

The upstream geomembrane allows reducing constraints in respect to the type and grading of fill material, to the type and thickness of drainage layer. The dam body can be constructed with a large range of materials locally available. The waterproofing system is “more forgiving” in respect to conventional systems as far as construction skills are concerned. Companies with limited previous experience in embankment dams can achieve construction of a safe dam, while a specialised contractor constructs the waterproofing system.

Construction can be staged to follow construction of the dam, so that the waterproofing system is completed concurrent with completion of civil works: construction of the embankment can start at the same time as construction of the cut-off or the grout curtain, while installation of the waterproofing system can start before the dam has completed most of the expected settlement, and can be carried out in stages, following the raising of the embankment, with no interference between placement of the fill and placement of the waterproofing layer. Since weather conditions to perform installation activities are temperatures sustainable for the workers, absence of heavy rain unless a shelter is provided, and wind speed allowing safe handling of geocomposite sheets, this system can accommodate practically any construction schedule, which would depend only on the rate of placement of the fill. The times for completion, and the costs of the entire dam construction, can be significantly reduced.

FIELD APPLICATION

The concept described has been applied in pioneer versions at Alento cofferdam in Italy, and at two pilot projects in Turkey and at Toulnustouc

CFRD in Canada (Scuero & Vaschetti 2006). Full field application has been made in two fill dams, as described in the following case histories, and is under consideration for several projects.

Kohrang head pond

Kohrang head pond is part of Kohrang hydropower plant on Kohrang River. The HPP has been designed with 3 units of 13.5 Mw installed capacity, that are fed from Kohrang head pond by three 1200m long steel penstocks. Water to the head pond comes from a 1700m diversion culvert from the outlet of Kohrang tunnel II. The head pond, which controls and regulates discharge and head of water for the generators, is an oval shaped embankment. The geometrical shape and layout was restricted on the east by a shallow valley, at the west by an access road, and south by the transmission tunnel. A free flow spillway inside the head pond body protects the embankment from overtopping, a concrete intake structure connects the head pond to the penstocks. During operation, several filling and emptying cycles may occur every day. The climate at site allows only 6-month outdoor construction activities, temperature in winter may go under -30°C and heavy snowfall up to 3 m is usual.

In preliminary studies for the dam forming the head pond, because of high risk of settlements arising from non-uniformity of soil layers in the foundations, the options of a rockfill dam with concrete face or with asphaltic face was eliminated (Jafarzadeh et al. 2005). Three alternatives were retained, based on the materials available for construction, on site conditions, and on construction times and costs: rockfill dam with clay core, homogenous impervious embankment, and homogenous embankment with geosynthetic liner. The three variants were fully designed and compared. The reasons for selecting the option of homogeneous embankment with exposed geomembrane (Yosefi et al. 2005) were

- the low permeability (almost zero) of the geomembrane
- the geomembrane system could grant watertightness even in the case of large settlements of the embankment
- the geomembrane system could establish a watertight connection to the concrete structures and sustain differential settlements
- the durability of the geomembrane in extremely cold and hot temperatures
- the overall lower cost, the higher speed of construction, and the simplification of construction works and quality control that the geomembrane system could provide.

Other very good advantages of the selection of the geomembrane option were the elimination of filter materials that not available naturally had to be prepared in a high cost procedure, the increased volume of the pond and the reduced volume of excavation and fills, as shown in Table 1. The main advantage from the client's point of view was a very short duration of construction works.

Table 1. Comparison of the three options (Jafarzadeh et al. 2005)

Item	Rock+clay core	Homogeneous	Homogeneous+GM
Max. height of embankment (from prepared foundation) (m)	16.5	14.5	14
Length of crest (m)	460	495	480
Upstream slope of embankment	1:2.1	1:2.5	1:1.7
Downstream slope of embankment	1:2.8	1:3	1:1.7
Effective volume of pond (m^3)	60000	60000	86000
Total volume of fills (m^3) (Rockfill, Clay, Random, Filter, Drain, Riprap)	216000	245000	99600
Volume of filter & Drain (m^3)	55000	68000	8930
Volume of excavation (m^3)	213000	165000	9000

Details on the reasons determining the selection of a PVC geocomposite as waterproofing liner are not in the scope of this paper and can be found elsewhere (Yosefi et al. 2005). The selected PVC geocomposite, a PVC geomembrane 2 mm thick on the slopes and 1.5 mm thick on the bottom, laminated during fabrication to a 500 g/m^2 geotextile, had tensile characteristics that allowed placing it directly over the compacted fill that forms the 14 m high dams and bottom of the head pond.

The face anchorage that is the most notable feature of the system is formed by PVC geomembrane strips embedded in the soil of the finished line of the earthworks. The trenches were designed to be filled with gravel, but the main contractor found it more convenient to fill them with porous concrete.

Figures 4 and 5. Kohrang: details of the PVC geomembrane strip embedded in a trench filled with porous concrete and of the PVC geocomposite being placed on it, and a view of the parallel anchor strips in embankments

The spacing between anchor strips, designed in function of the maximum wind velocity which is higher at the crest and reduces towards the bottom, is 8 m on the slopes and 16 m on the bottom. The PVC geocomposite sheets have been anchored by heat-welding onto the PVC anchor strips. Longitudinal anchorage was also provided at the base of the slope, by embedding the geocomposite in the peripheral drainage trench that runs all around the toe of the slopes and discharges by pipes downstream of the dam. The drainage system under the PVC liner has been designed to collect and discharge leakage and avoid that unbalanced backwater pressure can act on the liner.

Figures 6 and 7. Kohrang: the PVC geocomposite sheets already anchored on the PVC strips on part of the embankments, and ballasting the PVC geocomposite in the bottom peripheral trench

The PVC liner was anchored along the crest by embedment in a top trench. Connection to the concrete structures (spillway, intakes) was achieved with the well established and proven mechanical tie down method, with stainless steel batten strips bolted to the concrete and arranged so as to allow accepting large movements between the fill and the concrete structure.

Figures 8 and 9. Kohrang: backfilling of top trench for peripheral anchorage of the PVC geocomposite, and tie-down anchorage on concrete, with extra material to accommodate differential movements

Installation at Kohrang was completed in 35 days, for a total of 16500 m^2 of geomembrane placed. The settlements that had been foreseen during the design phase occurred after impounding; temperature went down to –37°C, with heavy snow and ice formation. As a consequence of the settlements, some fissures opened at the contact between the fill dam body and the concrete appurtenances. There was no damage to the exposed geomembrane.

Figures 10 and 11. Kohrang at completion, and during winter operation. Minimum temperature - 37°C, 3 m snow, 0.5 m ice

The owner (Yosefi et al., 2005) reports that the geomembrane system "is working effectively.... Because of the use of the geomembrane solution, the following benefits have been gained by the owner:

- A saving of 60 per cent was achieved in the volume of fill material.
- Construction was faster and easier.
- The volume of impounded water compared with the original design was increased by 25 per cent, which creates more energy generating capability.
- Construction costs for the head pond could be decreased by 50 per cent."

Sar Cheshmeh tailings dam

Sar Cheshmeh is a tailings dam owned by National Iranian Copper Industries Company (NICICO), with estimated ore reserve of about one billion tonnes. The upgrading project for tailings disposal and water management system carried out by NICICO includes strengthening and raising the Main Embankment and Saddle Dam No.1, from the existing crest at elevation 2175.5 m to elevation 2194 m (Stage IIB + Stage IIC). Maximum design earthquake is 0.8 g, temperatures at site vary from – 16°C to + 38 °C.

The designers, Australian Tailings Consultants and Middle East Water & Environment Joint Venture, have planned for Stage IIB and Stage IIC a zoned rockfill, with an upstream exposed PVC geocomposite system as the water-proofing component of the raised dam. The PVC geocomposite will be placed on a base layer made with extruded porous concrete curbs, inclination 1.5H:1V. A concrete plinth will be constructed at the rock abutments, while at the existing crest the geomembrane will be keyed to the existing clay core. Total length of the dam at Stage IIC will be 1000 m, total installed geocomposite 20500 m^2 for Stage IIB and 18000 m^2 for Stage IIC.

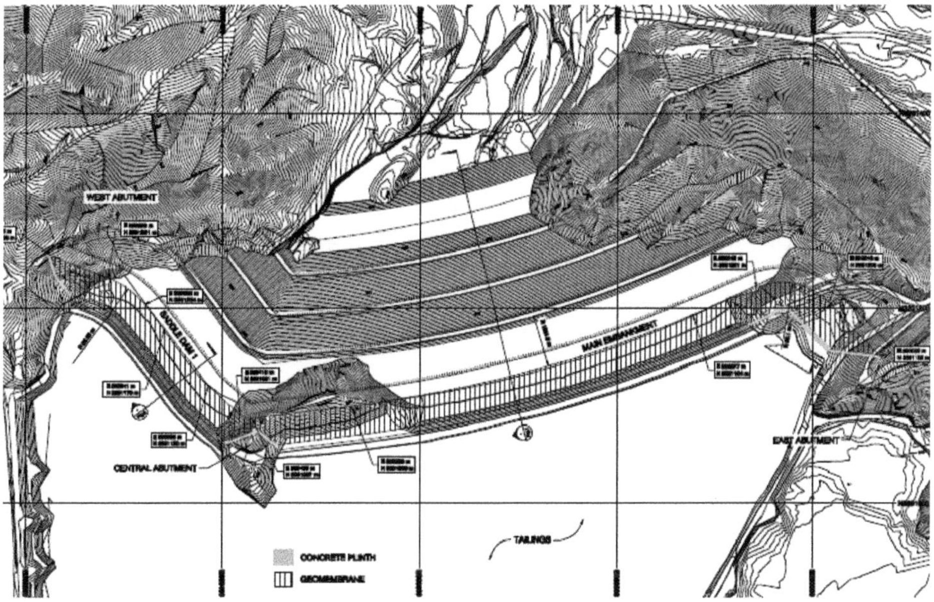

Figure 12. Layout of Sar Cheshmeh dam raising

Figures 13 and 14. The first curbs already placed at the main embankment: view from the central abutment (left) and from the east abutment

The waterproofing liner, Sibelon CNT 4400, is formed by a PVC geomembrane 3 mm thick laminated during fabrication to a 500 g/m^2

geotextile. Face anchorage is provided by anchor strips of the same material, placed over the extruded curbs and anchored by ballasting with the layers on top. To avoid that as the embankment is being raised the curbs are laterally displaced by the pressure of the fill and by the horizontal stresses locked-in as result of compaction, the curbs have been cast in such a way that they interlock with each other.

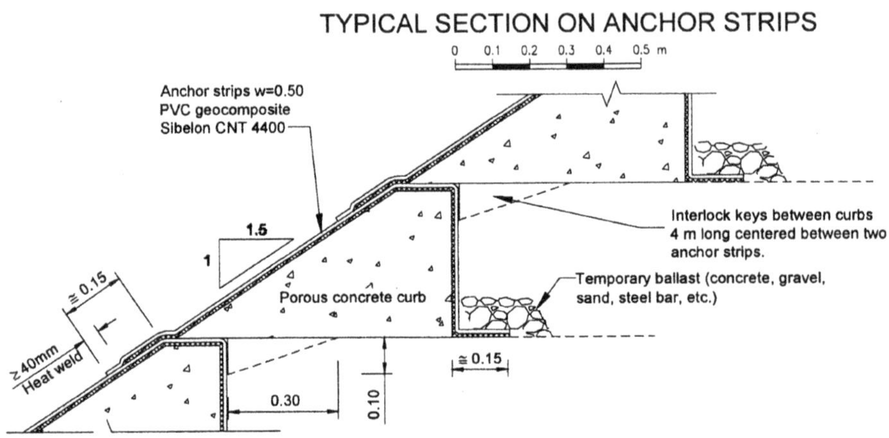

Figure 15. Face anchorage details at curbs

Figures 16 and 17. Detail and view of the PVC geomembrane wings placed over the curbs at Sar Cheshmeh in January 2008

IIB, bottom anchorage of Sibelon CNT 4400 at the rock abutments will be made by a watertight seal of the tie-down type placed on the concrete plinth, and consisting of a 60 x 6 mm flat stainless steel batten strip compressing the liner on the concrete, according to a well-known fastening system. Bottom anchorage at the existing crest will be made by connecting the PVC geocomposite to the existing clay core via a trench excavated in the existing crest and backfilled with clay. Top anchorage at elevation 2185 m will be made with a flat stainless steel batten strip.

The PVC geocomposite of Stage IIC will be deployed from the crest at elevation 2194 m, will overlap the top anchorage of Stage IIB at elevation 2185 m, and will be watertight welded to the PVC of Stage IIB. Top anchorage at Stage IIC will be made in a trench excavated at crest and backfilled with a mass concrete anchor beam.

Raising of Stage IIB started on October 2007. At the beginning of February, 9 lines of curbs have been placed that already incorporate the face anchorage system for the PVC liner. Expected time for installation and complete permanent anchorage of 20500 m^2 waterproofing system since completion of Stage IIB is 2 months.

REFERENCES

Scuero, A.M. and Vaschetti, G.L. (2004). *Watertightness and safety of dams by geomembranes.* Proc. British Dam Society Conference, Canterbury, pp 102-116.

Jafarzadeh, F., Javaheri, H., and Heidari, T. (2005). *Kouhrang HPP head pond, first hydraulic geomembrane project in Iran: comparison with other variants.* Proc. 73rd Annual Meeting of ICOLD, Tehran, paper 217-OT.

Yosefi M., Jafarzadeh F., Javaheri H., and Noori F. (2005). *Kohrang Headpond – the first geomembrane for a hydraulic structure in Iran.* The International Journal on Hydropower & Dams, Vol. 12, No. 2, pp 72-75.

Scuero, A.M. and Vaschetti, G.L. (2006). *Unconventional cross sections and materials in embankment dams.* Proc. 22nd International Congress on Large Dams, Vol. 1, Barcelona, pp 87-105.

Rutland Habitats - Design of Reservoirs for Environmental Habitat Benefits

J.R. FOSTER, Mott MacDonald Ltd, UK
A.M. KIRBY, Mott MacDonald Ltd, UK
S. SWAN, Anglian Water Services

SYNOPSIS. This Paper outlines a project involving the design of new reservoirs to create alternative habitats for wildlife at Rutland Water.

The work of Anglian Water and the Leicestershire and Rutland Wildlife Trust partnership since Rutland Water was completed in 1976 has been successful in providing new habitats for birds as well as other wildlife. However, with the need for Anglian Water to meet increased water demand there is a requirement to provide alternative habitats in the form of retained water lagoon areas.

The paper outlines the solution of Habitat Creation at Rutland Water.

INTRODUCTION

Rutland Water has been designated under the EU Habitats Directive as a Special Protection Area (SPA) and a RAMSAR site. An increase in water demand means that Anglian Water is required to alter the management of the reservoir which will result in greater drawdown of the reservoir. This will have an adverse effect on habitats for wading birds and wildfowl along the shoreline of the reservoir. The Habitats Directive requires mitigation in the form of additional habitat provision to be made to ensure that the integrity of the SPA is not adversely affected. Compensatory wetland areas are therefore to be created at the shallow western end of the reservoir close to Egleton village.

The Rutland Water Habitat Creation scheme will involve the impoundment of three existing areas of Rutland Water (Areas A, D1 & D2), the creation of water bodies away from the reservoir by excavating and forming water retaining embankments, referred to as terrestrial lagoons in a further five areas of land (Areas B & C1 to C4) and the creation of a wetland meadow (Area C5). The layout of the habitats creations is as illustrated in Figure 1.

The management of the new habitats will be carried out by the Leicestershire and Rutland Wildlife Trust who presently manage the existing nature reserve at Rutland Water. Anglian Water's Fisheries & Recreation department works with the trust and manages the remaining areas of Rutland Water.

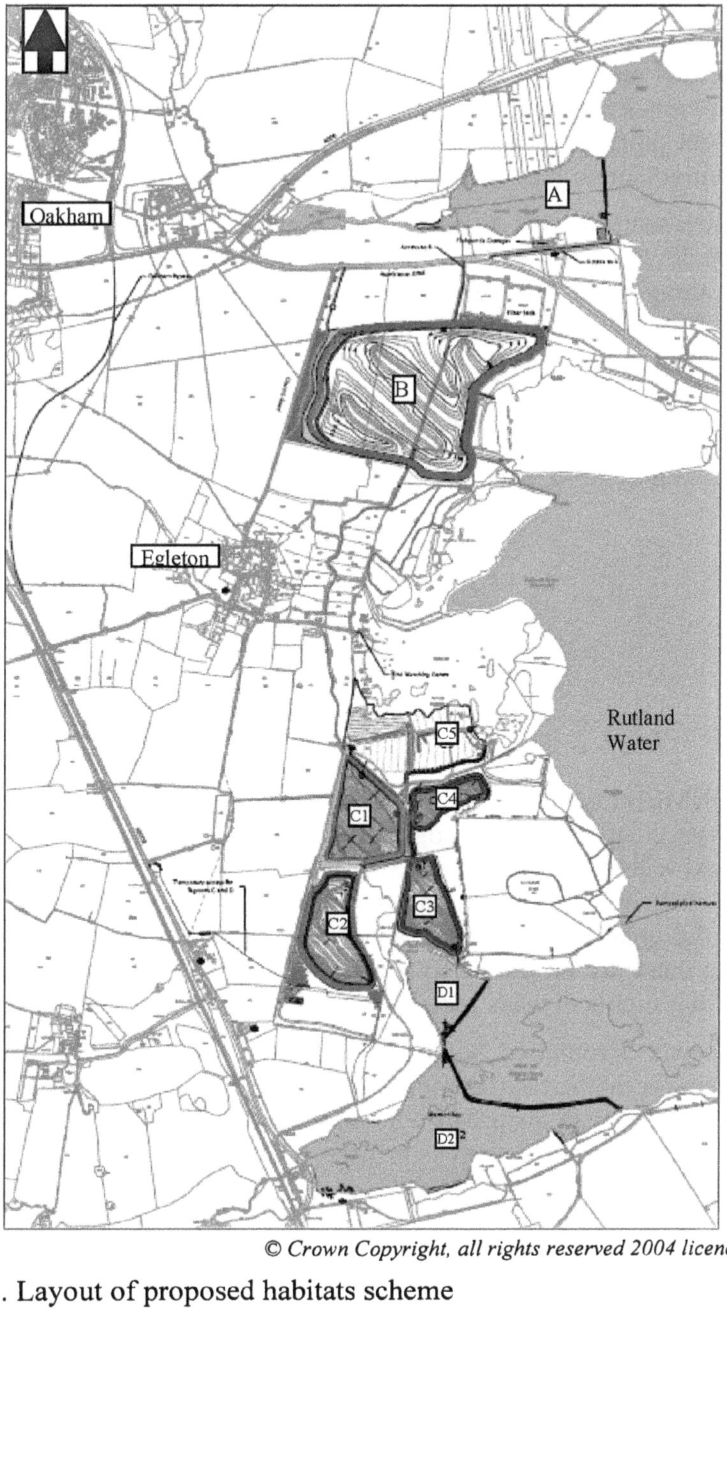

Figure 1. Layout of proposed habitats scheme

MULTI DISCIPLINARY APPROACH

Anglian Water is very aware of the importance Rutland Water has on wildfowl and the general environment and its own Fisheries & Recreation department has successfully worked with the Leicestershire & Rutland Wildlife Trust in developing and managing the reservoir to encourage this.

Knowing early on that a change in the management of the reservoir would have an adverse affect on habitat, Anglian Water commenced studies firstly into the effect on the habitats and then to design mitigation measures. The design approach for the provision of the mitigation habitats has been carrried out jointly between ecologists, landscape architects, engineers, client, contractor and the wildlife trust.

The requirements for the size, location, timing and nature of the lagoons have been developed in consultation with Natural England, Rutland County Council along with the Leicestershire & Rutland Wildlife Trust. In summary the following habitat creation areas are planned:

Table 1. Habitats to be created

Area	Size (Ha)	Description of proposal
A	6.0	Impoundment of reservoir spur
B	29.8	Terrestrial lagoon
C1	6.9	Terrestrial lagoon
C2	6.9	Terrestrial lagoon
C3	4.6	Terrestrial lagoon
C4	2.4	Terrestrial lagoon
C5	8.0	Wet meadow with ditches
D1	4.0	Impoundment of reservoir spur
D2	15.0	Impoundment of reservoir spur
Total	**83.8**	

ENVIRONMENTAL CONSTRAINTS

The habitats creation is to be constructed within and adjacent to the nature reserve and within the area of the SPA. Therefore, much preparatory work has been carried out on surveying the existing habitats, including badgers, bats and the great crested newts, all of which live within the construction areas. As these are all protected species under law, identifying where they live and their numbers leads onto how to safeguard them.

For the newts, alternative ponds were created for which they had to be collected and re-homed at the appropriate time of year. The badger colonies are to be avoided with disturbance monitored and tree clearance timed not to affect the bats. Wildlife surveys will be ongoing throughout and after the habitats creation work.

All this environmental preparatory work - mitigation works for the mitigation works - had to be conducted in agreement with Natural England to obtain the appropriate licences to allow the work to proceed. Anglian Water and the wildlife trust both assisted greatly in this process in allowing time and land.

Construction Sequence

The construction schedule is limited as far as possible to the late winter, spring and the summer months of each year to minimise adverse impacts on the SPA designated wildfowl populations and the operational regime of the reservoir. This construction schedule is over a 3 year period as follows:

- Year 1 - Lagoon area B commencing April 2008 and completion October 2008 with site set up and clearance from January 2008 and commissioning in November 2008.
- Year 2 – Dam area A commencing February 2009 and completion July 2009 with site set up in January 2009 and commissioning in August 2009.
- Year 3 – Dams areas D1 and D2, lagoons area C (C1 to C5) commencing January 2010 and completion October 2010.

The logic behind the three year schedule is:

- In disturbing one area an alternative area must be available as a refuge/habitat for wildfowl – logic: Cannot do A and D together;
- Before disturbing an area within the reservoir there should be an additional water area available as a refuge. Hence the construction of a terrestrial lagoon should be carried out before Lagoon A or D1/D2 is constructed – logic: Do B or C before A or D;
- Lagoons C and D are in a similar area so can be carried out in the same year to avoid disturbing that area twice – logic: Do C and D together.

SCOPE OF WORKS

Overall Scope Lagoon A

Lagoon A will be capable of retaining water at Rutland Water's 85 % full level by means of an over-toppable steel sheet pile dam. The length of the dam is approximately 250 m, with a maximum height of 2.7 m.

Along the downstream side of the sheet pile dam a rock bund is to be placed and a single tilting weir will control water levels in the lagoon when reservoir is below the 85% level. The lagoon will be supplied from the North Gwash stream inflow.

Overall Scope Lagoon B

Lagoon B will be a terrestrial lagoon, a water body formed by cut and fill, constructed on agricultural land away from Rutland Water. An earthfill dam will be constructed along its eastern perimeter to retain water. Along the remaining perimeters a screening embankment will be constructed to avoid disturbance to birds on the lagoon.

A tilting weir will be installed to provide water level control with a bottom outlet pipe for emergency drawdown. A small pump station will be constructed to supply water to the lagoon from Oakham's Sewage Treatment Works outfall and from the reservoir.

Overall Scope Lagoon C1 to C4

Lagoons C1 to C4 will also be terrestrial of a similar design to Lagoon B. The lagoons will be supplied by a pipeline fed from a pump station at Lax Hill. Each lagoon will be capable of being supplied with water separately.

Overall Scope Lagoon C5

Lagoon C5 will be created by modification of the existing grassland to create a wetland. This will primarily involve ditch excavation and construction of small, simple control structures. There is no separate water supply to Lagoon C5. It will receive water from existing drainage ditches, from excess flows from Lagoons C1 and C4 and from groundwater connections with Rutland Water. There will be three outfall structures from Lagoon C5 with one having a water level control sluice to manage water levels within the area.

Overall Scope Lagoon D1 and D2

Lagoons D1 and D2 will be capable of retaining water at the 90 % full level (Rutland Reservoir) which corresponds to an elevation of 82.78 mAOD.

The design of Lagoons D1 and D2 will be of the same general construction as Lagoon A. A culvert will connect the two lagoons to maintain a common water level. However, the cross-drainage structure will also have the capability of isolating Lagoon D1 from D2. The lagoons will be supplied by natural stream inflows from the River Gwash, however, it will be necessary to top-up during the summer and so is supplied by the same pipeline as for Lagoons C1 to C4.

TERRESTRIAL LAGOON PROFILING FOR HABITATS

The designs for the terrestrial lagoon topography have been developed with Roger Buisson of RPS and in consultation with Natural England, RSPB & others. The internal profiling and subsequent water depths are specific to develop habitat for particular wildfowl including the Shoveller and Gadwall. In summary the key design requirements for the terrestrial lagoons are:

1. Water levels in each lagoon managed independently varying 0.4 m between Winter Water Level (WWL) and Summer Water Level (SWL);
2. Sausage shaped islands linked to each other so that lagoons can be accessed for maintenance and wide enough in places for turning;
3. Lagoon B islands are set below the water level but that they are almost exposed during summer water levels; specifically for Gadwall
4. Lagoon C islands are permanently above the water mark;
5. The islands and channels orientated north-west to south east to reduce wave fetch from the south west and encourage seed trapping;
6. A deep water perimeter ditch within the lagoon to prevent access from predators and limit vegetation colonisation;
7. A causeway and bridge to the islands for maintenance access. Submerged at winter water level to prevent predator access;
8. A submerged bund across the perimeter ditch between inlet pipe and the control structure to encourage water circulation;
9. Shallow beaches;
10. Shallow water around islands gently sloping toward deepwater channels;
11. Clear sight lines for bird flight paths to connect habitat to existing lagoons/reservoir through bush clearance where necessary.

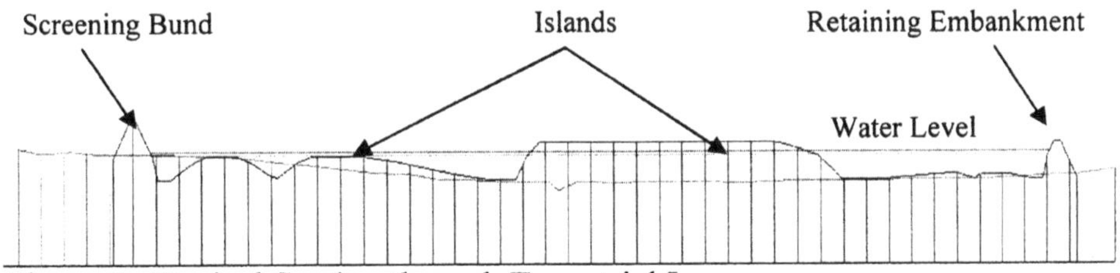

Figure 2. Typical Section through Terrestrial Lagoon

Island Profiling

The profiling of the islands is imperative to the success of the habitats as it is within the shallows where the invertebrates will prosper and provide the food for the wildfowl. The slopes of the island sides work in harmony with the proposed water level management to provide the optimum habitats for successful breeding. The deep perimeter ditches will also provide effective protection against land-based predators thus ensuring the ground based nesting will be successful in an uninterrupted environment.

The specification for the island material is to discourage vegetation and minimise releasing nutrients into the water. For this reason, topsoil will only be used in the island material mix within the core, with purely sub-soil used for the surfaces of the islands. The excavation and filling of the islands and channels shall require a high degree of accuracy, more than would be expected, to ensure the perfect habitats are created. It is expected that the Contractor will be using the latest surveying equipment to assist the process.

Water Level

The topography of the terrestrial lagoons has been developed to meet the design requirements for habitat creation. The ratios of deep to shallow water were developed to match the ratios of the existing nearby habitat lagoons based on bathymetric survey carried out in 2005.

The islands in Lagoons C are nominally 0.3 m above WWL when there exists a string of isolated islands, typically about 10 m wide. At SWL the receding water leaves a continuous path between islands with a typical width of 3 m.

Boundary Embankments

The terrestrial lagoon embankments can be categorised as:

- *Water retaining* – crest level dictated by the 150yr Return Period Rainfall event with wave freeboard.
- *Screening* – crest level is to provide 2 m screening above adjacent ground level. Winter water level is below natural ground level.
- *Combined* – embankment retains water above natural ground level but the crest level is dictated by the need to provide 2 m screening.

The construction and maintenance requirements differ between the different types:

- *Water retaining*: good compaction of high quality selected excavated material essential, good grass cover, no trees or shrubs, possible wave protection required;
- *Screening*: different grass seed mix to water retaining embankment, trees and shrubs permissible on downstream slope, requirements for compaction and selection of lower fill;
- *Combined*: generally as for water retaining embankment.

A crest width of 3.0m and side slopes on both inner and outer face of 1.0V:3.0H have been adopted. This is primarily dictated by maintenance requirements for access and to permit safe grass cutting. The stability of the embankments under a variety of load conditions was checked using Geoslope SLOPE/W software.

Access ramps onto the crest of the embankments have been provided for each lagoon at opposite ends of the lagoons. A single ramp into the lagoon has been provided for access to the islands. Turning circles suitable for standard 4WD vehicles have been located halfway between the ramps to reduce distances travelled on the crest.

Balanced Cut & Fill

An environmental constraint imposed by the client was to restrict lorry movements to and from the site, so that no earthwork material was to be brought onto or from the sites. Therefore, a digital terrain model was created for each terrestrial lagoon using Key Terrafirma software to carry out earthworks balancing. An earthworks balance was achieved by raising or lowering the lagoon surface which took into account the 'bulking' factor of excavating material and differing compaction rates for the various fill requirements.

Lagoon bed permeability

The Geotechnical Investigations suggested that permeable layers underlie some areas of the lagoons at a shallow depth. The risk of encountering these potential pathways for leakage has been minimised by orientating the islands so that the area of cut is reduced over where these potential pathways were identified in the geotechnical interpretation.

DESIGN OF IMPOUNDMENT AREAS

The Impoundment Areas, Lagoons A and D, are created within spurs of Rutland Water using a sheet pile cut-off wall with a rock bund to either side. A typical section is shown in Figure 3.

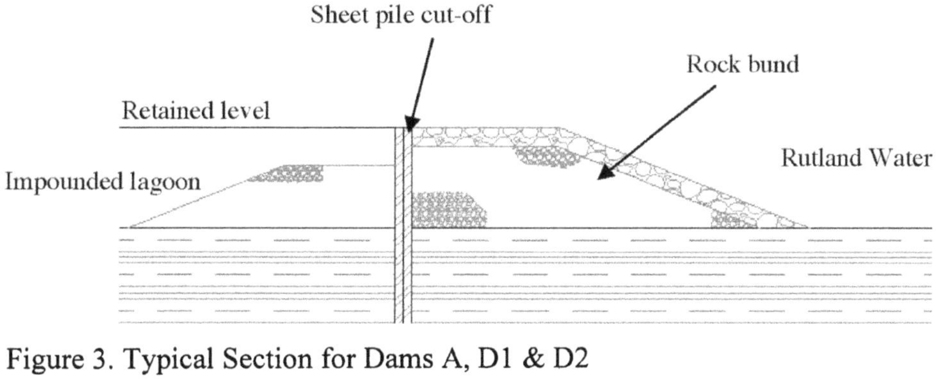

Figure 3. Typical Section for Dams A, D1 & D2

Sheet pile design

The integrity of the sheet pile dams were analysed using WALLUP software under a variety of load cases. To simplify the design and construction, three different lengths have been selected of the same pile section for all three dams. This shall provide the necessary pile section after a design life of 100 years with no protective coating. No protective coating has been used in order to avoid future maintenance of the piles and the risk of contaminating the reservoir.

Sealing between the pile joints is necessary to reduce water supply requirement of any during low rainfall periods. Welding of alternate piles and a non-bitumen sealant for the remaining joints has been adopted.

Rock bund design

The rock bund behind the sheet piles performs the following functions:

1. Access along the dam for maintenance, access to the control structure
2. Wave protection
3. Softens visual appearance

A crest width of 4 m allows maintenance access for heavy plant. Beyond the water level control structure, the width of the rock bund narrows to 1 m to reduce the volume of imported material and reduce costs.

The bund is formed with a granular core and a rock armour layer for wave protection. A modified Highways Specification Class 6F2 material has been adopted as the granular fill which is suitable for placing underwater. The rock armour, D_{50} of 0.2 m and 0.5 m thick, has been sized to withstand Rutland Water wave heights of 0.5 m predicted to occur during a 200 year wind speed event.

MANAGEMENT AND OPERATION OF HABITATS

The management and operation of the scheme is to be looked after by the Leicestershire and Rutland Wildlife Trust, so care has been taken in making it suitable for operation by non-mechanically skilled persons. The maintenance of the structures and pump stations shall be managed by Anglian Water.

Water Control Structures

The single water level controls in each lagoon have been designed to permit water level control from winter water level to 0.8 m below. The latter level is to allow draw down during periods of vegetation clearance and possible future changes to the water management regime.

A tilting weir gate has been adopted in preference to a simple stoplog or gate valve arrangement as it:

- Allows fine water level control for improved lagoon management
- Minimal leakage
- Easy operation
- Relatively low cost for small gates

The gates are manually operated via a gear box and crank handle mounted above the gate. For the water supply the pumps shall have a simple control system to provide pumping for a series of set times and a stop switch.

RESERVOIR SAFETY

Lagoons within the Reservoirs Act 1975.

The volumes to be stored in each of the terrestrial lagoons and impoundment areas are detailed in Table 2. All of the areas except Lagoon C4 fall within the Reservoirs Act.

Table 2. Lagoon Volumes Stored Above Natural GL

Lagoon	Total Volume Stored (m^3)	Total Volume Above GL (m^3)
A	44,000	44,000
B	349,000	349,000
C1	76,000	60,000
C2	79,000	60,000
C3	53,000	53,000
C4	29,000	16,000
D1	40,000	40,000
D2	245,000	245,000

Dam Categories

All of the lagoons are considered by the Construction Engineer to fall within Category D, according to the classification in Floods and Reservoir Safety (ICE, 1996). No loss of life can be foreseen as a result of a breach and very limited additional flood damage would be caused because the water released would be impounded in Rutland Water.

The reservoir design flood inflow for this category of dam is a 1 in 150 year flood. The initial reservoir condition for retained Lagoons A and D will be just overtopping, whilst terrestrial Lagoons B and C1 to C4 will be spilling with the design wave freeboard allowance.

Overflow arrangements

Design conditions for overflow are normal inflow, where they occur, plus the 150 year flood event. Inflows considered for Lagoons A, D1 and D2 are direct rainfall over the lagoon surface and design flood inflows from the North Gwash for Lagoon A and from the River Gwash for Lagoon D2. For the remaining lagoons, only direct rainfall over its surface was considered.

The impoundment dams are designed to allow over-topping of flood flow. The rock protection has been checked assuming the overflow is spread equally along the crest of Dams A and D2. The tilting weir and outlet structures have been analysed to study the consequences of weir failure at the highest winter water level, resulting in a head through the outlet of 0.8 m.

The terrestrial lagoon embankments have not been designed to withstand overtopping. Instead, the required wave freeboard has been provided and the tilting weir and spillway sized to accommodate design outflows. The spillway design concept is to use a composite channel whereby normal operating flows are conveyed down the embankment face in a pre-cast concrete channel. In the rare case of maximum outflow, additional flows are conveyed down a wider, concrete-reinforced-grass trapezoidal channel of which the low flow channel forms the invert.

At the embankment toe a stilling basin is required to contain the hydraulic jump and prevent undermining of the toe of the embankment. In keeping with the environmental aims of the project and given the infrequency of the design event, the use of concrete is the minimum necessary to prevent damage to the toe of the embankment. A permanent pool will be formed at the base of the spillway to dampen the visual impact of the concrete block and apron at the toe of the embankment.

Freeboard

The impoundment dams are designed to be over-toppable and therefore freeboard is not an issue. For the terrestrial lagoons, freeboard above winter water level will be required to account for flood surcharge and wave surcharge. The wave fetch has been minimised for Lagoons C by locating the islands perpendicular to the normal wind direction, however the islands are submerged in Lagoon B.

There are no direct inflows to the terrestrial lagoons; therefore a flood surcharge above winter water level would only be caused by direct rainfall over the lagoon surface. The following freeboard above winter water level has been adopted:

Table 3 Embankment Freeboard

Lagoon	Freeboard (m)	Comments
B	0.85	70 mm flood surcharge + 760 mm wave surcharge
C1 to C4	0.45	47 mm flood surcharge + 360 mm wave surcharge

WORKING PARTNERSHIPS

Working with the Leicestershire and Rutland Wildlife Trust throughout the design phase has developed into a mutually advantageous partnership. The design team worked closely with the wildlife trust gaining from their knowledge and expertise of the local area and habitats to develop an appropriate habitats creation scheme. As the wildlife trust will manage and operate the habitats they have been encouraged to input into the design to ensure the wildlife trust can operate the structures and maintain the lagoons safely.

This effective relationship also aided in communicating with Natural England and the local Planning Authority during the planning approval process. This was partially due to the good contacts and relationship the wildlife trust has with these statutory consultees. The wildlife trust's present interaction with neighbours has been valuable during the public consultations with the local residents.

A further recent example of this relationship was during the recent pre-work site clearance for Lagoon B where an area of woodland needed to be cleared. During initial discussions it was mentioned that the wildlife trust could undertake the work so that they could use the exercise as part of their chainsaw courses.

CONCLUSIONS

This Rutland project is primarily driven by environmental requirements as mitigation and compensation for loss of habitat. Habitat creation creates very different drivers in terms of design and construction than experienced in conventional dams and reservoir engineering. The habitat creation itself is constrained by the onerous environmental conditions within which the project is located which results in doing "mitigation works for the mitigation works".

Reservoir Repair and Upgrade: Still the same issues?

ANDREW DODD, Jacobs UK Ltd

SYNOPSIS. Data has been collated on remedial and upgrade works on over 200 dams and nearly 400 separate remedial works activities worldwide. An assessment of the works shows that the most common works required are stabilisation works, works to outlets and then grouting / cut-off works. Of the driving forces behind the works 65% come from reactive forces, i.e forces internal to the project such as deterioration and 35% from pro-active forces such as legislation. A comparison of the reactive forces shows that the breakdown of causes in this data set is essentially the same as that produced by ICOLD in 1983.

INTRODUCTION

Research was undertaken to:

- Assess and understand the activities that are undertaken on dams and hydropower facilities to maintain the structures in good condition.
- Determine what techniques are commonly used.
- Understand the reasons why the activities were undertaken.
- Assess what patterns and trends emerged.

The original research activities, gathered on papers published between 1995and 2004 were spilt into regular and "one-off" activities with a concentration on "one-off" activities. This paper only considers the causes for the "one-off" activities and does not discuss the techniques employed. Following completion of the assessment the data was compared with a study undertaken by ICOLD, published in 1983 and a comparison of the results of the two assessments was made.

DATA SET VARIABILITY

Data on over 200 dam/hydropower sites and approximately 400 cases of remedial works were recorded. All different types of dam were represented: Ages ranged from recently constructed structures through to dams dating back over 1000 years. Heights ranged from a few meters to over 150m with the majority of dams less than 40m high. Lengths ranged from less than 100m to over 10km. All regions of the world were represented with the majority of the dams in the dataset located in Western Europe and North America.

ACTIVITES UNDERTAKEN

The activities undertaken on the structures included: stabilization, reconstruction or abandonment, sealing works, membrane works, grouting or cut-off works, works to the reservoir (e.g. raising its level), outlet works (e.g. spillways, low level outlets, M&E works), monitoring activities and crest works. The distribution of the types of works is shown on Figure 1. Each of these areas of activities was further subdivided and the types of works assessed. Data on the subdivision of the activities are not provided due to space constraints but it is the intention to publish this data in the future.

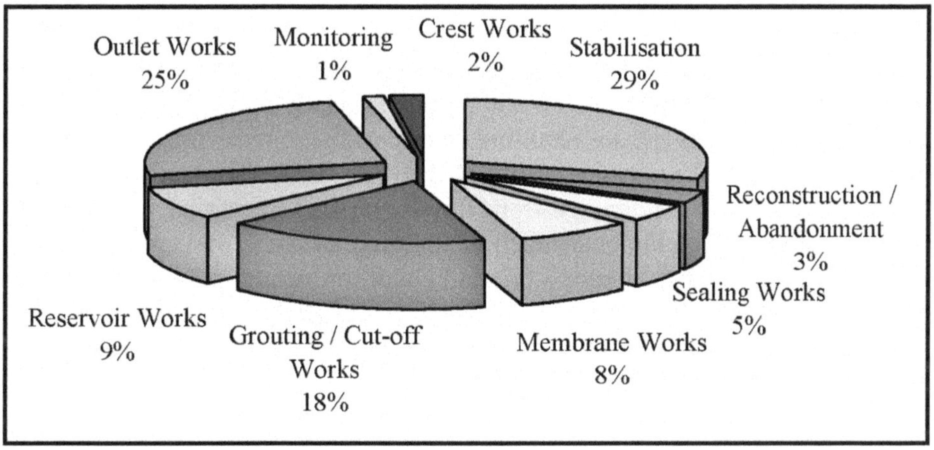

Figure 1: Types of Works Required

DRIVING FORCES

The activities identified were grouped in terms of the driving forces behind the activities. The driving forces were split into "Pro-active" and "Re-active" data sets i.e. those that are performed as a result of external requirements and those that are required to fix something. As can be seen from Figure 2 the majority of the works are Reactive.

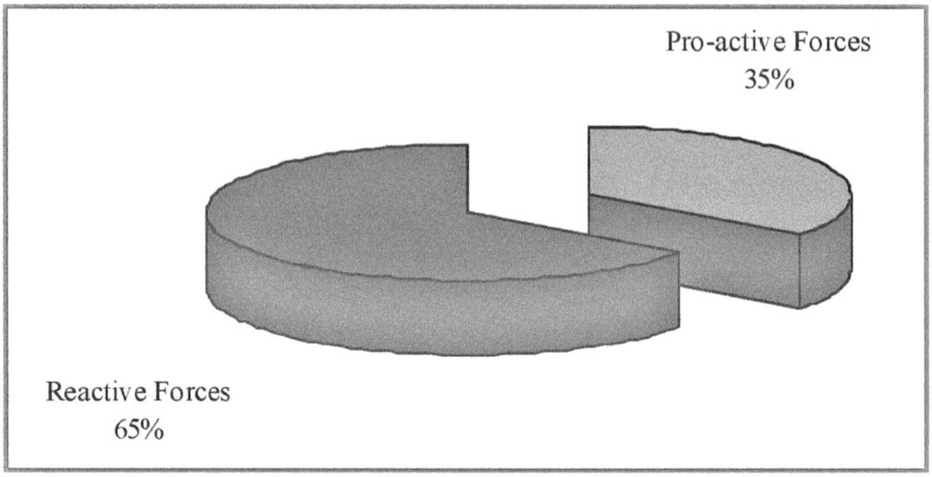

Figure 2: Remedial Works Driving Forces

Reactive Forces

Of the reactive forces, the data set demonstrates that the most common driving force behind the works is impermeability (leakage), closely followed by human influences (issues resulting from design, construction or operation errors). Together these make up over half of the causes for remedial works. Natural events include damage from floods, seismic events etc. Age related forces include age (elements reaching the end of their lifetime), siltation, corrosion and spillway erosion. Ground condition forces include settlement, abutment deterioration and landslides. Concrete expansion is self explanatory and includes alkali aggregate reaction. The break down of the Reactive Forces are shown in Figure 3.

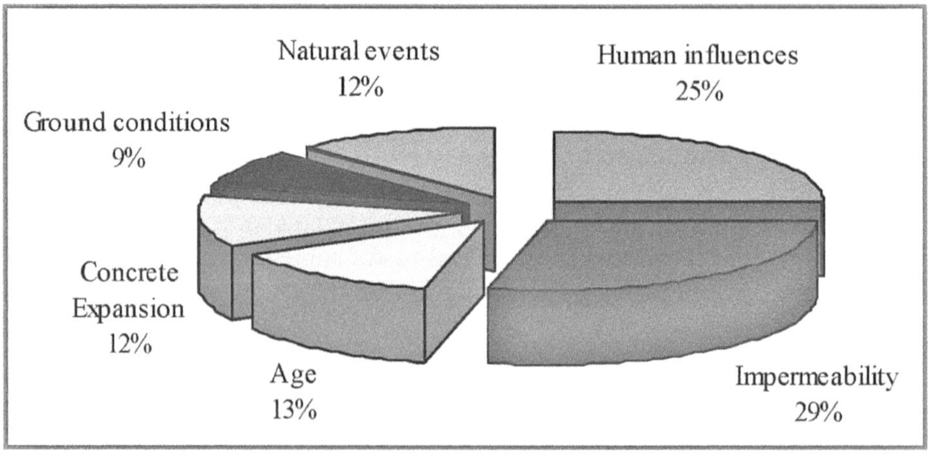

Figure 3: Reactive Forces

Pro-active Changes

The Pro-active changes were split into two sections, Legislational and Operational/Functional Changes and the split is shown on Figure 4.

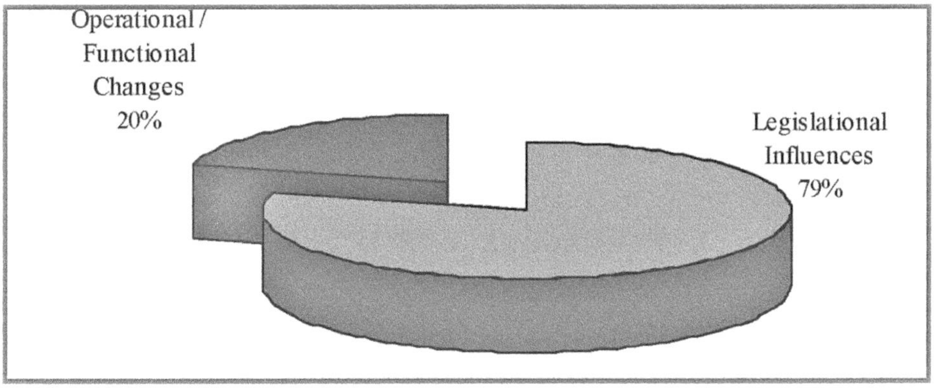

Figure 4: Breakdown of Pro-active Changes

Legislational Influences

Figure 5 shows a comparison between the different legislational influences. The data set includes all types and ages dams. With the exception of Old Aswan Dam, Egypt none of the dams is from Africa or South America, and only a few are located outside North America and Europe. This reflects the different levels of the development of dam safety legislation and guidance regulations worldwide. Of the influences New Flood relates to the revision of the design flood, crest details relates to the requirements for protection or freeboard, Stability relates to requirements for safety factors, and Instrumentation to requirements for instrumentation.

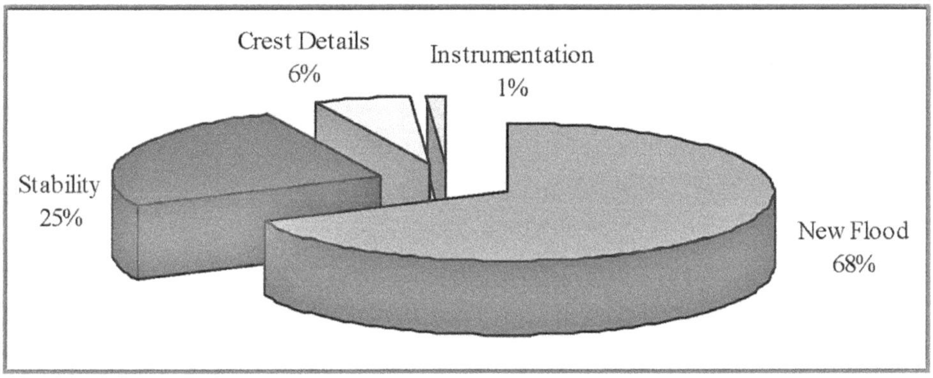

Figure 5 :Legislative Sub-Categories

Operational/Function Changes
The operational/Function Changes are shown in Figure 6. Most of the dams in the data set were constructed for water supply, with a few hydropower structures included. No flood dams or recreational structures are included. This is unsurprising given the nature of the dam functions and the economic return from raising or other operational changes. Or put another way there is insufficient income generated from a flood or a recreational structure to finance operational changes.

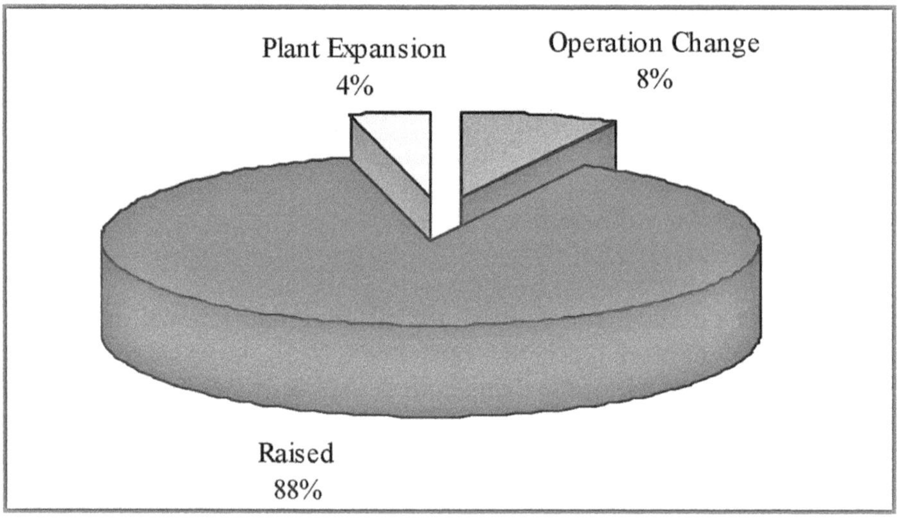

Figure 6: Operational/Functional Changes

Of the operational/function changes raising of the structures (resulting in increased reservoir volume) constitutes the vast majority. Operational changes such as to water supply occurred on a few structures and the expansion of the plant on even fewer. A considerable number of the dams were constructed in the period following the Second World War. This relates to the large number of dams built at this time and present in the data set, rather than to any other.

COMPARISON WITH THE ICOLD ASSESSMENT
A comparison between this data set and the ICOLD data set from 1983 is shown in Figure 7. This shows that despite the passage of time the proportions of the causes for remedial works as a result of reactive forces have essentially not changed. The one exception, where the new data set varies significantly from the ICOLD data set, is concrete expansion. The

difference in this instance is probably a question of time with more occurrences having been identified over the past few years.

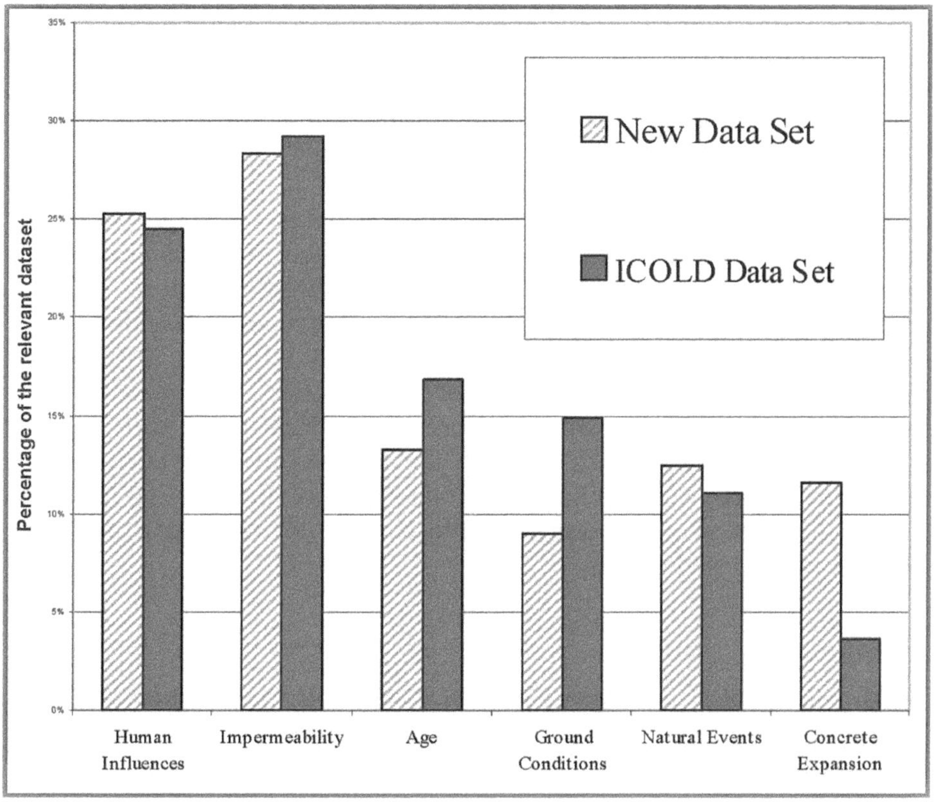

Figure 7: Comparison of ICOLD and New Data set for Reactive Forces – Grouped at the top level

The immediate logical conclusion to the general matching of the proportions of the reactive forces between the new and ICOLD data sets is that, despite publication of data, owners, designers and contractors are not eliminating problems that are known to be likely to occur. However, this is harsh as without comparative data sets of dams that have not required remedial works it is impossible to note whether the occurrence of these problems has reduced in the dams built since the publication of the ICOLD data set in 1983.

TRENDS

It has been difficult to establish specific trends from the data set. Ideally an evaluation of a controlled data set (e.g. all the dams in a region of country) would be required to properly assess trends, but this could not be accomplished. However, despite these shortcomings, a major trend that can reasonably be concluded from the works is the increase in the amount of works resulting from Pro-active Forces and in particular legislation. The other trend that can be concluded from the research is that the base causes for remedial works, as a proportion of the remedial works performed has not changed between the study performed by ICOLD in 1983 and this study.

SUMMARY & CONCLUSIONS

A dataset of reservoir activities, including remedial and upgrade works, has been collated and analysed. This data set shows the wide range of activities that have been performed on a wide variety of different structures. Of these activities the big three activities are stabilisation works, outlet works (including spillways and low level outlets) and works to improve the impermeability (reduce the leakage) of the structure. The data set also shows that the majority of works performed are "Reactive" i.e. required to fix something. However, "Pro-active" influences are still a significant proportion of the works required. Comparison of the data set with the ICOLD study shows that there is little change on the pattern of works between the two studies.

REFERENCES

Dodd, A.O., (2003) *Reservoir Remedial Works and Maintenance Activities,* Master Of Philosophy, City University School Of Engineering and Mathematical Sciences.

Decommissioning, discontinuation and abandonment of dams: is there a case for a national strategy?

C. S. MCCULLOCH, University of Oxford

SYNOPSIS. Dam removal has excited much interest in the United States in the last decade. Many small dams have been removed and a start on removing two large dams in Oregon is scheduled for autumn 2008. Valuable lessons relevant to England and Wales are drawn from the burgeoning scientific literature. Multi-disciplinary inspections are needed to assess dams for decommissioning and discontinuance based on aesthetic, ecological and recreational values as well as safety and utilitarian economic considerations. A case is made for a national strategy.

INTRODUCTION

Visions of the longevity of dams and reservoirs differ profoundly. To those impressed by those forces of nature which balance discharge, slope, sediment and channel shape, "Dams are unnatural, short-term features in the landscape" (Evans et al., 2000); "Dams have a finite life span, so dam age can be an important factor affecting removal decisions" (Poff and Hart, 2002). To proud engineers, "an artificial reservoir generally becomes a practically permanent part of the landscape" (Bonazzi, 1989), or "a well-designed and well-maintained dam can be operated indefinitely" (Back, 1989). Both views hold some truth. Some dams have not been well-designed or carefully constructed and some are not well-maintained. Reservoirs may become filled with sediment, concrete may disintegrate, moving parts may seize with rust, spillways may not route floods safely. The question is whether dam maintenance is a life sentence for society or can dams be removed and nature restored once costs outweigh benefits?

Physical changes are inevitable, needing attention and restoration. In addition, all dams are situated in a changing cultural context of values, economic, social and ecological. Economic rationale for a dam changes with time. Major water consumers, such as heavy manufacturing, may relocate overseas; small hydroelectric dams, water mills and canals may become replaced by more efficient power supplies and transport. Aspiration to higher standards of potability may condemn reservoirs with organic or industrial pollution. Conversely, new techniques may make use of brackish water more economical than transfer from distant sources. Economy of scale in water purification may make piping of water from fewer plants less costly than more local supply from small waterworks.

Social values also change appreciation of reservoirs. Post Second World War social idealism resulted in pressure on reservoir owners to open access to reservoirs. A new active group of stakeholders in reservoirs was created with interests in sailing, watersports and fishing. Expansion of population and of the extent of urban areas made many reservoirs local places where a form of 'nature' could be enjoyed. Even the most artificial reservoirs became appreciated for their contrast with noisy, crowded and dangerous urban streets congested with traffic. Some urban councils have purchased reservoirs no longer used for water supply for notional sums for recreation value. In other cases, the risk posed by dams situated upstream of vulnerable settlements coupled with costs of statutory maintenance and may tip decision-makers towards removal rather than change of use.

Ecological awareness and concern for the survival of other species, some of utility to humans, others of intrinsic value, has changed attitudes towards dams. Removal of dams for ecological reasons is advocated in the United States, particularly where damage to migrating fish caused by dam building has been obvious in living memory (Poff et al., 1997, Bednarek, 2001, Hart et al., 2002). In the Western states, salmon, shad and steelhead fish populations are declining so rapidly that 27 species, such as the coho salmon, have become endangered. Some salmon are even transported around dams by barge but such costly mitigation measures may not be sustainable. Whilst fewer than 500 small dams have been removed so far in the US, two high dams(33m and 64m with a joint capacity of $60Mm^3$) on the Elwha River in Oregon will be dismantled for ecological reasons, starting in autumn 2008 at a cost to the Federal government of $184 million, including $75m to provide water-treatment, flood protection and sanitation for settlements downstream (Tweit, 2006, Stokstad, 2006a). Other high dams, including four on the Snake River and the O'Shaughnessy Dam in Yosemite, are being reviewed for possible removal (Gray et al., 2006, Null and Lund, 2006, Gregory et al., 2002, Stokstad, 2006b). Methods of cost/benefit analysis to support such decisions are being refined (Whitelaw and Macmullan, 2002).

This paper will sample the burgeoning literature in the USA on dam redundancy and removal to identify transferable policies and science. Illustrative cases of recent dam breaching demonstrate the need for a strategic rather than an ad hoc approach to the decommissioning and discontinuance of dams in England and Wales.

DAM REMOVAL IN THE USA

The speed of transition of dams from benefits to liabilities pending decommissioning will vary from place to place. American dam redundancy has progressed more rapidly than in England and Wales and ambition for dam removal has grown. Every major US geoscientific and ecological society has held workshops on dam decommissioning/removal and two national panels (Aspen Institute and Heinz Centre) have produced reports (Pohl, 2002, Grant, 2001). Consequently, this presents an opportunity for UK dam stakeholders to learn from the large and growing literature describing US experience.

In dam removal and attempts to restore rivers to their natural state, the US has taken the lead, partly because of the political power of active NGOs, such as American Rivers (http://www.americanrivers.org) and Trout Unlimited (http://www.tu.org) which have demonstrated significant ecosystem damage caused by dams, and partly because of fears of loss of life and property caused by recent dam and levee failures. Pohl (2002) analysed 417 cases and concluded that environmental issues were the main cause of removal, followed by safety, then economic issues.

The hope to restore suitable conditions for migratory fish follows concern about drastic declines of salmon and trout. Canneries have closed and the loss has been devastating for native Indian tribes, such as the Oregon Elwha Klallam tribe, whose culture and nutrition depended on the salmon.

In addition to ecological concerns, the Federal Emergency Management Agency (FEMA) found that about 9200 dams in the U.S. are classed as high hazard because of inadequate spillway design, lack of emergency spillways, inadequate dam maintenance, and/or lack of active sediment management (FEMA, 2002). Accelerated soil erosion has filled many small reservoirs with sediment, reducing capacity and limiting attenuation of severe floods. Removal can be the cheapest option for a failing dam, "the cost of repairing a small dam can be as much as three times greater than the cost of removing it" (Born et al., 1998).

Ecological and geomorphological investigations of the effects of dam removal are in their infancy. Some introduced fish species may decline and native fish recuperate after removal. Sediments accumulated in the reservoir become mobilised and transported downstream changing channels, with consequences for fishing, spawning and flooding behaviour. More measurements of reservoirs, river channels and their biota are needed before and after dam removal to improve scientific understanding. "When understanding of environmental systems is lacking, scientific consensus, and thus policy, will often be years to decades away."(Doyle et al., 2002)

"Monitoring the site after the dam is removed is essential" and "ideally post removal monitoring should be included in the dam-removal budget (Baish et al., 2002). The aim is to develop dam removal as a science.

Dam removal may not always lead to ecological benefits particularly when contaminated sediments are mobilised. When transported from the disused dam site downstream, these may cause ecological damage. Shuman (1995) reports severe environmental damage caused by removal of the Fort Edward dam from the Upper Hudson:

> *'PCB contaminated sediment has settled over a 300 km length of the river, and commercial fishing of striped bass and other species has been banned due to the risk of bioaccumulation (EPA, 2002)'*

Others are concerned about the amount of carbon and nitrogen released as the sediments are mobilised (Riggsbee et al., 2007, Pacca, 2007).

Several papers make recommendations about good practice (e.g.(Bushaw-Newton et al., 2002, Bednarek, 2001, Babbit, 2002, American Rivers et al., 1999). Before dam removal, the history of the reservoir should be investigated to identify sources of pollution, past or current. The sediment should also be examined before any redistribution downstream is allowed. Precautions include careful control of the speed and timing of drawdown, also capping or dredging of polluted sediment. The dam removal should be considered on a catchment scale to anticipate effects on other parts of the system and to ensure that subsequent actions do not negate the beneficial environmental effects of restoring habitat connectivity.

The planning process for removal should be expected to be time-consuming because of the interests of many stakeholders. The decision to remove the Elwha dams was preceded by three decades of debate and only the decision of the Federal government to purchase the derelict dam and fund its removal resolved the issues.

RELEVANCE TO ENGLAND AND WALES

There are many emerging reasons why the whole life of dams, design, building, maintenance and removal, should be considered seriously in England and Wales. Most of current dams are old, (the average age of dams in the Building Research Establishment Register is 110 years) and reviews are needed of the relationship between the benefits of each dam and the cost of its safe maintenance. Embankment dams in small upland catchments risk damage with flashy flood responses to severe rain storms. Some dams have been poorly-designed with inadequate spillways, for example the Brushes Clough reservoir near Oldham where the existing spillway capacity was calculated to be able to pass only 40% of the PMF (Gardiner et al., 1996). 70% of categorised dams (A, B and C) pose a downstream risk (Hope,

2006) and liabilities have increased with housing expansion and growth in value of property. There is regulatory pressure from the Environment Agency to maintain dam safety despite the high costs involved. Economic change has meant that many dams no longer serve the purpose for which they were constructed. Privatisation of water supply into regional-scale companies has led to some rationalisation of reservoirs and many reservoir owners are faced with rising costs of maintenance without compensating income.

In contrast to American ambitions to restore rivers, pressure to breach dams in England and Wales has been driven by a pragmatic desire to diminish the capacity of the reservoir to below the Reservoirs Act 1975 limit of 25,000m^3 of water held above ground level. Economic drivers govern the choice between repair and rehabilitation on the one hand or breaching of the dam on the other. This difference is reflected in the vocabulary: in England and Wales, 'discontinuance' is used to indicate that the amount of water in the rump reservoir is less than 25,000 m^3 without implying restoration of ecological connectivity of the long profile of the river. The dam itself is usually only partially removed to effect the lowering of reservoir level. 'Decommissioning' is an ambiguous term used to indicate a significant change in the human use when a dam is taken out of the operation for which it was first designed but is sometimes used as if synonymous with removal. A third way, abandonment, postpones decision-making on the future of the dam but since the Water Act 2003, the Environment Agency, in taking responsibility for enforcement of regulations in England and Wales has identified owners of almost all reservoirs under the Act (Hope, 2006) and some will now be faced with decisions between costly repair or discontinuance.

<u>Reservoir inspection as a trigger for change</u> Inspection of dams under the Reservoirs Act 1975 provides an opportunity for review of both the safety and functionality of the dam and consideration of its future with the dam owner. Currently, the emphasis is on dam safety with the inspection being conducted by a sole Panel Engineer at least once every ten years. Ideally, where there is doubt about continuation of the reservoir, the inspection should be an opportunity also for consideration of ecological, recreational, aesthetic and other planning issues. In Wisconsin, interdisciplinary inspections and reviews are now possible and provide the basis for more considered evaluation of future options (Orr et al., 2004).

<u>Opportunities for research</u>

The zeal of the US scientific community reflects an appreciation that dam removal can be a positive environmental improvement and that each case presents valuable research opportunities (Hart et al., 2002, Pizzuto, 2002).

In the England and Wales, more co-ordination is needed between practising engineers and researchers to realise the value of discontinuation of dams as field experiments. Some recent examples illustrate potential for research:

Ecology. Even though ecological river restoration is not usually a stated aim of discontinuance in England and Wales, valuable research could be carried out on colonisation of the newly-exposed banks and on the changes in the biota left in the remnant pond after the lowering of the reservoir. Monitoring of the changes represents a wonderful educational opportunity for greater understanding of ecological principles and might suggest ways in which natural processes might be speeded up to restore landscapes more quickly.

At Boltby, a remnant pond still remains inaccessible to fish from the downstream river. A small fish ladder has been fashioned to link the much reduced pond to a feeder stream in the hope that the fish introduced by the local Fishing Club before the dam was breached might be tempted to migrate upstream to spawn. Research is needed on the long-term sustainability of the fish population confined to the smaller waterbody and on the functionality of the fish ladder.

Dam engineering. Breaching a dam could allow investigation into ageing processes as a cross-section of the dam is revealed. Recording the state of the clay core and other observations would be of use in predictions of behaviour of other old embankment dams.

Industrial archaeology and environmental history. Dams are historical features of the landscape, often part of our rich industrial archaeology, and with long lasting effects on fluvial behaviour. Before removal, records including maps, photographs and diagrams, of all the knowledge of the construction and usage of the dam should be deposited in local archives and in specialist libraries. Exposed on the bed may be traces of early settlements which were not fully recorded before the reservoir was filled[1] as well as artifacts thrown into the reservoir or accidentally lost. Consultation with local industrial archaeologists and historians should guide how many features of the breached dam and its structures should be preserved and advice sought on archiving of records. An example of good practice was co-operation of Northumbrian Water with Tees Archaeology and the Cleveland Industrial Archaeology Society in the removing of the Westworth dam built in 1874-75, near Guisborough. Interesting remnants of the structures were preserved and historical records published (Dixon, 1988, Johnson, 2008).

[1] During a low drawdown at Cow Green in Teesdale a Bronze Age Village was excavated.

Sediments. Sediment accumulation in reservoirs is not as problematic in Britain as in most of the US (White et al., 1997), even though rates of accelerated soil erosion in some places have been increasing with changing farming practices. Quality of sediments may be a problem, particularly in the Pennines and the Welsh uplands, where old lead mining may have released pollutants which accumulate in the reservoir sediments (Shotbolt et al., 2006) or where industrial waste waters have fed in to the reservoir. American experience suggests that before discontinuance, sampling of the sediment should be carried out to determine whether or not it should be removed physically before exposure to erosion and transport. Apart from such safety precautions, examination of the newly-accessible sediments in the reservoir allows calculation of rates of accretion and relationships between historical land use changes, soil erosion, pollution transport, algal growth, bacteriology and chemistry of sediments and reservoir water.

Legacy

Public access to reservoirs in England and Wales is the norm. Even those reservoirs supposedly closed to the public may be attractive to trespassers and perhaps even more prone to vandalism and illicit activity (Gardiner et al., 1996). American experience with tragic accidents at sites of former reservoirs left in dangerous states, gives rise to the advice that "Plans for dam removal should account for use of the site by a variety of people long after the dam has been removed" (Baish et al., 2002).

Discontinuance, like dam construction, needs aesthetic consideration to leave a landscape which will improve with time. Nature will soften disturbance caused by engineering work involved in dam breaching but care is needed over the appearance of the remnants left in situ for future generations. Whilst restoration to the pre-dam, natural state may be an unrealistic ambition, leaving a near-permanent blot on the landscape should not be an option, particularly when the site lies in an area of natural beauty.

This issue may be illustrated by the breaching of the Boltby dam and remedial work on the reservoir site completed in 2007 (Walker, 2008). Visitors to the notched Boltby dam in the North York Moors National Park will be faced with a disappointing vision. There is a striking contrast between the carefully-shaped masonry of local stone laid with appealing precision in the preserved remnants of the former dam and recent contributions of jumbled rough blocks captured in gabion wire netting, which is unlikely to last centuries. Boulders of Carboniferous limestone have been intruded rather than use of blocks of local calcareous grit and sandstone. The much lowered reservoir is now surrounded by oozes and mud which makes access to the shores of the pond hazardous and no standing for fishers has been provided. Fishing has not as yet resumed even though some of the fish introduced by the Fishing Club have survived in the remnant pond. The former Fishing Clubhouse fashioned from a hut built for the original 19th Century workmen stands abandoned.

Source: Author.

Figure 1: New overflow channel breaching the dam at Boltby Reservoir, September 2007, showing the contrast between new gabions on the left and old masonry on the right. The steep channel prevents migratory fish from entering the remnant reservoir. The stream has not been restored nor has the landscape been allowed to revert to nature.

Case for a national strategy

Dam removal is not a simple private matter for an individual dam owner because hydrological and ecological interdependencies involve many other stakeholders, both within the river catchment and nationally. The choice of which dam to remove may involve regional study to understand its relationship with other reservoirs and water sources in the region and on downstream flooding. Removal of a reservoir may throw added stress on other water sources whether nearby reservoirs, groundwater or places for recreation. The full ecological benefit of dam removal will be gained only if subsequent responses do not produce negative environmental effects, such as building a new reservoir or flood defence structures elsewhere. Hopes for the removal of dams from particularly pleasing landscapes may require a regional strategy so that an order of priority can be established. For example, the case for removal of the Hetch Hetchy dam from the stunning Yosemite landscape in California depends on the continuance of other dams in the region to provide alternative water supply. Ad hoc dam removal without wider consideration is unlikely to produce optimal results.

Small-scale, private interests have little incentive to fund removal and landscape restoration to a high standard. Many dams do not generate sufficient income to fund repairs (Hinks and Williams, 2004). Even though the cost of dam breaching maybe much less than cost of dam remediation, unless there is a serious safety incident some dam owners postpone decision-making on the future of their dams and they are abandoned, left as potential hazards littering the landscape. Financial assistance to encourage removal would be a useful lever both to raise standards and to encourage action. In the US, those states with a policy for removing dams, backed by grants, manage to remove the most. Support is justified because some objectives of dam removal such as restoring connectivity in the river, improving habitats, preserving industrial archaeology and improving recreational facilities are for the public good rather than private gain.

A national strategy would be based on long-term considerations and synergy with other national strategies. Climate change forecasts in England and Wales for more extreme weather events with wetter winters and drier summers may support building more reservoirs to even out water supply, if aquifer recharge is not sufficient. Another implication is that existing reservoirs are likely to be increasingly stressed and several will become liabilities. Spillways may become unable to cope with increased storm rainfalls and prolonged droughts may weaken embankment dams. Moves from use of fossil fuels will forfeit an era of cheap energy. As the water supply industry is a major consumer of energy for pumping water, this change will affect decisions about reservoir location, favouring reservoirs which can supply water by gravity. Ageing structures will need to be reassessed and some will be scheduled for removal when the cost of alteration to meet the new conditions exceeds anticipated benefits; others may become more worthy of restoration.

In the US, many early hydroelectric dams have been made redundant following the establishment of regional grids fuelled by cheap power. At the local level, continuation of such HEP supplies could not be justified but at the national level, attempts to meet carbon reduction targets might have justified continuation of such dams. Subsidised refurbishment might be in the national interest, if carbon reduction were given priority.

A national strategy could give due weight to non-economic factors especially merit goods, such as recreation and nature conservation, and priorities could be established after consultation with many stakeholders.

National support and technical advice would have benefits of data sharing. Cost-benefit analyses could be improved to take into account long-term, regional and even international strategies. Research programmes and standardised monitoring could be promoted to maximise scientific understanding of the fluvial and ecological responses to dam removal.

CONCLUSION

Dams are long-lasting even though they may have a limited economic life. Even when dams are removed, they leave lasting impressions on the landscape. American aspirations to remove dams and restore rivers to pre-dam nature with healthy populations of migratory wild fish have yet to inspire many followers overseas. In England and Wales, our romance with many bountiful rivers largely ended with the Industrial Revolution and permissive pollution of estuaries. Rather than enjoying rivers, cities turned their backs with disdain. In recent decades, better water treatment and pollution prevention is bringing back life to many rivers in their lower courses and ambitions for discontinuing dams and restoring connectivity for migratory fish may grow, as in the USA. Many lessons can be learnt from the American experiences in restoring habitats to hand on landscapes fit for future generations.

REFERENCES

Babbit, B. (2002) What goes up, may come down. Learning from our experiences with dam construction in the past can guide and improve dam removal in the future, *BioScience,* Vol.52, No.8, pp. 656-659.

Back, P. (1989) Reassessment of safety of existing dams In *World Bank Seminar on dam safety and environment.*

Baish, S. K., David, S. D. and Graf, W. L. (2002) The complex decisionmaking process for removing dams, *Environment,* Vol 44, No.4, pp.20-31.

Bednarek, A. (2001) Undamming rivers: a review of the ecological impacts of dam removal, *Environmental Management,* Vol.27, No.6, pp.803-814.

Bonazzi, D. (1989) Experience of the World Bank in implementation of dam safety. In *World Bank seminar on dam safety and environment.*

Born, S. M., Genskow, K. D., Filbert, T. L., Hernandez-Mora, N., Keefer, M. L. and White, K. A. (1998) Socioeconomic and institutional dimensions of dam removals: the Wisconsin experience., *Environmental Management,* Vol.22, No.3, pp.359-370.

Bushaw-Newton, K. L., Hart, D. D., Pizzuto, J. E., Thomson, J. R., Egan, J., Ashley, J. T., Johnson, T. E., Horwitz, R. J., Keeley, M., Lawrence, J., Charles, D., Gatenby, C., Kreeger, D. A., Nightengale, T., Thomas, R. L. and Velinsky, D. J. (2002)An integrative approach towards understanding ecological responses to dam removal: the Manatawny Creek study, *Journal of the American Water Resources Association*, Vol.38, No.6, pp.1581-1599.

Dixon, G. (1988) Guisborough's water supply: a century of development, *The Cleveland Industrial Archaeologist*, Vol.19, pp.1-14.

Doyle, M. W., Stanley, E. H. and Harbor, J. M. (2002)Geomorphic analogies for assessing probable channel response to dam removal, *Journal of the American Water Resources Association.*, Vol.38, No.6

EPA (2002) Responsiveness summary of Hudson River PCBs site record of decision. U.S. Environmental Protection Agency, U.S. Army Corps of Engineers, Washington.

Evans, J. E., Mackey, S. D., Gottgens, J. F. and Gill, W. M. (2000)Lessons from a dam failure, *Ohio Journal of Science*, Vol.100, pp.121-131.

FEMA (2002) Federal Emergency Management Agency National Dam Safety Inspection Program.

Gardiner, K., Sellars, P. and Baker, R. (1996) The discontinuance of Brushes Clough reservoir, *Dams and Reservoirs*.Vol.6, No.3, pp.23-28.

Grant, G. (2001) Dam removal: Panacea or Pandora for rivers?, *Hydrological Processes*, Vol.15, pp.1531-1532.

Gray, B., Hastings, U. C., Behar, D., Dempsey, H., Good, R. and McDevitt, R. (2006)Hetch Hetchy: To drain or not to drain, *Hastings Law Journal*, Vol. 57, pp.1261-1279.

Gregory, S., Li, H. and Li, J. (2002) The conceptual basis for ecological responses to dam removal, *BioScience*, Vol.52, No.8, pp.713-723

Hart, D. D., Johnson, T. E., Bushaw-Newton, K. L., Horwitz, R., Bednarek, A. T., Charles, D. F., Kreeger, D. A. and Velinsky, D. J. (2002)Dam removal: Challenges and opportunities for ecological research and river restoration, *BioScience*, Vol.52, No.8, pp.669-683.

Hinks, J. L. and Williams, P. (2004) Some problems at small dams in the United Kingdom. In *Long-term benefits and performance of dams* (Ed, Hewlett, H.) Thomas Telford, London pp.629-636.

Hope, I. (2006) Working together for the safety of our reservoirs, *Dams and Reservoirs*, Vol.16, pp.15-18.

Johnson, P. G. (2008) Archaeological Recording: Westworth Reservoir, North Yorkshire.Tees Archaeology and Northumbrian Water Ltd, Hartlepool, 21 pages

Null, S. E. and Lund, J. R. (2006) Reassembling Hetch Hetchy: Water supply without O'Shaughnessy Dam, *Journal of the American Water Resources Association*, Vol.42, No.2, pp.395-408.

Orr, C. H., Roth, B., Forshay K.J., Gonzales, J., Papenfus, M. and Wassell, R. (2004) Examination of physical and regulatory variables leading to small dam removal in Wisconsin, *Environmental Management*.

Pacca, S. (2007) Impacts from decommissioning of hydroelectric dams: a life cycle perspective., *Climatic Change,* Vol.84, pp.281-294.

Pizzuto, J. E. (2002) Effects of dam removal on river form and process, *BioScience,* Vol.52, No.8, pp.683-691.

Poff, N. L., Allan, J. D., Bain, M. B., Karr, J. R., Prestegaard, K. L., Richter, B. D., Sparks, R. E. and Stromberg, J. C. (1997) The natural flow regime, *BioScience,* Vol.47, pp.769-784.

Poff, N. L. and Hart, D. D. (2002) How dams vary and why it matters for the emerging science of dam removal, *BioScience,* Vol.52, No.8, pp.659-668.

Pohl, M. M. (2002) Bringing down our dams: trends in American dam removal rationales, *Journal of the American Water Resources Association,* Vol.38, No.6, pp.1511.

Riggsbee, J., Julian, J., Doyle, M. and Wetzel, R. (2007) Suspended sediment, dissolved organic carbon, and dissolved nitrogen export during the dam removal process, *Water Resources Research. American Geophysical Union, Washington, USA,* Vol.43, No.9, pp. 9414.

Shotbolt, L., Hutchinson, S. M. and Thomas, A. D. (2006) Sediment stratigraphy and heavy metal fluxes to reservoirs in the southern Pennine uplands, UK, *Journal of Paleolimnology,* Vol.35, No.2, pp.305-322.

Shuman, J. R. (1995) Environmental considerations for assessing dam removal alternatives for river restoration, *Regulated Rivers: Research and Management,* Vol.11, pp.249-261.

Stokstad, E. (2006a) Big dams ready for teardown *Science*, Vol. 314, No.5799 pp. 584.

Stokstad, E. (2006b) Environmental restoration: restoring Yosemite's twin, *Science,* Vol.314, No.5799, pp.582-584.

Tweit, S. (2006) Tearing down the Elwha River Dam, *Popular Mechanics*.

Walker, J. (2008) The discontinuance of Boltby Reservoir, *Dams and Reservoirs*. Forthcoming.

White, P., Butcher, D. P. and Labadz, J. C. (1997)Reservoir sedimentation and catchment sediment yield in the Strines catchment, U.K., *Physics and Chemistry of The Earth,* Vol.22, No. 3/4, pp.321-328.

Whitelaw, E. and Macmullan, E. (2002) A framework for estimating the costs and benefits of dam removal, *BioScience,* Vol.52, No.8, pp.724-730.

Clunie Dam Anchoring Works

A W GREGORY, Scottish and Southern Energy plc, UK
A E HENDERSON, Ritchies, UK
K C HENDERSON, Ritchies, UK
P NOWAK, Atkins, UK

SYNOPSIS. Following the statutory inspection of the Loch Tummel reservoir *"matters in the interest of safety"* were raised requiring a stability analysis be carried out to check the behaviour of the dam under load from a Probable Maximum Flood (PMF) event and also from a seismic event.

The paper describes the analysis; site investigation; subsequent design; and on-site construction works to implement a dam anchoring scheme which meets the requirements of the Qualified Civil Engineer.

INTRODUCTION

The analysis and subsequent works carried out to Clunie Dam were required following the 10-yearly statutory inspection under the Reservoirs Act 1975 which was carried out in June 1999 by Dr A K Hughes. Subsequent structural assessments demonstrated that the seismic stability of the dam was adequate, but that under worst case PMF conditions, the abutment sections of the dam were likely to crack becoming grossly unstable and were expected to fail. Further consequence studies and ALARP studies demonstrated justification for implementing strengthening works.

GENERAL DESCRIPTION OF DAM

Clunie Dam is located in a narrow valley at the eastern end of Loch Tummel and was constructed in the period 1949 – 1951 as one of a series of dams, of similar construction, to provide hydroelectric power. The dam is a concrete gravity structure some 21 metres high and 116m long retaining a reservoir with a capacity of 36,400,000 cubic metres.

The longitudinal axis of the dam runs in a generally north east/south west direction. The central portion contains the floating drum gates, each of which is 18m long. The left and right side gravity dams are 40.9m and 40.8m long respectively, with a maximum height of 23.8m. The gravity

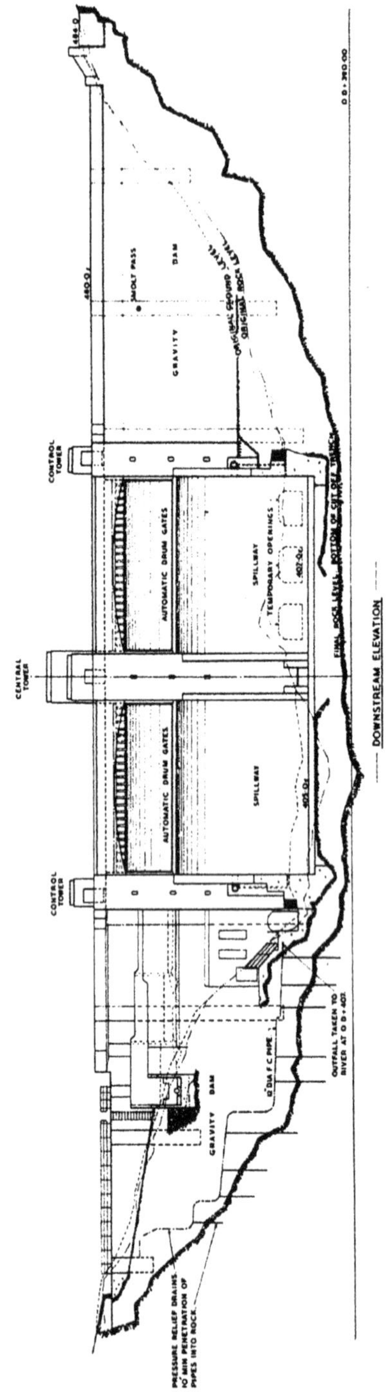

Figure 1 – Downstream Elevation on Dam

sections were both cast with a vertical upstream face and a downstream face at an inclination of 4V:3H.

The dam is founded on solid rock, with the gravity sections incorporating a cut off trench and a grout curtain drilled up to 24m into the rock.

The dam is categorised as falling within 'Category A' as defined in the publication *'Floods and Reservoir Safety'* (1996), and is placed in Category III as defined in the publication, *'An Engineering Guide to Seismic Risk to Dams in the United Kingdom'* (1991).

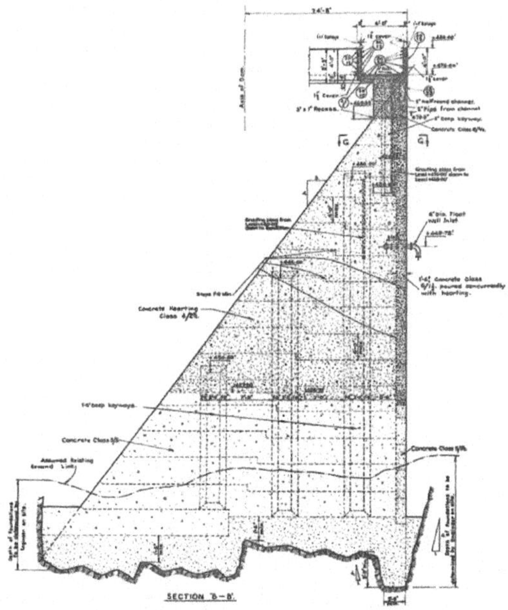

Figure 2 – Typical section through Left Bank gravity dam .

SECTION 10 INSPECTION REQUIREMENTS

The Section 10 inspection report recommended that a stability analysis be carried out to check the behaviour of the dam under the loading from a PMF event and also from a seismic event.

A number of studies were carried out and concluded that

- At PMF the reservoir would rise to a level such that the parapet wall would be overtopped by 2.9m for a duration of 6 days with unit discharge over the non-spillway sections of up to $16m^3/s/m$.
- The non-spillway sections of the dam have an inadequate safety factor against overturning under PMF loading

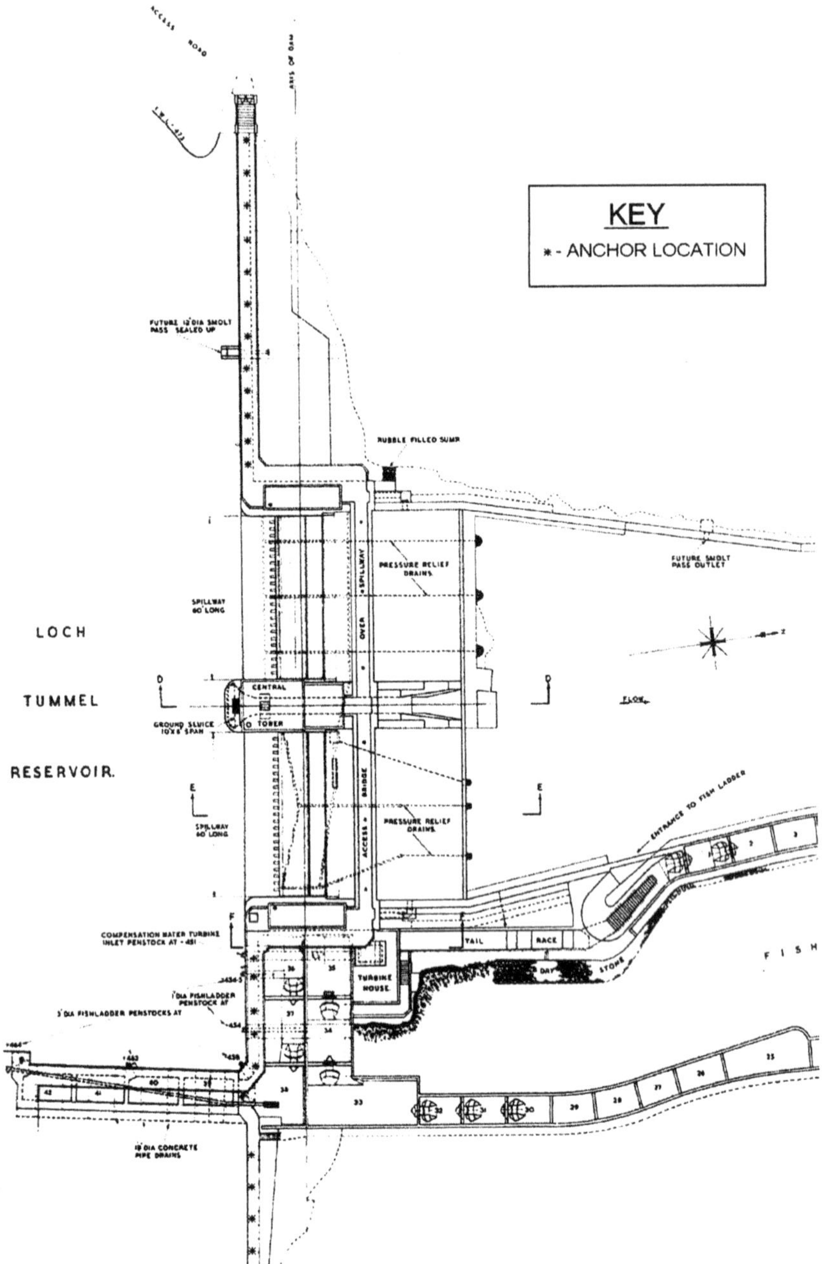

Figure 3 – Plan on Dam

- An additional failure mode, of sliding in the upper part of the rock foundation, was identified as being more critical than overturning. In this situation, the failure mode is brittle with a relatively rapid reduction in safety factor with an increase in water level, as the mode is determined by cracking at the heel, leading to both an increase in uplift pressures along the crack and reduction in shear strength along the cracked length as c' reduces to zero.
- The annual probability of failure was estimated as about 2×10^{-4}(1 in 5000 years), this being reduced to about 2×10^{-5} following anchor installation.
- The reports recommended the installation of post-tensioned anchors to prevent failure by either overturning or sliding, with an anchor load of 870kN/m required to give a factor of safety of 1.05 for the worst case PMF with a low tail water level.

Quantitative risk assessment (QRA) was used to assess whether the risk of dam failure would be reduced to "as low as reasonably practical" (ALARP) i.e. whether the cost of strengthening works would be grossly disproportionate to the reduction in risk achieved, and it was concluded that the strengthening works could be justified.

INVESTIGATION PHASE

In order to determine the nature of the ground acting as the dam foundation and confirm the nature of the dam construction a ground investigation and desk study were carried out.

Ground Investigation

Implementation

The investigation was carried out in June and July 2006, and comprised four rotary cored boreholes to a depth of between 35.24 metres and 45.0 metres below dam crest level. Subsequent to the completion of rotary coring an Optical Televiewer survey was carried out in all boreholes. On completion, all boreholes were filled with bentonite/cement grout. All rock/dam construction core was retained in core boxes of which photographs were taken and from which samples for laboratory testing were taken.

Results

The boreholes confirmed that the dam was founded on strong inter bedded schist and quartzite bedrock with little or no weathering below the concrete/bedrock interface..

The optical televiewer survey generally identified near vertical fracture orientation in the foundation bedrock. The televiewer survey showed the concrete within the dam construction to be essentially monolithic although one or two joints could be identified.

Laboratory Testing

Laboratory testing comprised the determination of unconfined compressive strength (UCS) and point load strength on cores of bedrock and dam concrete. A summary of strength test results is presented below.

Table 1. Rock Strength Summary

	UCS (MN/m2)			Point Load I_s50 (MN/m2)		
	Concrete	Schist	Quartzite	Concrete	Schist	Quartzite
No of Tests	13	8	4	17	13	3
Average	43.01	58.24	57.35	2.87	4.87	3.50
Maximum	59.99	85.69	91.51	4.96	9.69	4.39
Minimum	29.81	22.49	34.47	1.62	1.13	2.84

Desk Study

A comprehensive desk study of the construction drawings for the dam was carried out to allow the proposed anchors to be spaced such that they did not interfere with existing elements of the dam. Of particular interest was the location of penetrations through the dam for fish pass, compensation generation and control pipes, which would dictate the layout of the proposed anchors. Subsequently a full topographic survey of the gravity sections of the dam was carried out to confirm the key dimensions and physical layout.

ANCHOR DESIGN

From the inception of the project, it was intended that the successful anchoring contractor would undertake the works on a Design and Build basis. However, to allow a meaningful comparison of tenders, the number, spacing and load capacity of anchors was fixed by the Designer, and shown on the Contract Drawings. Anchors were positioned so as to avoid services and openings in the dam structure.

Design Philosophy

The philosophy of the design of rock anchors at Clunie Dam has followed the same principles as those adopted at Pitlochry dam in Sandilands *et al* (1994) with vertical anchors through the concrete gravity dams proposed to prevent failure of the abutments by sliding or overturning.

Forces for the design of anchor loads were derived from preliminary analyses taking into account changes in cross section variations across the dam. Six cross sections were analysed based on geometries provided from the As Constructed drawings. The analyses considered the beneficial effect required by anchors installed at the crest of the dam with regard, predominantly, to overturning of the dam about its downstream toe.

The width of the walkway on both the left and right abutments severely restricted the positioning of the anchor holes with respect to the geometry of the abutments. All anchors could only be positioned close to the upstream face with a clearance of some 750mm if they were to be installed in the existing walkway. This was the only practical location for installation of the anchors as alternative locations on the downstream face of the dam would involve significant temporary works and health and safety issues. The location chosen allows for verticality tolerances when drilling the anchor hole to prevent 'daylighting' of the anchor hole through the upstream face. Additionally, the downstream side of the walkway generally overhangs the downstream slope of the dam so the whole walkway width was not available for anchor installation.

The proposed location is advantageous when considering the contribution of the anchors with regard to overturning about the toe of the dam as it utilises the longest lever arm available to maximum the restoring action of the anchors.

Anchors were spaced at nominal 3.0 – 3.5 metre centres along the abutment crest. Anchor positions were constrained by the presence of pipework in the abutments and construction gaps from the original construction, and were been positioned to avoid these features assuming that the main concrete construction panels would act monolithically.

Proposed anchor locations are indicated in Figure 3.

Pre-Tender Anchor Design

The pre-tender design of rock anchors was carried out to BS 8081 (1989). Type A anchors were designed as advised by the Code. The anchor design requires a unit side friction which can be derived from the unconfined compressive strength (UCS) of the rock. Values of UCS were taken from laboratory testing of rock samples recovered from the ground investigation and a working friction of 1.2 N/mm2 was adopted for the pre-tender design.

All anchors were designed with a factor of safety of 3.0 as advised by BS8081. Design anchor loads have taken into account the constructed height of the abutment gravity walls and indicative design loads reduce from 2500kN per anchor close to the centre spillway to 1400kN per anchor close to the valley sides.

Additional analysis was carried out considering failure of a cone of rock around the anchor fixed length as the dam rotates in overturning after the approach after Tomlinson in BS8081, considering overlapping cones of failure for individual anchors.

This failure mechanism controls, and thus a minimum fixed anchor length of 5.0 metres into bedrock was stipulated. Individual anchors were staggered by 5.0 metre depth in order that failure cones did not significantly overlap i.e. 5 metres and 10 metres below the concrete/bedrock interface.

Load testing of anchors was specified after 9 months for three anchors to confirm that the anchors are fully stressed. A full round of load testing for anchors is scheduled for 2009 to align with other SSE dam anchor testing frequencies.

Contractor's Anchor Design

The Contractor-designed components comprised the tendon, the grout, the tendon/grout/ground bond interfaces, the bearing plate and the head assembly for each anchor. Design of these elements followed the requirements of BS8081 and was thus undertaken using a global factor of safety approach with a factor of 2 on the tendon ultimate capacity and a factor of 3 on all bond interfaces.

Tendon design was based on using multiples of 15.2mm dyform strand with an individual ultimate strand capacity of 300kN, supplied by Dywidag Systems International (DSI). The strands forming the tendon were noded and splayed within the fixed length to enhance the bond capacity by mechanical interlock. Double corrosion-protection (DCP) was specified and this was achieved by using DSI Type 7B anchors, comprising twin concentric corrugated plastic encapsulations over the bond length and a single corrugated plastic encapsulation enclosing individually greased and sheathed strands over the free length.

The grout design followed conventional practice for anchor grouts and used a neat CEM1 42.5N cement colloidally mixed with a water:cement ratio of 0.4:1 to give an homogenous liquid grout with a fluid density of approximately 19kN/m^3. The liquid grout displaces bore water without dilution when injected through a base-mounted tremie line and cures to give cube crushing strengths well in excess of 40N/mm^2 at 28 days. When properly mixed, the grout demonstrates 1hour bleed of between 3 and 4% and reaches initial set in approximately 45 minutes.

The bond interfaces were designed using the equivalent rectangular stress-distribution method described in the BS8081. Ultimate bond values for the grout/tendon and grout/encapsulation interfaces were drawn from the limiting value of 3N/mm^2 recommended in the code.

In common with much of Perthshire, the rock underlying the dam is high grade metamorphic, typically strong schists and quartzite. The schists demonstrate little fracturing and good mineralogical infill of the expressed schistosity. A grout/ground bond value was calculated using methods described by Rosenberg & Journeaux (1976) and Williams & Pells (1981), with a conservative unconfined compressive strength (UCS) of 35MPa (test values ranged from 29 to 87MPa) and a Rock Quality Designation (RQD) of 50. This yielded ultimate bond stresses of 1.75 and 2.275N/mm^2 respectively. A value of 1.5N/mm^2 was selected for design purposes; however, it is acknowledged that given the nature of the rock, substantially higher bond values could have been justified by intensive testing of a trial anchor.

In practice, the limiting bond interface was found to be the grout to inner encapsulation in all cases, leading to fixed lengths ranging from 5.0m to 7.3m.

Design of the head assembly bearing was based on a dam concrete strength of 35N/mm^2 (derived from UCS testing on samples recovered during the ground investigation) and an allowable bearing stress of $0.3f_{cu}$. Using circular bearing plates, this gave diameters of variously 500, 550 and 600mm. The bearing plate thicknesses were calculated using published formulae for simply-supported annular plates in biaxial bending and these gave thickness of between 70 and 85mm.

The head assemblies were standard DSI components, designed in accordance with the requirements of BS8081 and BS4447 (1973). The tendon strands are held using locking serrated wedges fitted into conical sockets machined into the solid headblock. The assemblies include a short transition sheath between the tendon outer encapsulation and the bearing plate to allow the strand bundle to splay slightly before entering the headblock. Each assembly was completed with a sealed steel cap filled with corrosion-inhibiting compound.

The anchor components (tendons, encapsulations, bearing plates and head assemblies) were manufactured in the DSI anchor manufacturing facility in Southam, UK and were brought to site as complete, quality-assured anchor units.

IMPLEMENTATION

Contract Strategy

The anchor installation contract was put out to tender to four contractors, with the Ritchies Division of Edmund Nuttall Ltd. being the successful tenderer. The contract was let under the NEC3 Engineering and Construction Contract under Option A, (Priced Contract with Activity Schedule). All works were successfully completed over a 13 week period in summer 2007 at a final cost of £498,000 (including all design and SI work).

Execution

The sequence of operations required to install the anchors can be broken down as follows:

1. Coring the anchor head recesses.
2. Drilling of the anchor boreholes.
3. Installation and grouting of the anchors.
4. Preparation of the head blocks and stressing of the anchors.
5. Capping of the anchors and installation of access covers.

Coring Works

To maintain access on the dam following completion of the project, each of the installed anchors was recessed into the crest walkway. The recess comprised a 700mm diameter x 500mm deep hole formed by coring through the concrete. Circular load bearing plates were used to optimise the available bearing contact area and 50mm diameter drainage holes were cored from the downstream face of the dam using roped access techniques.

Drilling Operations

Drilling operations were carried out using a Klemm KR701-01 restricted access drilling rig using a remote hydraulic power pack and 50m hydraulic hoses. This arrangement allowed the drilling rig to operate on the narrow dam walkway while the power pack remained on land. Drilling was undertaken using a 254mm diameter drill bit on an 8" down the hole hammer using air flush and water injection for dust suppression. A centraliser was used within the starter core hole together with stabilised drill rods to ensure verticality of the borehole during drilling. Spoil from the drilling process was contained using a well head and diverted into skips placed on the dam walkway which were replaced as required. Small diameter reinforcement bars were locally encountered within the dam during the drilling works. These obstructions were overcome using a tungsten-tipped core barrel in combination with the down the hole hammer.

Anchor Installation

Strand anchors supplied by Dywidag Systems International were used during the project with installed capacities of between 1400kN and 2500kN. Anchor lengths ranged from 11.5m to 39.8m and used between 12 and 17 strands depending on the required working load.

The dam is accessed by a narrow single track road which precluded the use of traditional craneage for installation of the anchors. During the tender period a number of options were considered, including the use of helicopters. To overcome the access restrictions an alternative anchor installation method was developed which minimised the environmental impact of the project and reduced the projected project cost.

The installation method employed used a de-rated 360-degree telehandler mounted on a work barge operating on the impounded loch. Anchors were delivered to site pre-coiled and with a 1m section of the bond length grouted. Each anchor was loaded onto a bespoke anchor wheel which could be attached to the telehandler. The work barge then transported the anchor to the borehole location, where it was uncoiled down the borehole using a hydraulic winch on the telehandler mounted anchor wheel. This method was successfully and safely utilised for all of the anchors installed over water.

Figure 4 – Anchor installation using barge mounted telehandler

Anchor Stressing and Capping

Grouting of the anchors was undertaken using a neat Portland cement grout mixed in a high shear colloidal grout mixer. Grout was injected into the borehole through three internal grout lines in the anchor. Once the grout had achieved sufficient compressive strength, cyclic load testing of the installed anchors was undertaken in accordance with BS EN 1537 (1999). All anchors performed within the expected limits both during the suitability and acceptance tests that were undertaken. On completion of the stressing operations removable access covers were installed over the stressing recess to allow the walkway to be re-opened for normal use.

CONCLUSION

The anchoring works have been successfully carried out, using an innovative method of installation giving environmental, cost and programme benefits.

The dam structure now complies with current UK safety standards, and has been issued with a 10(6) certificate.

REFERENCES

HMSO, (1975). *Reservoirs Act.*

British Standards Institution, (1989). *BS8081, British Standard Code of Practice for Ground Anchorages.*

British Standards Institution, (1973). *BS4447, Specification for the performance of prestressing anchorages for post-tensioned construction.*

British Standards Institution, (1999). *BS EN1537, Execution of special geotechnical work – Ground Anchors.*

Building Research Establishment, (1991). *An Engineering Guide to Seismic Risk to Dams in the United Kingdom.*

Institution of Civil Engineers, (1996). *Floods and Reservoir Safety - Third Edition* , Thomas Telford, London.

Rosenberg, P. and Journeaux, N.L., (1976). Friction and end-bearing tests on bedrock for high capacity socket design, *Canadian Geotechnical Journal,* Vol. 13, pp. 324-333.

Sandilands, N.M., Cameron, A.T., Bryce, W.M., (1994), Pitlochry Dam, the use of post-tensioned ground anchors to increase stability, *Reservoir Safety and the Environment,* Thomas Telford, London.

Williams, A.F and Pells, P.J.N, (1981). Side resistance rock sockets in sandstone mudstone and shale, *Canadian Geotechnical Journal,* Vol. 18, pp. 502-513.

Authors index